Bosch Professional Automotive Information

Bosch Professional Automotive Information is a definitive reference for automotive engineers. The series is compiled by one of the world´s largest automotive equipment suppliers. All topics are covered in a concise but descriptive way backed up by diagrams, graphs, photographs and tables enabling the reader to better comprehend the subject.

There is now greater detail on electronics and their application in the motor vehicle, including electrical energy management (EEM) and discusses the topic of intersystem networking within vehicle. The series will benefit automotive engineers and design engineers, automotive technicians in training and mechanics and technicians in garages.

Konrad Reif
Editor

Gasoline Engine Management

Systems and Components

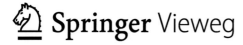 Springer Vieweg

Editor
Prof. Dr.-Ing. Konrad Reif
Duale Hochschule Baden-Württemberg
Friedrichshafen, Germany
reif@dhbw-ravensburg.de

ISBN 978-3-658-03963-9   ISBN 978-3-658-03964-6 (eBook)
DOI 10.1007/978-3-658-03964-6

Library of Congress Control Number: 2014945106

Springer Vieweg
© Springer Fachmedien Wiesbaden 2015

Springer is part of Springer Science+Business Media
www.springer.com

▶ Foreword

The call for environmentally compatible and economical vehicles necessitates immense efforts to develop innovative engine concepts. Technical concepts such as gasoline direct injection helped to save fuel up to 20 % and reduce CO_2-emissions. Descriptions of the cylinder-charge control, fuel injection, ignition and catalytic emission-control systems provides comprehensive overview of today´s gasoline engines. This book also describes emission-control systems and explains the diagnostic systems. The publication provides information on engine-management-systems and emission-control regulations.

Complex technology of modern motor vehicles and increasing functions need a reliable source of information to understand the components or systems. The rapid and secure access to these informations in the field of Automotive Electrics and Electronics provides the book in the series "Bosch Professional Automotive Information" which contains necessary fundamentals, data and explanations clearly, systematically, currently and application-oriented. The series is intended for automotive professionals in practice and study which need to understand issues in their area of work. It provides simultaneously the theoretical tools for understanding as well as the applications.

Contents

History of the automobile
Dipl.-Ing. Karl-Heinz Dietsche,
Dietrich Kuhlgatz.

Basics of the gasoline (SI) engine
Dr. rer. nat. Dirk Hofmann,
Dipl.-Ing. Bernhard Mencher,
Dipl.-Ing. Werner Häming,
Dipl.-Ing. Werner Hess.

Fuels
Dr. rer. nat. Jörg Ullmann,
Dipl.-Ing. (FH) Thorsten Allgeier.

Cylinder-charge control systems
Dr. rer. nat. Heinz Fuchs,
Dipl.-Ing. (FH) Bernhard Bauer,
Dipl.-Phys. Torsten Schulz,
Dipl.-Ing. Michael Bäuerle,
Dipl.-Ing. Kristina Milos.

Gasoline injection systems over the years
Dipl.-Ing. Karl-Heinz Dietsche.

Fuel supply
Dipl.-Ing. Jens Wolber,
Ing. grad. Peter Schelhas,
Dipl.-Ing. Uwe Müller,
Dipl.-Ing. (FH) Andreas Baumann,
Dipl.-Betriebsw. Meike Keller.

Manifold injection
Dipl.-Ing. Anja Melsheimer,
Dipl.-Ing. Rainer Ecker,
Dipl.-Ing. Ferdinand Reiter,
Dipl.-Ing. Markus Gesk.

Gasoline direct injection
Dipl.-Ing. Andreas Binder,
Dipl.-Ing. Rainer Ecker,
Dipl.-Ing. Andreas Glaser,
Dr.-Ing. Klaus Müller.

Operation of gasoline engines on natural gas
Dipl.-Ing. (FH) Thorsten Allgeier,
Dipl.-Ing. (FH) Martin Haug,
Dipl.-Ing. Roger Frehoff,
Dipl.-Ing. Michael Weikert,
Dipl.-Ing. (FH) Kai Kröger,

Dr. rer. nat. Winfried Langer,
Dr.-Ing. habil. Jürgen Förster,
Dr.-Ing. Jens Thurso,
Jürgen Wörsinger.

Ignition systems over the years
Dipl.-Ing. Karl-Heinz Dietsche.

Inductive ignition system
Dipl.-Ing. Walter Gollin.

Ignition coils
Dipl.-Ing. (FH) Klaus Lerchenmüller,
Dipl.-Ing. (FH) Markus Weimert,
Dipl.-Ing. Tim Skowronek.

Spark plugs
Dipl.-Ing. Erich Breuser.

Electronic Control
Dipl.-Ing. Bernhard Mencher,
Dipl.-Ing. (FH) Thorsten Allgeier,
Dipl.-Ing. (FH) Klaus Joos,
Dipl.-Ing. (BA) Andreas Blumenstock,
Dipl.-Red. Ulrich Michelt.

Sensors
Dr.-Ing. Wolfgang-Michael Müller,
Dr.-Ing. Uwe Konzelmann,
Dipl.-Ing. Roger Frehoff,
Dipl.-Ing. Martin Mast,
Dr.-Ing. Johann Riegel.

Electronic control unit (ECU)
Dipl.-Ing. Martin Kaiser.

Exhaust emissions
Dipl.-Ing. Christian Köhler,
Dipl.-Ing. (FH) Thorsten Allgeier.

Catalytic emission control
Dr.-Ing. Jörg Frauhammer,
Dr. rer. nat. Alexander Schenck zu Schweinsberg,
Dipl.-Ing. Klaus Winkler.

Emission-control legislation
Dipl.-Ing. Bernd Kesch,
Dipl.-Ing Ramon Amirpour,
Dr. Michael Eggers.

Exhaust-gas measuring techniques
Dipl.-Phys. Martin-Andreas Drühe.

Diagnosis
Dr.-Ing. Matthias Knirsch,
Dipl.-Ing. Bernd Kesch,
Dr.-Ing. Matthias Tappe,
Dr.-Ing. Günter Driedger,
Dr. rer. nat. Walter Lehle.

ECU development
Dipl.-Ing. Martin Kaiser,
Dipl.-Phys. Lutz Reuschenbach,
Dipl.-Ing. (FH) Bert Scheible,
Dipl.-Ing. Eberhard Frech.

and the editorial team in cooperation with the
responsible in-house specialist departments.

History of the automobile

Mobility has always played a crucial role in the course of human development. In almost every era, man has attempted to find the means to allow him to transport people over long distances at the highest possible speed. It took the development of reliable internal-combustion engines that were operated on liquid fuels to turn the vision of a self-propelling "automobile" into reality (combination of Greek: autos = self and Latin: mobilis = mobile).

Development history

It would be hard to imagine life in our modern day without the motor car. Its emergence required the existence of many conditions without which an undertaking of this kind would not have been possible. At this point, some development landmarks may be worthy of note. They represent an essential contribution to the development of the automobile:

- About 3500 B.C.
  The development of the wheel is attributed to the Sumerians
- About 1300
  Further refinement of the carriage with elements such as steering, wheel suspension and carriage springs
- 1770
  Steam buggy by Joseph Cugnot
- 1801
  Étienne Lenoir develops the gas engine
- 1870
  Nikolaus Otto builds the first four-stroke internal-combustion engine

In 1885 Carl Benz enters the annals of history as the inventor of the first automobile. His patent marks the beginning of the rapid development of the automobile

powered by the internal-combustion engine. Public opinion remained divided, however. While the proponents of the new age lauded the automobile as the epitome of progress, the majority of the population protested against the increasing annoyances of dust, noise, accident hazard, and inconsiderate motorists. Despite all of this, the progress of the automobile proved unstoppable.

In the beginning, the acquisition of an automobile represented a serious challenge.
A road network was virtually nonexistent; repair shops were unknown, fuel was purchased at the drugstore, and spare parts were produced on demand by the local blacksmith. The prevailing circumstances made the first long-distance journey by Bertha Benz in 1888 an even more astonishing accomplishment. She is thought to have been the first woman behind the wheel of a motorized vehicle. She also demonstrated the reliability of the automobile by journeying the then enormous distance of more than 100 kilometers (about 60 miles) between Mannheim and Pforzheim in south-western Germany.

In the early days, however, few entrepreneurs – with the exception of Benz – considered the significance of the engine-powered vehicle on a worldwide scale. It was the French who were to help the automobile to greatness. Panhard & Levassor used licenses for Daimler engines to build their own automobiles. Panhard pioneered construction features such as the steering wheel, inclined steering column, clutch pedal, pneumatic tires, and tube-type radiator.
  In the years that followed, the industry mushroomed with the arrival of companies such as Peugeot, Citroën, Renault, Fiat, Ford, Rolls-Royce, Austin, and others. The influence of Gottlieb Daimler, who was selling his engines almost all over the world, added significant impetus to these developments.

Daimler Motorized Carriage, 1894 (Source: DaimlerChrysler Classic, Corporate Archives)

The first journey with an engine-powered vehicle is attributed to Joseph Cugnot (in 1770). His lumbering, steam-powered, wooden three-wheeled vehicle was able to travel for all of 12 minutes on a single tankful of water.

The patent issued to Benz on January 29, 1886 was not based on a converted carriage. Instead, it was a totally new, independent construction (Source: DaimlerChrysler Classic, Corporate Archives)

1st stroke: Induction

Referred to Top Dead Center (TDC), the piston is moving downwards and increases the volume of the combustion chamber (7) so that fresh air (gasoline direct injection) or fresh air/fuel mixture (manifold injection) is drawn into the combustion chamber past the opened intake valve (5).

The combustion chamber reaches maximum volume ($V_h + V_c$) at Bottom Dead Center (BDC).

2nd stroke: Compression

The gas-exchange valves are closed, and the piston is moving upwards in the cylinder. In doing so it reduces the combustion-chamber volume and compresses the air/fuel mixture. On manifold-injection engines the air/fuel mixture has already entered the combustion chamber at the end of the induction stroke. With a direct-injection engine on the other hand, depending upon the operating mode, the fuel is first injected towards the end of the compression stroke.

At Top Dead Center (TDC) the combustion-chamber volume is at minimum (compression volume V_c).

3rd stroke: Power (or combustion)

Before the piston reaches Top Dead Center (TDC), the spark plug (2) initiates the combustion of the air/fuel mixture at a given ignition point (ignition angle). This form of ignition is known as externally supplied ignition. The piston has already passed its TDC point before the mixture has combusted completely.

The gas-exchange valves remain closed and the combustion heat increases the pressure in the cylinder to such an extent that the piston is forced downward.

4th stroke: Exhaust

The exhaust valve (6) opens shortly before Bottom Dead Center (BDC). The hot (exhaust) gases are under high pressure and leave the cylinder through the exhaust valve. The remaining exhaust gas is forced out by the upwards-moving piston.

A new operating cycle starts again with the induction stroke after every two revolutions of the crankshaft.

Valve timing

The gas-exchange valves are opened and closed by the cams on the intake and exhaust camshafts (3 and 1 respectively). On engines with only 1 camshaft, a lever mechanism transfers the cam lift to the gas-exchange valves.

The valve timing defines the opening and closing times of the gas-exchange valves. Since it is referred to the crankshaft position, timing is given in "degrees crankshaft". Gas flow and gas-column vibration effects are applied to improve the filling of the combustion chamber with air/fuel mixture and to remove the exhaust gases. This is the reason for the valve opening and closing times overlapping in a given crankshaft angular-position range.

The camshaft is driven from the crankshaft through a toothed belt (or a chain or gear pair). On 4-stroke engines, a complete working cycle takes two rotations of the crankshaft. In other words, the camshaft only turns at half crankshaft speed, so that the step-down ratio between crankshaft and camshaft is 2:1.

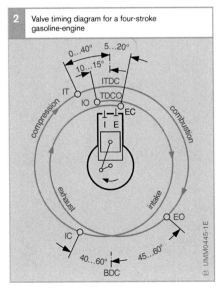

2 Valve timing diagram for a four-stroke gasoline-engine

Fig. 2
I   Intake valve
IO   Intake valve opens
IC   Intake valve closes
E   Exhaust valve
EO   Exhaust valve opens
EC   Exhaust valve closes
TDC   Top Dead Center
TDCO Overlap at TDC
ITDC Ignition at TDC
BDC   Bottom Dead Center
IT   Ignition point

Compression

The difference between the maximum piston displacement V_h and the compression volume V_c is the compression ratio

$$\varepsilon = (V_h + V_c)/V_c.$$

The engine's compression ratio is a vital factor in determining
● Torque generation
● Power generation
● Fuel economy and
● Emissions of harmful pollutants

The gasoline-engine's compression ratio ε varies according to design configuration and the selected form of fuel injection (manifold or direct injection $\varepsilon = 7...13$). Extreme compression ratios of the kind employed in diesel powerplants ($\varepsilon = 14...24$) are not suitable for use in gasoline engines. Because the knock resistance of the fuel is limited, the extreme compression pressures and the high combustion-chamber temperatures resulting from such compression ratios must be avoided in order to prevent spontaneous and uncontrolled detonation of the air/fuel mixture. The resulting knock can damage the engine.

Air/fuel ratio

Complete combustion of the air/fuel mixture relies on a stoichiometric mixture ratio. A stoichiometric ratio is defined as 14.7 kg of air for 1 kg of fuel, that is, a 14.7 to 1 mixture ratio.

The air/fuel ratio λ (lambda) indicates the extent to which the instantaneous monitored air/fuel ratio deviates from the theoretical ideal:

$$\lambda = \frac{\text{induction air mass}}{\text{theoretical air requirement}}$$

The lambda factor for a stoichiometric ratio is λ 1.0. λ is also referred to as the excess-air factor.

Richer fuel mixtures result in λ figures of less than 1. Leaning out the fuel produces mixtures with excess air: λ then exceeds 1. Beyond a certain point the mixture encounters the lean-burn limit, beyond which ignition is no longer possible. The excess-air factor has a decisive effect on the specific fuel consumption (Fig. 3) and untreated pollutant emissions (Fig. 4).

Induction-mixture distribution in the combustion chamber

Homogeneous distribution
The induction systems on engines with manifold injection distribute a homogeneous air/fuel mixture throughout the combustion chamber. The entire induction charge has a single excess-air factor λ (Fig. 5a). Lean-burn engines, which operate on excess air under

Fig. 3
a   Rich air/fuel mixture
    (air deficiency)
b   Lean air/fuel mixture
    (excess air)

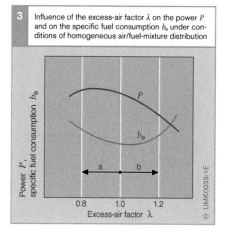

3   Influence of the excess-air factor λ on the power P and on the specific fuel consumption b_e under conditions of homogeneous air/fuel-mixture distribution

Power P, specific fuel consumption b_e

0.8   1.0   1.2
Excess-air factor λ

UMK0033-1E

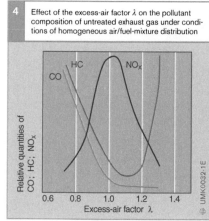

4   Effect of the excess-air factor λ on the pollutant composition of untreated exhaust gas under conditions of homogeneous air/fuel-mixture distribution

Relative quantities of CO; HC; NO_x

CO   HC NO_x

0.6   0.8   1.0   1.2   1.4
Excess-air factor λ

UMK0032-1E

specific operating conditions, also rely on homogeneous mixture distribution.

Stratified-charge concept
A combustible mixture cloud with $\lambda \approx 1$ surrounds the tip of the spark plug at the instant ignition is triggered. At this point the remainder of the combustion chamber contains either non-combustible gas with no fuel, or an extremely lean air/fuel charge. The corresponding strategy, in which the ignitable mixture cloud is present only in one portion of the combustion chamber, is the stratified-charge concept (Fig. 5b). With this concept, the overall mixture – meaning the average mixture ratio within the entire combustion chamber – is extremely lean (up to $\lambda \approx 10$). This type of lean operation fosters extremely high levels of fuel economy.

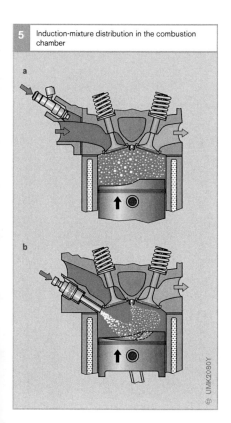

| 5 | Induction-mixture distribution in the combustion chamber |

a

b

Efficient implementation of the stratified-charge concept is impossible without direct fuel injection, as the entire induction strategy depends on the ability to inject fuel directly into the combustion chamber just before ignition.

Ignition and flame propagation
The spark plug ignites the air/fuel mixture by discharging a spark across a gap. The extent to which ignition will result in reliable flame propagation and secure combustion depends in large part on the air/fuel mixture λ, which should be in a range extending from $\lambda = 0.75...1.3$. Suitable flow patterns in the area immediately adjacent to the spark-plug electrodes can be employed to ignite mixtures as lean as $\lambda \leq 1.7$.

The initial ignition event is followed by formation of a flame-front. The flame front's propagation rate rises as a function of combustion pressure before dropping off again toward the end of the combustion process. The mean flame front propagation rate is on the order of 15...25 m/s.

The flame front's propagation rate is the combination of mixture transport and combustion rates, and one of its defining factors is the air/fuel ratio λ. The combustion rate peaks at slightly rich mixtures on the order of $\lambda = 0.8...0.9$. In this range it is possible to approach the conditions coinciding with an ideal constant-volume combustion process (refer to section on "Engine efficiency"). Rapid combustion rates provide highly satisfactory full-throttle, full-load performance at high engine speeds.

Good thermodynamic efficiency is produced by the high combustion temperatures achieved with air/fuel mixtures of $\lambda = 1.05...1.1$. However, high combustion temperatures and lean mixtures also promote generation of nitrous oxides (NO_X), which are subject to strict limitations under official emissions standards.

Fig. 5
a Homogeneous mixture distribution
b Stratified charge

Cylinder charge

An air/fuel mixture is required for the combustion process in the cylinder. The engine draws in air through the intake manifolds (Fig. 1, Pos. 14), the throttle valve (13) ensuring that the air quantity is metered. The fuel is metered through fuel injectors. Furthermore, usually part of the burnt mixture (exhaust gas) from the last combustion is retained as residual gas (9) in the cylinder or exhaust gas is returned specifically to increase the residual-gas content in the cylinder (4).

Components of the cylinder charge

The gas mixture trapped in the combustion chamber when the intake valve closes is referred to as the cylinder charge. This is comprised of the fresh gas and the residual gas.

The term "relative air charge rac" has been introduced in order to have a quantity which is independent of the engine's displacement. It describes the air content in the cylinder and is defined as the ratio of the current air quantity in the cylinder to the air quantity that would be contained in the engine displacement under standard conditions ($p_0 = 1013$ hPa, $T_0 = 273$ K). Accordingly, there is a relative fuel quantity rfq; this is defined in such a way that identical values for rac and rfq result in $\lambda = 1$, i.e., $\lambda = rac/rfq$, or with specified λ : $rfq = rac/\lambda$.

Fresh gas

The freshly introduced gas mixture in the cylinder is comprised of the fresh air drawn in and the fuel entrained with it. In a manifold-injection engine, all the fuel has already been mixed with the fresh air upstream of the intake valve. On direct-injection systems, on the other hand, the fuel is injected directly into the combustion chamber.

The majority of the fresh air enters the cylinder with the air-mass flow (Fig. 1, Pos. 6, 7) via the throttle valve (13). Additional fresh gas, comprising fresh air and fuel vapor, is directed to the cylinder via the evaporative-emissions control system (3, 2).

For homogeneous operation at $\lambda \leq 1$, the air in the cylinder directed via the throttle valve after the intake valve (11) has closed is the decisive quantity for the work at the piston during the combustion stroke and therefore for the engine's output torque. In this case, the air charge corresponds to the torque and the engine load. Here, changing the throttle-valve angle only indirectly leads to a change in the air charge. First of all, the pressure in the intake manifold must rise so that a greater air mass flows into the cylinder via the intake valves. Fuel can, on the other hand, be injected more contemporaneously with the combustion process and metered precisely to the individual cylinder. Therefore the injected fuel quantity is dependent on the current air quantity, and the gasoline engine is an air-directed system in "conventional" homogeneous mode at $\lambda \leq 1$.

During lean-burn operation (stratified charge), however, the torque (engine load) – on account of the excess air – is a direct product of the injected fuel mass. The air mass can thus differ for the same torque. The gasoline engine is therefore fuel-directed during lean-burn operation.

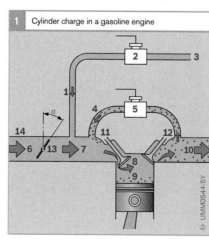

1   Cylinder charge in a gasoline engine

UMM0544-5Y

Almost always, measures aimed at increasing the engine's maximum torque and maximum output power necessitate an increase in the maximum possible fresh-gas charge. This can be achieved by increasing the engine displacement but also by supercharging (see section entitled "Supercharging").

Residual gas

The residual-gas share of the cylinder charge comprises that portion of the cylinder charge which has already taken part in the combustion process. In principle, one differentiates between internal and external residual gas. Internal residual gas is the exhaust gas which remains in the upper clearance volume of the cylinder after combustion or which, while the intake and exhaust valves are simultaneously open (valve overlap, see section entitled "Gas exchange"), is drawn from the exhaust port back into the intake manifold (internal exhaust-gas recirculation).

External residual gas is exhaust gas which is introduced via an exhaust-gas recirculation valve (Fig. 1, Pos. 4, 5) into the intake manifold (external exhaust-gas recirculation).

The residual gas is made up of inert gas [1] and – in the event of excess air, i.e., during lean-burn operation – of unburnt air. The amount of inert gas in the residual gas is particularly important. This no longer contains any oxygen and therefore does not participate in combustion during the following power cycle. However, it does delay ignition and slows down the course of combustion, which results in slightly lower efficiency but also in lower peak pressures and temperatures. In this way, a specifically used amount of residual gas can reduce the emission of nitrogen oxides (NO_X). This then is the benefit of inert gas in lean-burn operation in that the three-way catalytic converter is unable to reduce the nitrogen oxides in the event of excess air.

In homogeneous engine mode, the fresh-gas charge displaced by the residual gas (consisting in this case of inert gas only) is compensated by means of a greater opening of the throttle valve. With a constant fresh-gas charge, this increases the intake-manifold pressure, therefore reduces the throttling losses (see section entitled "Gas exchange"), and in all results in reduced fuel consumption.

Gas exchange

The process of replacing the consumed cylinder charge (exhaust gas, also referred to in the above as residual gas) with fresh gas is known as gas exchange or the charge cycle. It is controlled by the opening and closing of the intake and exhaust valves in combination with the piston stroke. The shape and position of the camshaft cams determine the progression of the valve lift and thereby influence the cylinder charge.

The opening and closing times of the valves are called valve timing and the maximum distance a valve is lifted from its seat is known as the valve lift or valve stroke. The characteristic variables are Exhaust Opens (EO), Exhaust Closes (EC), Intake Opens (IO), Intake Closes (IC) and the valve lift. There are engines with fixed and others with variable timing and valve lifts (see chapter entitled "Cylinder-charge control systems").

The amount of residual gas for the following power cycle can be significantly influenced by a valve overlap. During the valve overlap, intake and exhaust valves are simultaneously open for a certain amount of time, i.e., the intake valve opens before the exhaust valve closes. If in the overlap phase the pressure in the intake manifold is lower than that in the exhaust train, the residual gas flows back into the intake manifold; because the residual gas drawn back in this way is drawn in again after Exhaust Closes, this results in an increase in the residual-gas content.

---

[1] Components in the combustion chamber which behave inertly, that is, do not participate in the combustion process.

In the case of supercharging, the pressure before the intake valve can also be higher during the overlap phase; in this event, the residual gas flows in the direction of the exhaust train such that it is properly cleared away ("scavenging") and it is also possible for the air to flow through into the exhaust train.

When the residual gas is successfully scavenged, its volume is then available for an increased fresh-gas charge. The scavenging effect is therefore used to increase torque in the lower speed range (up to approx. 2000 rpm), either in combination with dynamic supercharging in naturally aspirated engines or with turbocharging.

Volumetric efficiency and air consumption

The success of the gas-exchange process is measured in the variables volumetric efficiency, air consumption and retention rate. The volumetric efficiency is the ratio of the fresh-gas charge actually remaining in the cylinder to the theoretically maximum possible charge. It differs from the relative air charge in that the volumetric efficiency is referred to the external conditions at the time of measurement and not to standard conditions.

The air consumption describes the total air-mass throughput during the gas-exchange process, likewise referred to the theoretically maximum possible charge. The air consumption can also include the air mass which is transferred directly into the exhaust train during the valve overlap. The retention rate, the ratio of volumetric efficiency to air consumption, specifies the proportion of the air-mass throughput which remains in the cylinder at the end of the gas-exchange process.

The maximum volumetric efficiency for naturally aspirated engines is 0.6...0.9. It depends on the combustion-chamber shape, the opened cross-sections of the gas-exchange valves, and the valve timing.

Pumping losses

Work is expended in the form of pumping losses or gas-exchange losses in order to replace the exhaust gas with fresh gas in the gas-exchange process. These losses use up part of the mechanical work generated and therefore reduce the effective efficiency of the engine. In the intake phase, i.e., during the downward stroke of the piston, the intake-manifold pressure in throttled mode is less than the ambient pressure and in particular the pressure in the piston return chamber. The piston must work against this pressure differential (throttling losses).

A dynamic pressure occurs in the combustion chamber during the upward stroke of the piston when the burnt gas is emitted, particularly at high engine speeds and loads; the piston must expend energy in order to overcome this pressure (push-out losses).

If with gasoline direct injection stratified-charge operation is used with the throttle valve fully opened or high exhaust-gas recirculation is used in homogeneous operation ($\lambda \leq 1$), this increases the intake-manifold pressure and reduces the pressure differential above the piston. In this way, the engine's throttling losses can be reduced, which in turn improves the effective efficiency.

Supercharging

The torque which can be achieved during homogenous operation at $\lambda \leq 1$ is proportional to the fresh-gas charge. This means that maximum torque can be increased by compressing the air before it enters the cylinder (supercharging). This leads to an increase in volumetric efficiency to values above 1.

Dynamic supercharging

Supercharging can be achieved simply by taking advantage of the dynamic effects inside the intake manifold. The supercharging level depends on the intake manifold's design and on its operating point (for the most part, on engine speed, but also on cylinder charge). The possibility of changing the intake-manifold geometry while the engine is running (variable intake-manifold geometry) means that dynamic supercharging can be applied across a wide operating range to increase the maximum cylinder charge.

Mechanical supercharging

The intake-air density can be further increased by compressors which are driven mechanically from the engine's crankshaft. The compressed air is forced through the intake manifold and into the engine's cylinders.

Exhaust-gas turbocharging

In contrast mechanical supercharging, the compressor of the exhaust-gas turbocharger is driven by an exhaust-gas turbine located in the exhaust-gas flow, and not by the engine's crankshaft. This enables recovery of some of the energy in the exhaust gas.

Charge recording

In a gasoline engine with homogeneous $\lambda = 1$ operation, the injected fuel quantity is dependent on the air quantity. This is necessary because after a change to the throttle-valve angle the air charge changes only gradually while the fuel quantity can be varied from injection to injection.

For this reason, the current available air charge must be determined for each combustion in the engine-management system (charge recording). There are essentially three systems which can be used to record the charge:

- A hot-film air-mass meter (HFM) measures the air-mass flow into the intake manifold.

- A model is used to calculate the air-mass flow from the temperature before the throttle valve, the pressure before and after the throttle valve, and the throttle-valve angle (throttle-valve model, α/n system[1])).

- A model is used to calculate the charge drawn in by the cylinder from the engine speed (n), the pressure (p) in the intake manifold (i.e., before the intake valve), the temperature in the intake passage and further additional information (e.g., camshaft/valve-lift adjustment, intake-manifold changeover, position of the swirl control valve) $(p/n$ system). Sophisticated models may be necessary, depending on the complexity of the engine, particularly with regard to the variabilities of the valve gear.

Because only the mass flow passing into the intake manifold can be determined with a hot-film air-mass meter or a throttle-valve model, both these systems only provide a cylinder-charge value during stationary engine operation. Stationary means at constant intake-manifold pressure; because then the mass flows flowing into the intake manifold and off into the engine are identical.

In the event of a sudden load change (change in the throttle-valve angle), the inflowing mass flow changes spontaneously, while the off-flowing mass flow and with it the cylinder charge only change if the intake-manifold pressure has increased or reduced. The accumulator behavior of the intake manifold must therefore also be imitated (intake-manifold model).

---

[1]) The designation α/n system is historically conditioned since originally the pressure after the throttle valve was not taken into account and the mass flow was stored in a program map covering throttle-valve angle and engine speed. This simplified approach is sometimes still used today.

Torque and power

Torques at the drivetrain

The power P delivered by a gasoline engine is defined by the available clutch torque M and the engine speed n. The clutch torque is the torque developed by the combustion process less friction torque (friction losses in the engine), pumping losses, and the torque needed to drive the auxiliary equipment (Fig. 1). The drive torque is derived from the clutch torque plus the losses arising at the clutch and transmission.

The combustion torque is generated in the power cycle and is determined in engines with manifold injection by the following variables:
- The air mass which is available for combustion when the intake valves close
- The fuel mass which is available at the same moment, and
- The moment in time when the ignition spark initiates the combustion of the air/fuel mixture

Direct-injection gasoline engines function at certain operating points with excess air (lean-burn operation). The cylinder thus contains air, which has no effect on the generated torque. Here, it is the fuel mass which has the most effect.

Generation of torque

The physical quantity torque M is the product of force F times lever arm s:

$$M = F \cdot s$$

The connecting rod utilizes the throw of the crankshaft to convert the piston's linear travel into rotary motion. The force with which the expanding air/fuel mixture drives the piston down the cylinder is converted into torque by the lever arm generated by the throw.

The lever arm l which is effective for the torque is the lever component vertical to the force (Fig. 2). The force and the leverage angle are parallel at Top Dead Center (TDC).

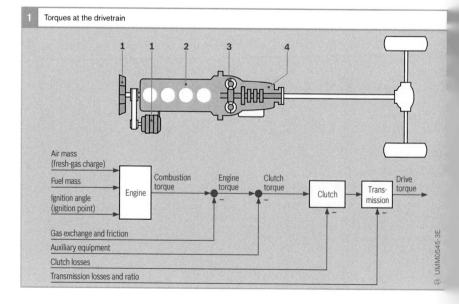

1   Torques at the drivetrain

Air mass (fresh-gas charge)

Fuel mass

Ignition angle (ignition point)

Engine

Combustion torque

Engine torque

Clutch torque

Clutch

Trans-mission

Drive torque

Gas exchange and friction

Auxiliary equipment

Clutch losses

Transmission losses and ratio

Fig. 1
1   Auxiliary equipment (A/C compressor, alternator, etc.)
2   Engine
3   Clutch
4   Transmission

This results in an effective lever arm of zero. The ignition angle must be selected in such a way as to trigger mixture ignition while the crankshaft is rotating through a phase of increasing lever arm (0...90°crankshaft). This enables the engine to generate the maximum possible torque. The engine's design (for instance, piston displacement, combustion-chamber geometry, volumetric efficiency, charge) determines the maximum possible torque M that it can generate.

Essentially, the torque is adapted to the requirements of actual driving by adjusting the quality and quantity of the air/fuel mixture and the ignition angle. Fig. 3 shows the typical torque and power curves, plotted against engine speed, for a manifold-injection gasoline engine. As engine speed increases, full-load torque initially increases to its maximum M_{max}. At higher engine speeds, torque falls off again as the shorter opening times of the intake valves limits the cylinder charge.

Engine designers focus on attempting to obtain maximum torque at low engine speeds of around 2000 rpm. This rpm range coincides with optimal fuel economy. Engines with exhaust-gas turbochargers are able to meet these requirements.

Relationship between torque and power

The engine's power output P climbs along with increasing torque M and engine speed n. The following applies:

$$P = 2 \cdot \pi \cdot M \cdot n$$

Engine power increases until it reaches its peak value at rated speed n_{rat} with rated power P_{rat}. Owing to the substantial decrease in torque, power generation drops again at extremely high engine speeds.

A transmission to vary conversion ratios is needed to adapt the gasoline engine's torque and power curves to meet the requirements of vehicle operation.

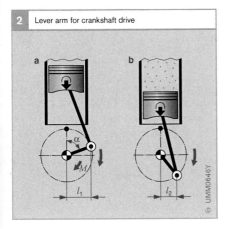

2 Lever arm for crankshaft drive

a   b

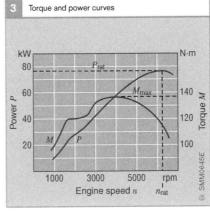

3 Torque and power curves

Fig. 2
Changing the effective lever arm during the power cycle
a   Increasing lever arm l_1
b   Decreasing lever arm l_2

Fig. 3
Typical curves for a manifold-injection gasoline engine

Engine efficiency

Thermal efficiency

The internal-combustion engine does not convert all the energy which is chemically available in the fuel into mechanical work, and some of the added energy is lost. This means that an engine's efficiency is less than 100% (Fig. 1). Thermal efficiency is one of the important links in the engine's efficiency chain.

Pressure-volume diagram (p-V diagram)
The p-V diagram is used to display the pressure and volume conditions during a complete working cycle of the 4-stroke IC engine.

The ideal cycle
Figure 2 (curve A) shows the compression and power strokes of an ideal process as defined by the laws of Boyle/Mariotte and Gay-Lussac. The piston travels from BDC to TDC (point 1 to point 2), and the air/fuel mixture is compressed without the addition of heat (Boyle/Mariotte). Subsequently, the mixture burns accompanied by a pressure rise (point 2 to point 3) while volume remains constant (Gay-Lussac).

From TDC (point 3), the piston travels towards BDC (point 4), and the combustion-chamber volume increases. The pressure of the burnt gases drops whereby no heat is released (Boyle/Mariotte). Finally, the burnt mixture cools off again with the volume remaining constant (Gay-Lussac) until the initial status (point 1) is reached again.

The area inside the points $1 - 2 - 3 - 4$ shows the work gained during a complete working cycle. The exhaust valve opens at point 4 and the gas, which is still under pressure, escapes from the cylinder. If it were possible for the gas to expand completely by the time point 5 is reached, the area described by $1 - 4 - 5$ would represent usable energy. On an exhaust-gas-turbocharged engine, the part above the atmospheric line (1 bar) can to some extent be utilized ($1 - 4 - 5'$).

Real p-V diagram
Since it is impossible during normal engine operation to maintain the basic conditions for the ideal cycle, the actual p-V diagram (Fig. 2, curve B) differs from the ideal p-V diagram.

Measures for increasing thermal efficiency
The thermal efficiency rises along with increasing air/fuel-mixture compression. The higher the compression, the higher the pressure in the cylinder at the end of the compression phase, and the larger is the enclosed area in the p-V diagram. This area is an indication of the energy generated during the combustion process. When selecting the compression ratio, the fuel's antiknock qualities must be taken into account.

Manifold-injection engines inject the fuel into the intake manifold onto the closed intake valve, where it is stored until drawn into the cylinder. During the formation of the air/fuel mixture, the fine fuel droplets vaporize. The energy needed for this process is in the form of heat and is taken from the air and the intake-manifold walls. On direct-injection engines the fuel is injected into the combustion chamber, and the energy needed for fuel-droplet vaporization is taken from the air trapped in the cylinder which cools off as a result. This means that the compressed air/fuel mixture is at a lower temperature than is the case with a manifold-injection engine, so that a higher compression ratio can be chosen.

Thermal losses

The heat generated during combustion heats up the cylinder walls. Part of this thermal energy is radiated and lost. In the case of gasoline direct injection, the stratified-charge air/fuel mixture cloud is surrounded by a jacket of gases which do not participate in the combustion process. This gas jacket hinders the transfer of heat to the cylinder walls and therefore reduces the thermal losses.

Further losses stem from the incomplete combustion of the fuel which has condensed onto the cylinder walls. Thanks to the insulating effects of the gas jacket, these losses are reduced in stratified-charge operation. Further thermal losses result from the residual heat of the exhaust gases.

Losses at $\lambda = 1$

The efficiency of the constant-volume cycle climbs along with increasing excess-air factor (λ). Due to the reduced flame-propagation velocity common to lean air/fuel mixtures, at $\lambda > 1.1$ combustion is increasingly sluggish, a fact which has a negative effect upon the SI engine's efficiency curve. In the final analysis, efficiency is the highest in the range $\lambda = 1.1...1.3$. Efficiency is therefore less for a homogeneous air/fuel-mixture formation with $\lambda = 1$ than it is for an air/fuel mixture featuring excess air. When a 3-way catalytic converter is used for emissions control, an air/fuel mixture with $\lambda = 1$ is absolutely imperative for efficient operation.

Pumping losses

During the exhaust and refill cycle, the engine draws in fresh gas during the 1st (induction) stroke. The desired quantity of gas is controlled by the throttle-valve opening.
A vacuum is generated in the intake manifold which opposes engine operation (throttling losses). Since with a gasoline direct-injection engine the throttle valve is wide open at idle and part load, and the torque is determined by the injected fuel mass, the pumping losses (throttling losses) are lower.

In the 4th stroke, work is also involved in forcing the remaining exhaust gases out of the cylinder.

Frictional losses

The frictional losses are the total of all the friction between moving parts in the engine itself and in its auxiliary equipment. For instance, due to the piston-ring friction at the cylinder walls, the bearing friction, and the friction of the alternator drive.

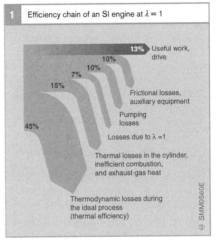

1   Efficiency chain of an SI engine at $\lambda = 1$

45%
15%
7%
10%
10%
13% ▸ Useful work, drive

Frictional losses, auxiliary equipment

Pumping losses

Losses due to $\lambda = 1$

Thermal losses in the cylinder, inefficient combustion, and exhaust-gas heat

Thermodynamic losses during the ideal process (thermal efficiency)

SMM0560E

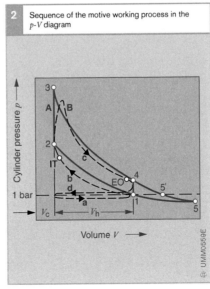

2   Sequence of the motive working process in the p-V diagram

Cylinder pressure p →

Volume V →

V_c V_h

1 bar

UMM0559E

Fig. 2
A   Ideal constant-volume cycle
B   Real p-V diagram

a   Induction
b   Compression
c   Work (combustion)
d   Exhaust

IT   Ignition point
EO   Exhaust valve opens

Specific fuel consumption

Specific fuel consumption b_e is defined as the mass of the fuel (in grams) that the internal-combustion engine requires to perform a specified amount of work (kW·h, kilowatt hours). This parameter thus provides a more accurate measure of the energy extracted from each unit of fuel than the terms liters per hour, litres per 100 kilometers or miles per gallon.

Effects of excess-air factor

Homogeneous mixture distribution
When engines operate on homogeneous induction mixtures, specific fuel consumption initially responds to increases in excess-air factor λ by falling (Fig. 1). The progressive reductions in the range extending to $\lambda = 1.0$ are explained by the incomplete combustion that results when a rich air/fuel mixture burns with inadequate air.

The throttle plate must be opened to wider apertures to obtain a given torque during operation in the lean range ($\lambda > 1$). The resulting reduction in throttling losses combines with enhanced thermodynamic efficiency to furnish lower rates of specific fuel consumption.

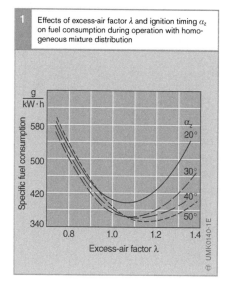

1 Effects of excess-air factor λ and ignition timing α_z on fuel consumption during operation with homogeneous mixture distribution

$\frac{g}{kW \cdot h}$

Specific fuel consumption

580
500
420
340

α_z
20°
30°
40°
50°

0.8   1.0   1.2   1.4

Excess-air factor λ

UMK0140-1E

As the excess-air factor is increased, the flame front's propagation rate falls in the resulting, progressively leaner mixtures.
The ignition timing must be further advanced to compensate for the resulting lag in ignition of the combustion mixture.

As the excess-air factor continues to rise, the engine approaches the lean-burn limit, where incomplete combustion takes place (combustion miss). This results in a radical increase in fuel consumption. The excess-air factor that coincides with the lean-burn limit varies according to engine design.

Stratified-charge concept
Engines featuring direct gasoline injection can operate with high excess-air factors in their stratified-charge mode. The only fuel in the combustion chamber is found in the stratification layer immediately adjacent to the tip of the spark plug. The excess-air factor within this layer is approximately $\lambda = 1$.

The remainder of the combustion chamber is filled with air and inert gases (exhaust-gas recirculation). The large throttle-plate apertures available in this mode lead to a reduction in pumping losses. This combines with the thermodynamic benefits to provide a substantial reduction in specific fuel consumption.

Effects of ignition timing

Homogeneous mixture distribution
Each point in the cycle corresponds to an optimal phase in the combustion process with its own defined ignition timing (Fig. 1). Any deviation from this ignition timing will have negative effects on specific fuel consumption.

Stratified-charge concept
The range of possibilities for varying the ignition angle is limited on direct-injection gasoline engines operating in the stratified-charge mode. Because the ignition spark must be triggered as soon as the mixture cloud reaches the spark plug, the ideal ignition point is largely determined by injection timing.

Achieving ideal fuel consumption

During operation on homogeneous induction mixtures, gasoline engines must operate on a stoichiometric air/fuel ratio of $\lambda = 1$ to create an optimal operating environment for the 3-way catalytic converter. Under these conditions using the excess-air factor to manipulate specific fuel consumption is not an option. Instead, the only available recourse is to vary the ignition timing. Defining ignition timing always equates with finding the best compromise between maximum fuel economy and minimal levels of raw exhaust emissions. Because the catalytic converter's treatment of toxic emissions is very effective once it is hot, the aspects related to fuel economy are the primary considerations once the engine has warmed to normal operating temperature.

Fuel-consumption map

Testing on an engine dynamometer can be used to determine specific fuel consumption in its relation to brake mean effective pressure and to engine speed. The monitored data are then entered in the fuel consumption map (Fig. 2). The points representing levels of specific fuel consumption are joined to form curves. Because the resulting graphic portrayal resembles a sea shell, the lines are also known as shell or conchoid curves.

As the diagram indicates, the point of minimum specific fuel consumption coincides with a high level of brake mean effective pressure p_{me} at an engine speed of roughly 2600 rpm.

Because the brake mean effective pressure also serves as an index of torque generation M, curves representing power output P can also be entered in the chart. Each curve assumes the form of a hyperbola. Although the chart indicates identical power at different engine speeds and torques (operating points A and B), the specific fuel consumption rates at these operating points are not the same. At Point B the engine speed is lower and the torque is higher than at Point A. Engine operation can be shifted toward Point A by using the transmission to select a gear with a higher conversion ratio.

Fig. 2
Engine data:
  4-cylinder gasoline
  engine
Displacement:
  $V_H = 2.3$ litres
Power:
  $P = 110\,kW$ at
  5400 rpm
Torque peak:
  $M = 220\,N \cdot m$ at
  3700...4500 rpm
Brake mean effective
pressure:
  $p_{me} = 12\,bar$ (100 %)

Calculating torque M
and power P with
numerical value
equations:
  $M = V_H \cdot p_{me}/0.12566$
  $P = M \cdot n/9549$

M in N·m
V_H in dm³
p_{me} in bar
n in rpm
P in kW

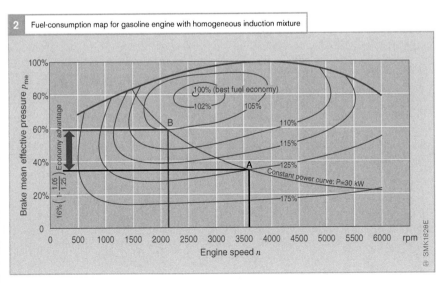

2 Fuel-consumption map for gasoline engine with homogeneous induction mixture

Combustion knock

Among the factors imposing limits on the latitude for enhancing an engine's thermodynamic efficiency and increasing power-plant performance are spontaneous pre-ignition and detonation. This highly undesirable phenomenon is frequently accompanied by an audible "pinging" noise, which is why the generally applicable term for this condition is "knock". Knock occurs when portions of the mixture ignite spontaneously before being reached by the flame front. The intense heat and immense pressure peaks produced by combustion knock subject pistons, bearings, cylinder head and head gasket to enormous mechanical and thermal loads. Extended periods of knock can produce blown head gaskets, holed piston crowns and engine seizure, and leads to destruction of the engine.

The sources of combustion knock

The spark plug ignites the air/fuel mixture toward the end of the compression stroke, just before the piston reaches Top Dead Centre (TDC). Because several milliseconds can elapse until the entire air/fuel mixture can ignite (the precise ignition lag varies according to engine speed), the actual combustion peak occurs after TDC.

The flame front extends outward from the spark plug. After being compressed during the compression stroke, the induction mixture is heated and pressurized as it burns within the combustion chamber. This further compresses any unburned air/fuel mixture within the chamber. As a result, some portions of the compressed air/fuel mixture can attain temperatures high enough to induce spontaneous auto-ignition (Fig. 1). Sudden detonation and uncontrolled combustion are the results.

When this type of detonation occurs it produces a flame front with a propagation rate 10 to 100 times that associated with the normal combustion triggered by the spark plug (approximately 20 m/s). This uncontrolled combustion generates pressure pulses which spread out in circular patterns from the core of the process. It is when these pulsations impact against the walls of the cylinder that they generate the metallic pinging sound typically associated with combustion knock.

Other flame fronts can be initiated at hot spots within the combustion chamber. Among the potential sources of this hot-spot ignition are spark plugs which during operation heat up excessively due to their heat range being too low. This type of pre-ignition produces engine knock by initiating combustion before the ignition spark is triggered.

Engine knock can occur throughout the engine's speed range. However, it is not possible to hear it at extremely high rpm, when its sound is obscured by the noise from general engine operation.

Factors affecting tendency to knock

Substantial ignition advance: Advancing the timing to ignite the mixture earlier produces progressively higher combustion-chamber temperatures and correspondingly extreme pressure rises.

High cylinder-charge density: The charge density must increase as torque demand rises (engine load factor). This leads to high temperatures during compression.

Fuel grade: Because fuels with low octane ratings furnish only limited resistance to knock, compliance with manufacturer's specifications for fuel grade(s) is vital.

Excessively high compression ratio: One potential source of excessively high compression would be a cylinder head gasket of less than the specified thickness. This leads to higher pressures and temperatures in the air/fuel mixture during compression. Deposits and residue in the combustion chamber (from ageing, etc.) can also produce a slight increase in the effective compression ratio.

Cooling: Ineffective heat dissipation within the engine can lead to high mixture temperatures within the combustion chamber.

Geometry: The engine's knock tendency can be aggravated by unfavorable combustion-chamber geometry. Poor turbulence and swirl characteristics caused by unsatisfactory intake-manifold tract configurations are yet another potential problem source.

Engine knock with direct gasoline injection

With regard to engine knock, when operating with homogeneous air/fuel mixtures direct-injection gasoline engines behave the same as manifold-injected power plants. One major difference is the cooling effect exerted by the evaporating fuel during direct injection, which reduces the temperature of the air within the cylinder to levels lower than those encountered with manifold injection.

During operation in the stratified-charge mode it is only in the area immediately adjacent to the spark plug tip that an ignitable mixture is present. When the remainder of the combustion chamber is filled with air or inert gases, there is no danger of spontaneous ignition and engine knock. Nor is there any danger of detonation when an extremely lean air/fuel mixture is present within these outlying sections of the combustion chamber. The ignition energy required to generate a flame in this kind of lean mixture would be substantially higher than that needed to spark a stoichiometric combustion mixture. This is why stratified-charge operation effectively banishes the danger of engine knock.

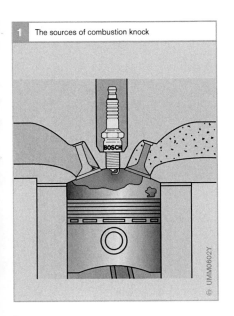

1   The sources of combustion knock

UMM0602Y

Avoiding consistent engine knock

To effectively avoid pre-ignition and detonation, ignition systems not equipped with knock detection rely on ignition timing with a safety margin of 5...8 degrees (crankshaft) relative to the knock limit.

Ignition systems featuring knock detection employ one or several knock sensors to monitor acoustic waves in the engine. The engine-management ECU detects knock in individual combustion cycles by analysing the electrical signals relayed by these sensors. The ECU then responds by retarding the ignition timing for the affected cylinder to prevent continuous knock. The system then gradually advances the ignition timing back toward its original position. This progressive advance process continues until the ignition timing is either back at the initial reference point programmed into the engine's software map, or until the system starts to detect knock again. The engine management regulates the timing advance for each cylinder individually.

The limited number of combustion events with mild knock of the kind that also occur with knock control are not injurious to the health of the engine. On the contrary: They help dissolve deposits formed by oil and fuel additives within the combustion chamber (on intake and exhaust valves, etc.), allowing them to be combusted and/or discharged with the exhaust gases.

Advantages of knock control

Thanks to reliable knock recognition, engines with knock control can use higher compression ratios. Co-ordinated control of the ignition's timing advance also makes it possible to do without the safety margin between the timing point and the knock threshold; the ignition timing can be selected for the "best case" instead of the "worst case" scenario. This provides benefits in terms of thermodynamic efficiency. Knock control

- reduces fuel consumption,
- enhances torque and power, and
- allows engine operation on different fuels within an extended range of octane ratings (both premium and regular unleaded, etc.).

Fuels

The most important energy source from which fuels are extracted is petroleum or crude oil. Crude oil was formed over millions of years from the remains of decomposed living organisms and is made up of many different hydrocarbons. High-quality fuels make an important contribution to trouble-free vehicle operation and to low exhaust-gas emissions. The composition and properties of fuels are therefore governed by legal provisions.

Fuels for spark-ignition engines (gasolines)

Components of gasolines

Gasolines are composed primarily of paraffins and aromatics. Their basic properties can be improved by the use of organic components containing oxygen and additives.

Paraffins with a pure chain structure (standard paraffins, Fig. 1) demonstrate very good ignition performance but also very low knock resistance. A chain molecule structure with additional side chains (iso paraffins) or a benzene ring as the skeletal structure (aromatics) give rise to fuel components with high knock resistance.

The following different fuels are sold in Germany: Normal, Super and Super Plus. "Regular" and "Premium" are sold in the United States; these fuels are roughly comparable with Normal and Super respectively in Germany. Super or Premium gasolines, thanks to their higher aromatic content and the addition of components containing oxygen, demonstrate higher knock resistance and are preferred for use in higher-compression engines.

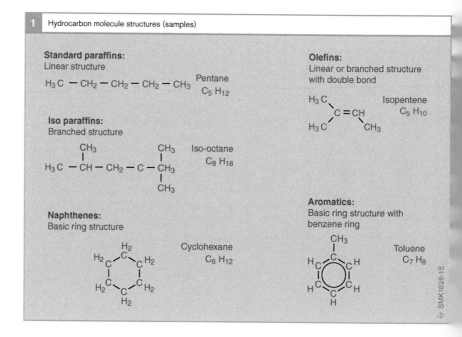

1 Hydrocarbon molecule structures (samples)

Processing of gasolines

Crude oil cannot be used directly in gasoline engines, but must be processed first in refineries. Primarily the following processes are used to refine crude oil:

- Hydrocarbon mixtures are separated by distillation in accordance with the boiling characteristics into fractions, i.e., groups of different molecule sizes
- Larger hydrocarbon molecules deriving from distillation are split up by cracking into smaller molecules
- The molecular structure of hydrocarbons is altered by reforming, e.g., paraffins can be converted into higher-octane aromatics
- Unwanted constituents are removed from hydrocarbons in the refining process (e.g., desulfurization of sulfurous components)

Fuel standards

The European standard EN 228 (Table 1) defines the requirements for unleaded gasoline for use in spark-ignition engines. Further, country-specific characteristic values are set out in the national appendices to this standard. Leaded gasolines are now prohibited in Europe.

The US specifications defining fuels for spark-ignition engines are contained in ASTM D 4814 (American Society for Testing and Materials).

Characteristic quantities

Calorific value

Normally the net calorific value H_n is specified for the energy content of fuels; it corresponds to the usable heat quantity released during full combustion. The gross calorific value H_g, on the other hand, specifies the total reaction heat released and therefore comprises as well as the usable heat the latent heat in the water vapor created. However, this component is not used in the vehicle. The net calorific value of gasoline is 40.1...41.8 MJ/kg.

Oxygenates, i.e. fuels or fuel constituents containing oxygen, such as alcohol fuels, ether, or fatty-acid methyl ester, have a lower calorific value than pure hydrocarbons because the oxygen bonded in them does not contribute to the combustion process. Comparable engine power with conventional fuels therefore results in higher fuel consumption.

Calorific value of air/fuel mixture

The calorific value of the combustible air/fuel mixture determines engine power output. With a stoichiometric air/fuel ratio, this is roughly 3.5...3.7 MJ/m³ for all liquid fuels and liquified petroleum gases.

Density

European standard EN 228 limits the density of gasolines to 720...775 kg/m³. Because premium fuels generally include a higher proportion of aromatic compounds, they are denser than regular gasoline, and also have a slightly higher calorific value.

Knock resistance (octane number)

The octane number defines the knock resistance of a particular gasoline. The higher the octane number, the greater the resistance to engine knock. Iso-octane (trimethyl pentane), which is extremely knock-resistant, is assigned the octane number 100, while n-heptane, which is extremely knock-susceptible, is assigned the number 0.

The octane number of a fuel is determined in a standardized test engine: The numerical value corresponds to the proportion (percent by volume) of iso-octane in an iso-octane/ n-heptane mixture which demonstrates the same knock characteristics as the fuel to be tested.

RON, MON
The number determined in testing using the Research Method is the Research Octane Number, or RON. It serves as the essential index of acceleration knock. The Motor Octane Number, or MON, is derived from testing according to the Motor Method. The MON basically provides an indication of the tendency to knock at high speeds.

The Motor Method differs from the Research Method by using preheated mixtures, higher engine speeds and variable ignition timing, thereby placing more stringent thermal demands on the fuel under examination. MON figures are lower than those for RON.

Enhancing knock resistance
Normal (untreated) straight-run gasoline has only limited resistance to knock. Only by mixing such gasoline with different knock-resistant refinery components (rexformed components, isomerisates) is it possible to produce high-octane fuels for modern engines. It is also important to maintain the

highest possible octane level throughout the entire boiling range. It is possible to increase knock resistance to good effect by adding components containing oxygen. Ethers (e.g., MTBE: methyl tertiary butyl ether, ETBE: ethyl tertiary butyl ether, 3...15%) and alcohols (methanol, ethanol) are used. The following are permitted by way of example: in Europe E5 (max. 5% ethanol), in the USA E10 and in Brazil E22...E26.

However, adding alcohols can also give rise to difficulties. Alcohols increase volatility, can damage the materials used in the fuel-injection equipment, and may cause, for example, elastomer swelling and corrosion.

Octane-number improvers (e.g., MMT: methylcyclopentadienyl manganese tricarbonyl) form ashes during combustion and are therefore used only occasionally (e.g., in Canada).

Volatility
The volatility of gasoline has both upper and lower limits. On the one hand, they must contain an adequate proportion of highly

1	Essential properties of gasolines, EN 228 (March 2004)		
Requirements		**Unit**	**Parameter**
Knock resistance			
Super/Premium, min.		RON/MON	95/85
Normal/Regular, min. [1]		RON/MON	91/82.5
Super Plus [1]		RON/MON	98/88
Density		kg/m³	720...775
Sulfur, max.		mg/kg	50
Benzene, max.		% vol.	1
Lead, max.		mg/l	5
Volatility			
Summer vapor pressure, min./max.		kPa	45/60
Winter vapor pressure, min./max.		kPa	60/90 [1]
Evaporated volume at 70°C in summer, min./max.		% vol.	20/48
Evaporated volume at 70°C in winter, min./max.		% vol.	22/50
Evaporated volume at 100°C, min./max.		% vol.	46/71
Evaporated volume at 150°C, min./max.		% vol.	75/–
Final boiling point, max.		°C	210
VLI transition time [3], max. [2]			1150 [1]

[1] National values for Germany,
[2] VLI = Vapor-Lock Index,
[3] Spring and fall.

volatile components to ensure reliable cold starting. At the same time, volatility should not be so high as to lead to starting and performance problems during operation in high-temperature environments ("vapor lock"). Still another factor is environmental protection, which demands that evaporative losses be kept low.

Fuel volatility is defined by different characteristic quantities. EN 228 defines 10 different volatility classes distinguished by various levels of boiling curve, vapor pressure and VLI (Vapor-Lock Index). To meet special requirements stemming from variations in climatic conditions, countries can incorporate specific individual classes into their own national appendices to the standard. Different values are laid down for summer and winter.

Boiling curve
In order to assess the fuel with regard to its performance, it is necessary to view the individual areas of the boiling curve separately. EN 228 therefore contains limit values laid down for the volume of fuel that vaporizes at 70 °C, at 100 °C and at 150 °C. The volume of fuel that vaporizes up to 70 °C must achieve a minimum volume in order to ensure that the engine starts easily when cold (previously important for vehicles with carburetors). However, the volume of fuel that vaporizes must not be too great either because otherwise vapor bubbles may be formed when the engine is hot. While the volume of fuel that vaporizes up to 100 °C determines the engine's warm-up characteristics, this factor's most pronounced effects are reflected in the acceleration and response provided by the engine once it warms to normal operating temperature. The volume of fuel that vaporizes up to 150 °C should be high enough to minimize dilution of the engine oil. Especially when the engine is cold, the non-volatile gasoline components find it difficult to vaporize and can pass from the combustion chamber via the cylinder walls into the engine oil.

Vapor pressure
Fuel vapor pressure as measured at 37.8 °C (100 °F) in accordance with EN 13016-1 is primarily an index of the safety with which the fuel can be pumped into and out of the vehicle's tank. Vapor pressure has upper and lower limits in all specifications. In Germany, for example, it is max. 60 kPa in summer and max. 90 kPa in winter.

In order to configure a fuel-injection system, it is also important to know the vapor pressure at higher temperatures (80...100 °C) because a rise in the vapor pressure due to the admixture of alcohol, for example, becomes apparent particularly at higher temperatures. If the fuel vapor pressure rises above the fuel-injection system pressure for example during vehicle operation due to the effect of the engine temperature, this may result in malfunctions caused by vapor-bubble formation.

Vapor/liquid ratio
The vapor/liquid ratio is a measure of a fuel's tendency to form bubbles. It refers to the volume of vapor generated by a specific quantity of fuel at a defined back pressure and a defined temperature.

If the back pressure drops (e.g., when driving over a mountain pass) and/or the temperature rises, this will raise the vapor/liquid ratio and with it the probability of operating problems. ASTM D 4814 lays down for example for each volatility class a temperature at which a vapor/liquid ratio of 20 must not be exceeded.

Vapor-lock index (VLI)
The vapor-lock index is the mathematically calculated sum total of ten times the vapor pressure (in kPa at 37.8 °C) and seven times the volume of fuel that vaporizes up to 70 °C. The properties of the fuel in terms of hot-starting and hot-running performance are described better with the additional quantity VLI than by the characteristic vapor-pressure and boiling values alone.

Sulfur content

In the interests of reducing SO_2 emissions and protecting the catalytic converters for exhaust-gas treatment, the sulfur content of gasolines will be limited on a Europe-wide basis to 10 mg/kg as from 2009. Fuels which adhere to this limit value are known as "sulfur-free fuels". These have already been introduced in Germany because a penalty tax has been levied on sour fuels in that country since 2003.

Sulfur-free fuels are being introduced gradually throughout Europe. Since 1.1.2005 only low-sulfur fuels (sulfur content < 50 mg/kg) have been permitted to be brought onto the market.

In the USA, the limit value for the sulfur content of gasolines commercially available to the end user is currently set at max. 300 mg/kg, although an upper average value of 30 mg/kg for the total amount of sold and imported fuel has been in place since the start of 2005. As of 2006, while the average value of 30 mg/kg is maintained, the limit will be reduced to max. 80 mg/kg. Individual states, California for example, have laid down lower limits. The sulfur content of certification fuels is subject to separate regulation.

Additives

Additives can be added to improve fuel quality in order to counteract deteriorations in engine performance and in the exhaust-gas composition during vehicle operation. The packages generally used combine individual components with various attributes.

Extreme care and precision are required both when testing additives and in determining their optimal concentrations. Undesirable side-effects must be avoided. They are usually added to the individually branded fuels at the refinery's filling stations when the road tankers are filled (end-point dosing). Vehicle operators should refrain from adding supplementary additives of their own into the vehicle fuel tanks as this would invalidate the manufacturer's warranty.

Detergent additives

The entire intake system (fuel injectors, intake valves) should remain free of contamination and deposits for several reasons. A clean intake tract is essential for maintaining the factory-defined air/fuel ratios, as well as for trouble-free operation and minimal exhaust emissions. To achieve this end, effective detergent additives should be added to the fuel.

Corrosion inhibitors

The ingress of water/moisture may lead to corrosion in fuel-system components. Corrosion is effectively eliminated by the addition of corrosion inhibitors, which form a thin protective film on the metal surface.

Oxidation stabilizers

Anti-aging agents (antioxidants) are added to fuels to improve their stability during storage. They prevent oxidation caused by oxygen in the air. Metal deactivators prevent a catalytic influence by metal ions on fuel aging.

Reformulated gasoline

Reformulated gasoline is the term used to describe gasoline which, through its altered composition, generates fewer evaporative and pollutant emissions than conventional gasoline. The demands placed on reformulated gasoline are laid down in the US Clean Air Act of 1990. This legislation prescribes, for example, lower values for vapor pressure, aromatic and benzene content, and final boiling point. It also prescribes the use of additives to keep the intake system free of contamination and deposits.

Alternative fuels

As well as the processes for producing gasolines and diesel fuels, there are different
technical formulations for producing alternative fuels from different sources of energy.
The main creation and conversion processes
are shown in Figure 2. The complete journey
fuel takes in the course of its production and
provision – from primary-energy extraction
through to its introduction in the vehicle's
fuel tank – is known as the "well to tank"
path. In order to evaluate the different fuel
options with regard to CO_2 emissions and
energy balance, it is necessary not only to
include this entire path but also to take into
account the efficiency of the respective vehicle drive system as it is this latter factor
which determines fuel consumption. It is
not enough simply to evaluate the combustion of the fuel.

In this context, a distinction is made between fossil fuels, which are produced on the
basis of crude oil or natural gas, and regenerative fuels, which are created from renewable sources of energy, such as biomass,
wind power or solar power.

Alternative fossil fuels include liquified
petroleum gas, natural gas, synfuels (synthetic liquid fuels) created from natural gas,
and hydrogen produced from natural gas.

Regenerative fuels include methane,
methanol and ethanol, provided these fuels
are created from biomass. Further biomass-
based regenerative fuels are sunfuels (synthetic liquid fuels) and biodiesel. Hydrogen
extracted by electrolysis is then classed as
regenerative if the current used comes from
renewable sources (wind energy, solar energy). Biomass-based regenerative hydrogen
can also be produced.

With the sole exception of hydrogen, all
regenerative and fossil fuels contain carbon
and therefore release CO_2 during combustion. In the case of fuels produced from
biomass, however, the CO_2 absorbed by
the plants as they grow is offset against the
emissions produced during combustion.
The CO_2 emissions to be attributed to
combustion are thereby reduced.

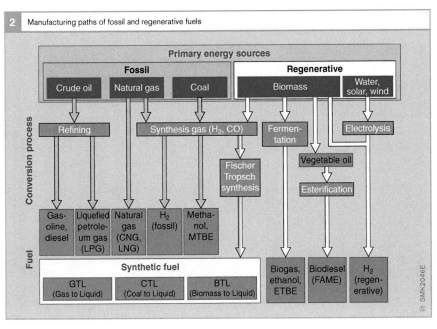

2   Manufacturing paths of fossil and regenerative fuels

Alternative fuels for spark-ignition engines

Natural gas and liquified petroleum gas are primarily used as alternative fuels in spark-ignition engines. Spark-ignition engines that run on hydrogen are currently restricted to test vehicles. Alcohols are mainly used in Europe and the US as gasoline additives. In Brazil, pure ethanol is also used as a fuel.

Synthetic fuels are used exclusively in diesel engines.

To enable engines to run on many of the alternative fuels mentioned, it may be necessary to adapt the fuel-injection components and where required the vehicle engine and the fuel tank. Today, more and more vehicle manufacturers are offering natural-gas vehicles straight off the production lines. Bivalent vehicles are primarily used here, i.e., the driver can switch between gasoline and gas operation.

Natural gas (CNG, LNG)

The primary component of natural gas is methane (CH_4), which is present in proportions of 80...99%. Further components are inert gases, such as carbon dioxide, nitrogen and low-chain hydrocarbons.

Natural gas is stored either in gas form as Compressed Natural Gas (CNG) at a pressure of 200 bar or as a liquified gas (LNG: Liquefied Natural Gas) at −162 °C in a cold-resistant tank. LNG requires only one third of the storage volume of CNG, however the storage of LNG requires a high expenditure of energy for cooling. For this reason, natural gas is offered almost exclusively as CNG at the roughly 550 natural-gas filling stations in Germany today.

Natural-gas vehicles are characterized by low CO_2 emissions, due to the lower proportion of carbon in natural gas. The hydrogen-carbon ratio of natural gas stands at approx. 4:1, that of gasoline, on the other hand, is 2.3:1. Thus, the process of burning natural gas produces less CO_2 and more H_2O. A spark-ignition engine converted to natural gas – without any further optimizations – already produces roughly 25% fewer CO_2 emissions than a gasoline engine.

Because of the extremely high knock resistance of natural gas of up to 130 RON (gasoline 91...100 RON), the natural-gas engine is ideally suited for turbocharging and enables the compression ratio to be increased. In this way, it is possible in conjunction with a downsizing concept (reduction of displacement) to improve engine efficiency and further reduce CO_2 emissions.

Liquefied petroleum gas (LPG)

Liquefied Petroleum Gas (LPG) is primarily a mixture of propane and butane and is used to a limited extent as a fuel for motor vehicles. It is a by-product of the crude-oil-refining process and can be liquified under pressure.

The demands placed on LPG for use in motor vehicles are laid down in the European standard EN 589. The octane number MON is at least 89.

CO_2 emissions from an LPG engine are roughly 10% lower than from a gasoline engine.

Alcohol fuels

Specially adapted spark-ignition engines can be run on pure methanol (M100) or ethanol (E100). These alcohols are, however, mostly used as fuel components for increasing the octane number (e.g., E24 in Brazil and E10, E85, M85 in the USA). Even the ethers that can be manufactured from these alcohols MTBE (methyl tertiary butyl ether) and ETBE (ethyl tertiary butyl ether) are important octane-number improvers.

Ethanol, because of its biogenous origin, has become a highly significant alternative fuel in some countries, above all in Brazil (manufactured by fermentation of sugar cane) and the USA (from wheat).

Methanol can be manufactured from readily available natural hydrocarbons found in plentiful substances such as coal, natural gas, heavy oils, etc.

Compared with petroleum-based fuels, alcohols have different material properties (calorific value, vapor pressure, material resistance, corrosivity, etc.), which must be taken into consideration with respect to design.

Engines with can burn gasolines and alcohols in any mixture ratio without the driver having to intervene are used in "flexible fuel" vehicles.

Hydrogen

Hydrogen can be used both in fuel-cell drives and directly in internal-combustion engines. CO_2 advantages are enjoyed, particularly when the hydrogen is created regeneratively by electrolysis from water or from biomass. Today, however, hydrogen is predominantly obtained on a major industrial scale by means of steam reforming from natural gas, in the course of which CO_2 is released.

Even the distribution and storage of hydrogen is still technically complex and expensive today. Because of its low density, hydrogen is mainly stored in one of two ways:
- Pressure storage at 350 bar or 700 bar; at 350 bar, the storage volume referred to the energy content is 10 times greater than with gasoline.
- Liquid storage at a temperature of −253 °C (cryogenic storage); this gives rise to four times the tank volume of gasoline.

Electric drive with fuel-cell power supply
The fuel cell converts hydrogen with oxygen in the air in a cold-combustion process into electrical current; the only by-product of this process is water vapor. The currents serves to power an electric motor acting as the vehicle drive.

Polymer-electrolyte fuel cells (PEM fuel cells), which operate at relatively low temperatures of 60…100 °C, are primarily used for the vehicle drive. The system efficiency of a hydrogen-fueled PEM fuel cell including electric motor is in the range of 30…40 % (referred to the New European Driving Cycle NEDC) and thus clearly surpasses the typical efficiency of an internal-combustion engine of 18...24 %.

Hydrogen in a spark-ignition engine
Hydrogen is an extremely ignitable fuel. Its very high ignition performance permits a strong leaning of the hydrogen/air mixture up to approx. $\lambda = 4...5$ and thus extensive dethrottling of the engine. The extended ignition limits compared with gasoline, however, also increase the risk of backfiring.

The efficiency of a hydrogen combustion engine is generally higher than that of a gasoline engine, but lower than that of a fuel-cell drive.

The process of burning hydrogen produces water and no CO_2.

Cylinder-charge control systems

In the case of a homogeneously operated gasoline engine with a defined air/fuel ratio λ, the output torque and thus the power is determined by the intake-air mass and the injected fuel quantity. The air mass must be proportioned exactly so that λ can be adhered to precisely.

Electronic throttle control (ETC)

For it to burn, fuel needs oxygen, which the engine takes from the intake air. In engines with external mixture formation (manifold injection), as well as in direct-injection engines operating on a homogeneous mixture, the output torque is directly dependent on the intake-air mass. The engine must therefore be throttled for the purpose of setting a defined air charge.

Function and method of operation

The torque requested by the driver is derived from the position of the accelerator pedal. In the case of the ETC system (Electronic Throttle Control), a position sensor in the accelerator-pedal module (Fig. 1, Pos. 1) records this variable. Further torque requests are derived from functional requests, such as, for example, an additional torque when the air-conditioning system is switched on or a torque reduction during a gearshift.

The Motronic ECU (2) – ME-Motronic for systems with manifold injection or DI-Motronic for gasoline direct injection – calculates the required air mass from the torque to be set and generates the triggering signals for the electrically actuated throttle valve (5). In this way, the opening cross-section and thus the air-mass flow inducted by the gasoline engine are set. Using the feedback information from the throttle-valve-angle sensor (3) regarding the current position of the throttle valve, it then becomes possible to adjust the throttle valve precisely to the required setting.

A cruise-control function can also be easily integrated with ETC. The ECU adjusts the torque in such a way that the vehicle speed preselected at the control element for cruise control is maintained. There is no need to press the accelerator pedal.

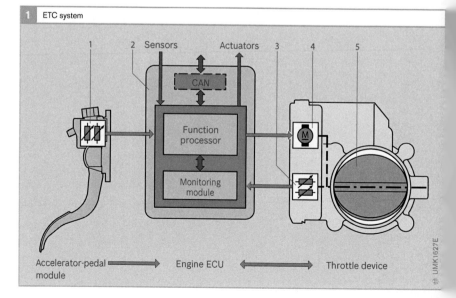

1   ETC system

Fig. 1
1   Pedal-travel sensor
2   Motronic ECU
3   Throttle-valve-angle sensor
4   Throttle-valve drive
5   Throttle valve

Throttle device

The throttle device (Fig. 2) consists of a housing (1), in which the rotating throttle valve (2) is mounted. The DC motor (3) drives the shaft of the throttle valve via a gear unit (5). The shaft connects two throttle-valve-angle sensors (ETC monitoring concept). Potentiometers are used in the DV-E5 throttle device. Alternatively, proximity-type sensors are also used in the DV-E8 version. All the connections are connected by way of a plug to the vehicle wiring harness.

The throttle device is assembled in accordance with a modular principle. This modular design enables it to be easily adapted to the relevant requirements – for example, the air requirement dependent on the swept volume of the engine cylinders.

The use of plastic in the DV-E8 offers the following advantages over the aluminum housing of the DV-E5:

- Weight saving
- Optimal throttle-valve geometry
- Corrosion resistance
- Low wear
- Less sensitivity to temperature influences
- Less tendency to icing (omission of water heater)

Accelerator-pedal sensors

In Motronic systems with Electronic Throttle Control (ETC), the pedal-travel sensor records the travel or the angular position of the accelerator pedal. For this purpose, potentiometers are used in addition to proximity-type sensors.

The pedal-travel sensor is integrated together with the accelerator pedal in the accelerator-pedal module. These ready-to-install units make adjustments on the vehicle a thing of the past.

Potentiometric pedal-travel sensor

The engine ECU receives the measured value picked off at the potentiometer wiper as a voltage. The ECU uses a stored sensor curve to convert this voltage into the relative pedal travel or the angular position of the accelerator pedal (Fig. 3).

A second (redundant) sensor is incorporated for diagnosis purposes and for use in case of malfunctions. It is a component part of the monitoring system. One sensor version operates with a second potentiometer, which always delivers half the voltage of the first potentiometer at all operating points. Thus, two independent signals are available for fault-detection purposes (Fig. 3). Instead of the second potentiometer, another version uses a low-idle switch, which signals the

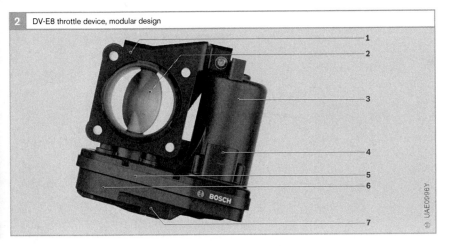

2   DV-E8 throttle device, modular design

UAE0996Y

Fig. 2
1   Pneumatic housing
2   Throttle valve
3   DC motor
4   Plug module
5   Gear-unit housing
6   Integrated throttle-valve-angle sensor
7   Cover module

idle position to the ECU. The status of this switch and the potentiometer voltage must be plausible.

For vehicles with automatic transmissions, a further switch can be incorporated for an electrical kickdown signal. Alternatively, this information can also be derived from the rate of change of the potentiometer voltage. A further possibility is to trigger the kickdown function by means of a defined voltage value of the sensor curve; here, the driver receives feedback on a jump in force in a mechanical kickdown cell. This is the most frequently used solution.

Hall-effect angle-of-rotation sensors

Hall-effect sensors are used to measure the movement of the accelerator pedal on a non-contact basis. In the case of the Type ARS1 Hall-effect angle-of-rotation sensor, the magnetic flux of a roughly semicircular, permanent-magnetic disk is fed back via a pole shoe, two further conductive elements and the similarly ferromagnetic shaft to the magnet (Fig. 4). Depending upon the angular setting, the flux is led to a greater or lesser degree through the two conductive elements, in the magnetic path of which a Hall-effect sensor is also situated. Using this principle, it is possible to achieve a practically linear characteristic in the measuring range of 90°.

The Type ARS2 is a simplified version which does without soft magnetic conductive elements (Fig. 5). In this version, the magnet moves around the Hall-effect sensor in a circular arc. Only a relatively small section of the resulting sinusoidal characteristic curve features good linearity. If the Hall-effect sensor is located slightly outside the center of the circular arc, the characteristic curve increasingly deviates from the sinusoidal, and now features a short measuring range of almost 90°, and a longer measuring range of more than 180° with good linearity.

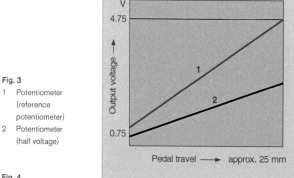

3   Characteristic curve of a pedal-travel sensor

Output voltage → (V)
4.75
0.75
Pedal travel → approx. 25 mm

1
2

UAE0724E

Fig. 3
1 Potentiometer
   (reference
   potentiometer)
2 Potentiometer
   (half voltage)

Fig. 4
1 Rotor disc
   (permanent-
   magnetic)
2 Pole shoe
3 Conductive element
4 Air gap
5 Hall-effect sensor
6 Shaft (soft magnetic)

Fig. 5
a Principle of operation
b Characteristic curve

1 Hall IC positioned
   in the mid-point of
   the circular path
2 Hall IC located
   outside the mid-
   point (linearization)
3 Magnet

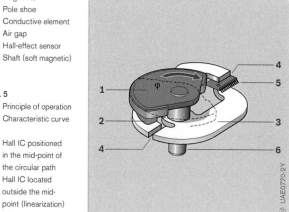

4   ARS1 Hall-effect angle-of-rotation sensor

1
2
4

4
5
3
6

UAE0770-2Y

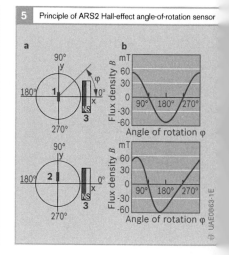

5   Principle of ARS2 Hall-effect angle-of-rotation sensor

a

90°
y
180°   0°
x
270°
1
3

b

Flux density B (mT)
60
30
0
-30
-60
90°   180°   270°
Angle of rotation φ

90°
y
180°   0°
x
270°
2
3

Flux density B (mT)
60
30
0
-30
-60
90°   180°   270°
Angle of rotation φ

UAE0863-1E

A great disadvantage though is the low level of shielding against external fields, as well as the remaining dependence on the geometric tolerances of the magnetic circuit, and the intensity fluctuations of the magnetic flux in the permanent magnet as a function of temperature and age.

In the case of the FPM2.3 Type Hall-effect angle-of-rotation sensor, it is not the field strength but rather the direction of the magnetic field which is used to generate the output signal. The field lines are recorded by four radially arranged measuring elements lying in one plane in the x- and y-directions (Fig. 6). The output signals are derived in the ASIC from the raw data (cos and sin signals) using the arctan function. The sensor is positioned between two magnets to generate a homogenous magnetic field. The sensor is therefore insensitive to component tolerances and temperature-resistant.

As with the accelerator-pedal module with a potentiometric sensor, these proximity-type systems also contain two sensors in order to receive two redundant voltage signals.

ETC monitoring concept

The ETC system is classified as a safety-related system. The engine-management system therefore contains the facility for diagnosing the individual components. Input information representing the power-determining driver command (accelerator-pedal position) or the engine status (throttle-valve position) is directed to the ECU by two sensors (redundancy). The two sensors in the accelerator-pedal module and the two sensors in the throttle device supply signals that are independent of each other to such an extent that, if one signal should fail, the other signal supplies a valid value. Different characteristic curves ensure that a short circuit between the two signals is detected.

Fig. 6
a   Design
b   Principle
c   Measurement
    signals

1   Integrated Circuit
    (IC) with Hall-effect
    elements
2   Magnet (opposing
    magnet not shown
    here)
3   Conductive element
4   Hall-effect elements
    (for recording
    x-component of B)
5   Hall-effect elements
    (for recording
    y-component of B)

B_x Homogenous
    magnetic field
    (x-component)
B_y Homogenous
    magnetic field
    (y-component)

Fig. 7
1   Pedal
2   Cover
3   Spacer sleeve
4   Sensor block with
    housing and plug
5   Bearing block
6   Shaft with two mag-
    nets and hysteresis
    elements (round
    magnets not visible)
7   Kickdown (optional)
8   Two springs
9   Stop damper
10 Thrust member
11 Floor cover

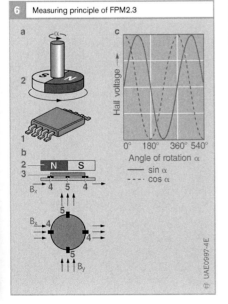

6   Measuring principle of FPM2.3

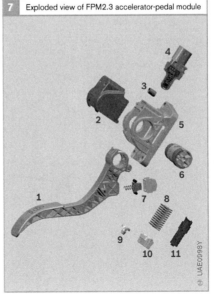

7   Exploded view of FPM2.3 accelerator-pedal module

Variable valve timing

In addition to controlling the fresh-gas flow inducted by the engine via the throttle valve, it is possible to adjust the cylinder charge with the aid of variable valve timing. It is possible to use the variability of valve timing and valve lifts to influence the ratio of fresh gas to residual gas and mixture formation.

Periodic opening and closing of the valves causes the fresh gas to be inducted spasmodically into the combustion chamber and the burnt gases to be forced out of the chamber. This stimulates the gas columns flowing in and out to vibrate. The frequency and amplitude of the vibrations are very much dependent on the geometry of the air duct and the exhaust-system branch, the engine speed, and the throttle-valve position. Valve timing and valve lifts can thus only be adapted to the best possible extent for the gas-exchange process in a specific operating range. An invariant design of the valve gear for a production engine will therefore always constitute a compromise over the entire load and engine-speed range. Variable valve timing facilitates extended adaptation to different operating conditions, from which the following benefits are derived:
● Higher rated power
● Favorable torque curve over the entire engine-speed range
● Reduction of toxic emissions
● Reduced fuel consumption, and
● Improved smooth running at low engine speeds

Camshaft phase adjustment

In today's production engines, the camshafts are driven from the crankshaft via toothed belts, timing chains or gearwheels. Adjusting the camshafts relative to the crankshaft position is becoming increasingly more important. In many applications, only the intake camshaft is designed to be rotated in relation to the crankshaft position. However, the tendency is now towards engines with additionally adjustable exhaust camshafts.

Hydraulically driven actuators, which permit continuous adjustment, are used predominantly in this context.

Rotating the intake camshaft, for example, determines the time of "Intake Opens". Via the fixed cam-lobe profile the same lifting curve and thus also "Intake Closes" are displaced in parallel. Figure 1 shows the displacement of the lifting curve of an intake valve with regard to the top dead center of the piston when the intake camshaft is rotated in the "Retard" or "Advance" direction.

Timing retardation of the intake camshaft
Timing retardation at low engine speeds
Timing retardation of the intake camshaft leads to the intake valve opening later so that valve overlap is reduced. This reduces the amount of burnt gases flowing back via the intake valve into the intake manifold. The lower residual-gas content of the mixture then inducted results at low engine speeds (< 2000 rpm) in more stable combustion and smoother engine operation. The idle speed can be lowered, which in particular reduces the fuel consumption.

Timing retardation at high engine speeds
The camshaft is also retarded at high engine speeds (> 5000 rpm). Because of the associated later closing of the intake valve after BDC, the fresh gas has as much time to flow into the cylinder until the pressure differential is compensated via the intake valve by the continuing upwards stroke of the piston. This boost effect helps to increase the cylinder charge significantly.

Timing advance of the intake camshaft
Timing advance at middle engine speeds
In the middle speed range, there is no boost effect, due to the changed gas dynamics. The early closing of the intake valve shortly after BDC prevents the upward-moving piston under these conditions from forcing the inducted fresh gas back into the intake manifold. This results in the best possible charge and with it a good torque curve.

Timing advance of the intake camshaft also means a larger valve overlap. The early opening of the intake valve before TDC causes burnt residual gas to be forced by the upward stroke of the piston via the open valve into the intake manifold and then inducted again. The upshot of this is an increased proportion of residual gas in the cylinder charge (internal exhaust-gas recirculation). This in turn lowers the peak combustion temperatures and reduces NO_X formation. The higher proportion of inert gas also permits further opening of the throttle valve (dethrottling) and thereby results in a reduction of the fuel consumption associated with throttling losses.

Adjustment of the exhaust camshaft
In addition, it is possible with systems in which the exhaust camshaft is also adjustable to adjust the proportion of residual gas by choosing the point at which the exhaust valve closes. This facilitates a largely isolated adjustment of optimal charge with fresh gas (Intake Closes) and residual-gas content.

Camshaft-lobe control

Camshaft-lobe control (Fig. 2) involves alternating between different cam-lobe profiles. In this way, both the valve-lifting curve and the timing can be varied. Generally, a first cam determines the optimal timing and the valve lift for the lower and middle engine-speed ranges, while at high engine speeds and loads the arrangement switches to a second cam with a greater valve lift and longer opening duration.

Alternation between the different cams is achieved, for example, by bucket tappets or by cutting in an otherwise freely vibrating rocker arm on an additional cam.

Alternation in the lower and middle engine-speed ranges
In the lower and middle engine-speed ranges, the narrower valve gap associated with a small valve lift results in a higher admission speed and better swirling of the charge gas (in the case of gasoline direct injection) or of the air/fuel mixture (in the case of manifold injection). This ensures good mixture formation even at part load.

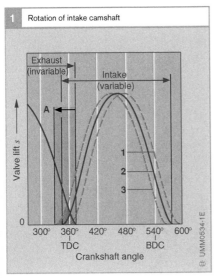

1   Rotation of intake camshaft

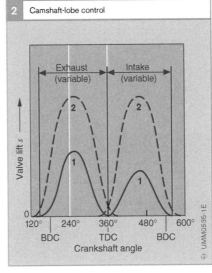

2   Camshaft-lobe control

Fig. 1
1   Rotation in "Retard" direction
2   Normal position
3   Rotation in "Advance" direction

A   Valve overlap

Fig. 2
1   Standard cam
2   Additional cam

Alternation in the upper engine-speed range
A high torque request (full load) calls for
maximum charge. The opening cross-section widened by the large valve lift allows
fresh gas to flow in relatively unhindered
and reduces the pumping losses.

Fully variable valve timing with camshaft

Valve timing in which both the valve lift and
the timing can be continuously varied is
known as fully variable valve gear. Spatial
cam-lobe profiles in conjunction with an axially adjustable camshaft (Fig. 3) or variable
lever-arm geometries can adjust the valve lift
and the opening duration. Additional phase
adjusters determine the phase position. This
fully variable valve timing facilitates extensive dethrottling through specific opening
and closing of the intake valve. This enables
fuel consumption to be further reduced in
comparison with simple variation of the
timing.

Fully variable valve timing without camshaft

The greatest degree of freedom for valve
timing and the greatest potential for reducing consumption are afforded by camshaft-
free systems. Here the valves are operated,
for example, by electromagnetic or electrohydraulic actuators. A supplementary ECU
is responsible for triggering. The purpose of
these camshaft-free fully variable valve gears
is to achieve the best possible dethrottling,
coupled with very low pumping losses.
The extensive variability of the valve gear
provides for the best possible charge and
with it maximum torque as well as improved
mixture preparation. Additional fuel savings
can be achieved by incorporating valve and
cylinder shutoff. However, technical risks
and high costs stand in the way of the general introduction of such systems in the
short term.

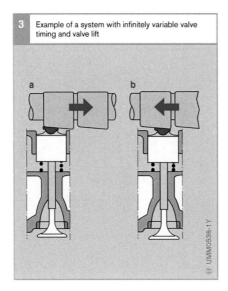

3   Example of a system with infinitely variable valve
timing and valve lift

Fig. 3
a   Minimum lift
b   Maximum lift

Dynamic supercharging

Approximately speaking, the achievable engine torque is proportional to the fresh-gas content in the cylinder charge. This means that the maximum torque can be increased to a certain extent by compressing the air before it enters the cylinder.

The gas-exchange processes are not only influenced by the valve timing, but also by the intake and exhaust lines. The piston's induction work causes the open intake valve to trigger a return pressure wave. At the open end of the intake manifold, the pressure wave encounters the quiescent ambient air from which it is reflected back again so that it returns in the direction of the intake valve. The resulting pressure fluctuations at the intake valve can be utilized to increase the fresh-gas charge and thus achieve the highest possible torque.

This supercharging effect thus depends on utilization of the incoming air's dynamic response. In the intake manifold, the dynamic effects depend upon the geometrical relationships in the intake manifold and on the engine speed.

For the even distribution of the air/fuel mixture, the intake manifolds for carburetor engines and single-point injection must have short pipes which as far as possible must be of the same length for all cylinders. In the case of multi-point injection, the fuel is either injected into the intake manifold onto the intake valve (manifold injection), or it is injected directly into the combustion chamber (gasoline direct injection). The intake manifolds essentially carry air only. This opens up more varied possibilities for intake-manifold layout because practically no fuel can be precipitated on the intake manifolds. This is the reason for there being no problems with multi-point injection systems regarding the even distribution of fuel.

Ram-tube supercharging

The intake manifolds for multi-point injection systems are composed of the individual resonance tubes or runners and the manifold chamber.

In the case of ram-tube supercharging (Fig. 1), each cylinder is allocated its own individual resonance tube (2) of specific length which is usually attached to the manifold chamber (3). The pressure waves are able to propagate in the individual resonance tubes independently.

The supercharging effect depends upon the intake-manifold geometry and the engine speed. For this reason, the length and diameter of the individual resonance tubes are matched to the valve timing so that in the required speed range a pressure wave reflected at the end of the resonance tube is able to enter the cylinder (1) through the open intake valve and improve the cylinder charge. Long, narrow tubes result in a marked supercharging effect at low engine speeds. On the other hand, short, large-diameter tubes have a positive effect on the torque curve at higher engine speeds.

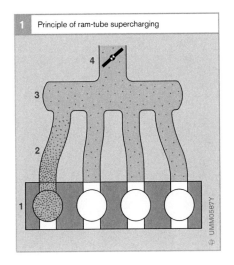

1 Principle of ram-tube supercharging

Fig. 1
1   Cylinder
2   Individual resonance tube
3   Manifold chamber
4   Throttle valve

Tuned-intake pressure charging

At a given engine speed, the periodic piston movement causes the intake-manifold gas column to vibrate at resonant frequency. This results in a further increase of pressure and leads to an additional supercharging effect.

In the tuned-intake-tube system (Fig. 2), groups of cylinders (1) with identical angular ignition spacing are each connected to a resonance chamber (3) through short tubes (2). The chambers, in turn, are connected through tuned intake tubes (4) with either atmosphere or with the manifold chamber (5) and function as Helmholtz resonators. The subdivision into two groups of cylinders each with its own tuned intake tube prevents the overlapping of the flow processes of two neighboring cylinders which are adjacent to each other in the firing sequence.

The length of the tuned intake tubes and the size of the resonance chamber are a function of the speed range in which the supercharging effect due to resonance is required to be at maximum. Due to the accumulator effect of the considerable chamber volumes which are sometimes needed, dynamic-response errors can occur in some cases when the load is changed abruptly.

Variable-geometry intake-manifolds

The supplementary cylinder charge resulting from dynamic supercharging depends on the engine's operating point. The two systems just dealt with increase the achievable maximum charge (volumetric efficiency), above all in the low engine-speed range (Fig. 3).

Practically ideal torque characteristics can be achieved with variable-geometry intake-manifolds in which, as a function of the engine operating point, flaps are used to implement a variety of different adjustments such as:

- Adjustment of the ram-tube length
- Switchover between different ram-tube lengths or different tube diameters
- Selected switchoff of one of the cylinder's intake tubes on multiple-tube systems
- Switchover to different chamber volumes

Electrically or electropneumatically actuated flaps are used for changeover operations in these variable-geometry systems.

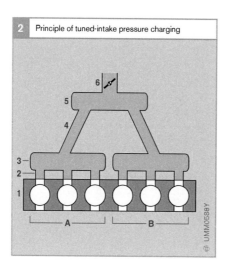

2   Principle of tuned-intake pressure charging

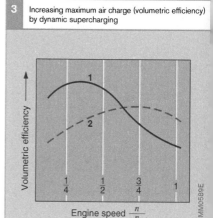

3   Increasing maximum air charge (volumetric efficiency) by dynamic supercharging

Volumetric efficiency →

Engine speed $\frac{n}{n_{rat}}$

Ram-tube systems

The manifold system shown in Fig. 4 can switch between two different ram tubes. In the lower speed range, the changeover flap (1) is closed and the intake air flows to the cylinders through the long ram tube (3). At higher speeds and with the changeover flap open, the intake air flows through the short, wide diameter ram tube (4), and thus contributes to improved cylinder charge at high engine speeds.

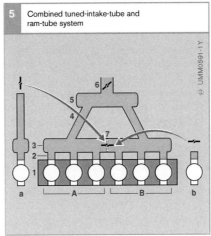

5 Combined tuned-intake-tube and ram-tube system

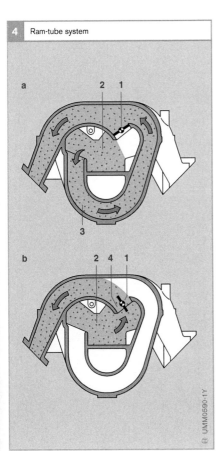

4 Ram-tube system

Tuned-intake-tube systems

Opening the resonance flap switches in a second tuned intake tube. The changed geometry of this configuration has an effect upon the resonant frequency of the intake system. Cylinder charge in the lower speed range is improved by the higher effective volume resulting from the second tuned intake pipe.

Combined tuned-intake-tube and ram-tube system

When design permits the open changeover flap (Fig. 5, Pos. 7) to combine both the resonance chambers (3) to form a single volume, one speaks of a combined tuned-intake-tube and ram-tube system. A single intake-air chamber with a high resonant frequency is then formed for the short ram tubes (2).

The changeover flap is closed at low and middle speeds. The system acts as a tuned-intake-tube system. The low resonant frequency is then defined by the long tuned intake tube (4).

Fig. 5
1   Cylinder
2   Ram tube (short intake tube)
3   Resonance chamber
4   Tuned intake tube
5   Manifold chamber
6   Throttle valve
7   Changeover flap

A   Cylinder group A
B   Cylinder group B

a   Intake-tube conditions with changeover flap closed
b   Intake-tube conditions with changeover flap open

Fig. 4
a   Manifold geometry with changeover flap closed
b   Manifold geometry with changeover flap open

1   Changeover flap
2   Manifold chamber
3   Changeover flap closed: long, narrow-diameter ram tube
4   Changeover flap opened: short, wide-diameter ram tube

Mechanical supercharging

Design and method of operation

The application of supercharging units leads to increased cylinder charge and therefore to increased torque. Mechanical supercharging uses a compressor which is driven directly by the IC engine. Mechanically driven compressors are either positive-displacement superchargers with different types of construction (e.g. Roots supercharger, sliding-vane supercharger, spiral-type supercharger, screw-type supercharger), or they are centrifugal turbo-compressors (e.g. radial-flow compressor). Fig. 1 shows the principle of functioning of the rotary-screw supercharger with the two counter-rotating screw elements. As a rule, engine and compressor speeds are directly coupled to one another through a belt drive.

Boost-pressure control

On the mechanical supercharger, a bypass can be applied to control the boost pressure. A portion of the compressd air is directed into the cylinder and the remainder is returned to the supercharger input via the bypass. The engine management is responsible for controlling the bypass valve.

Advantages and disadvantages

On the mechanical supercharger, the direct coupling between compressor and engine crankshaft means that when engine speed increases there is no delay in supercharger acceleration. This means therefore, that compared to exhaust-gas turbocharging engine torque is higher and dynamic response is better.

Since the power required to drive the compressor is not available as effective engine power, the above advantage is counteracted by a slightly higher fuel-consumption figure compared to the exhaust-gas turbocharger. This disadvantage though is somewhat alleviated when the engine management is able to switch off the compressor via a clutch at low engine loading.

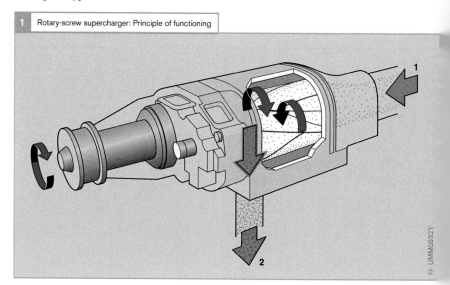

1 Rotary-screw supercharger: Principle of functioning

Fig. 1
1   Intake air
2   Compressed air

At the 1921 Berlin Motor Show, the "Daimler Motoren Gesellschaft" introduced the passenger cars designated Type 6/20 HP and 10/35 HP, each featuring a four-cylinder compressor engine.

1922 marked the first time a vehicle powered by a compressor engine was entered in a car race. At the Targa Florio on the island of Sicily, Max Sailer drove a 28/95 HP Mercedes equipped with a compressor engine to victory in the production-car category.

In 1924 the new Mercedes 15/70/100 HP and 24/100/140 HP passenger cars with six-cylinder compressor engines were introduced at the Berlin Motor Show. On the Avus race track in Berlin that same year, NSU set spectacular speed records and victories with their 5/15 compressor racing car.

In 1927 the Type "S" Mercedes-Benz captured a triple victory at the inaugural race at the new Nuerburgring race track. The championship was won by Rudolf Caracciola. With its 6.8-litre displacement, the 6-cylinder compressor engine of the Type "S" delivered 180 HP and 120 HP without compressor. In 1932 Manfred von Brauchitsch, driving an "SSKL" (abbreviation for the German "super-sport-short-light") won the Avus race in Berlin, establishing a world record for this category at 200 km/h. The "SSKL" represented the final stage in the development of the "S" series.

The era of the Silver Arrows began in 1934. The very first time they were seen was at the International Eifel Race on the Nuerburgring. It ended with pilot von Brauchitsch winning and establishing a new speed record. In 1938 it was a streamlined Silver Arrow that established a speed record on public motorways that remains unbroken to this day: 426.6 km/h at 1 km, with a flying start. Silver Arrows with compressor engines were also built by Auto-Union; they engaged in riveting duels with the Mercedes-built Silver Arrows.

1934 saw the introduction of the Mercedes Type 500K featuring an 8-cylinder compressor engine. The successor model launched in 1936 featured an even more powerful 5.4-l engine. In 1938 the Type 770 "Grand Mercedes" (W150) was introduced at the Berlin Motor and Motorcycle Show. It featured a 7.7-liter, eight-cylinder in-line engine with compressor. This engine developed a power output of 230 HP.

Compared with today's compressor engines, the power output per engine size then was rather modest. The Mercedes SLK, launched in the late 1990s, develops 192 HP from a 2.3-l engine at 5300 rpm.

1922
Mercedes 28/95 HP

1924
Mercedes Type 24/100
6 cylinders

1927
Mercedes Type "S"
6 cylinders, 6.8 l, 180 HP

1934
Mercedes Type 500K
8 cylinders

1938
"Grand Mercedes"
Type 770 (W150),
7.7 l, 230 HP

(All photos:
DaimlerChrysler Classic,
Corporate Archives)

Exhaust-gas turbocharging

Of all the possible methods for supercharging the internal-combustion engine, exhaust-gas turbocharging is the most widely used. Even on engines with low swept volumes, exhaust-gas supercharging leads to high torques and power outputs together with high levels of engine efficiency.

Whereas, in the past, exhaust-gas turbocharging was utilized in the quest for increased power output, it is today mostly used in order to increase the maximum torque at very low engine speeds. On the other hand, downsizing concepts are possible; while maintaining comparable engine power, it is possible to reduce engine displacement and lower fuel consumption. This applies in particular to turbocharged engines with gasoline direct injection.

Design and method of operation

The main components of the exhaust-gas turbocharger (Fig. 1) are the exhaust-gas turbine (3) and the compressor (1). The compressor impeller and the turbine rotor are mounted on a common shaft (2). This component is located in the exhaust-gas system so that the flow of exhaust gas can drive the turbine (Fig. 2, Pos. 14). The compressor (12) compresses the inducted air and thereby increases the cylinder charge. In the process of being compressed, the intake air is heated up. As a result, the intake air passes through an intercooler (5) to reduce its temperature again.

The energy for driving the exhaust-gas turbine is for the most part taken from the exhaust gas. The energy utilized here is contained in the hot and pressurized exhaust gas. On the other hand, energy must be also used to "dam" the exhaust gas when it leaves the engine so as to generate the required compressor power.

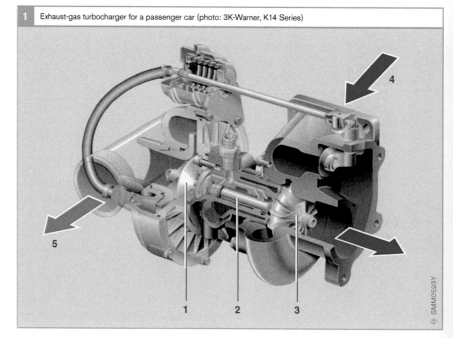

1   Exhaust-gas turbocharger for a passenger car (photo: 3K-Warner, K14 Series)

Fig. 1
1 Compressor impeller
2 Shaft
3 Exhaust-gas turbine
4 Inlet for exhaust-gas
   mass flow
5 Outlet for
   compressed air

The hot exhaust gas (Fig. 3, Pos. 7) flows into the exhaust-gas turbine (4) in a radial direction and, by so doing, forces it to rotate at high speeds (up to approx. 250,000 rpm). The turbine-rotor blades direct the exhaust gas to the center, from where it then exits axially.

The compressor (3) also turns along with the turbine, but here the flow conditions are reversed. The fresh incoming air (5) enters axially at the center of the compressor and is forced radially to the outside by the blades and compressed in the process.

In order to prevent the compressor from pumping and generating unwanted additional noise or even suffering damage, the divert-air valve (Fig. 2, Pos. 9) in the compressor bypass passage is opened when the throttle valve (2) is closed.

The turbine of the exhaust-gas turbocharger is situated in the hot exhaust-gas system. It therefore has to be made of highly heat-resistant materials.

Exhaust-gas turbochargers: designs

Wastegate supercharger

The objective is for engines to develop high torques at low engine speeds. For this reason, the turbine housing is designed for low exhaust-gas mass flow rates, e.g., full load at

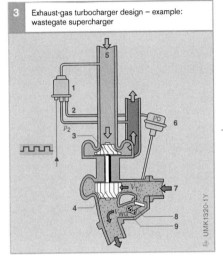

3 Exhaust-gas turbocharger design – example: wastegate supercharger

Fig. 3
1 Pulse valve
2 Pneumatic control line
3 Compressor
4 Exhaust-gas turbine
5 Fresh incoming air
6 Boost-pressure control valve
7 Exhaust gas
8 Wastegate
9 Bypass duct

ЛЛЛ Triggering signal for pulse valve
V_T Volume flow through the turbine
V_{WG} Volume flow through the wastegate
p_2 Boost/charge-air pressure
p_D Pressure on the valve diaphragm

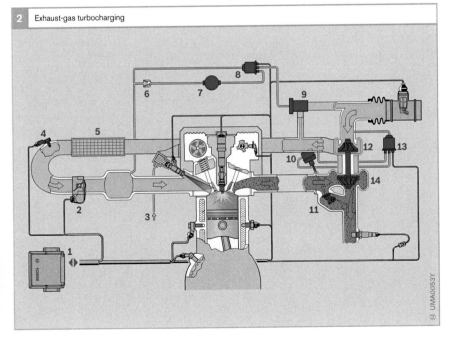

2 Exhaust-gas turbocharging

Fig. 2
1 Engine ECU
2 Throttle device
3 Fuel supply
4 Charge-air pressure and charge-air temperature sensor
5 Intercooler
6 Non-return valve
7 Vacuum reservoir
8 Solenoid valve (pulse valve)
9 Divert-air valve (dump valve)
10 Boost-pressure control valve
11 Wastegate (bypass valve)
12 Turbocharger compressor
13 Solenoid valve (pulse valve)
14 Exhaust-gas turbine

$n \leq 2000$ rpm. In order to prevent the turbocharger from overloading the engine at higher exhaust-gas mass flow rates, it is necessary in this range for some of the exhaust flow to be discharged via a bypass valve, the wastegate (Fig. 3, Pos. 8), past the turbine into the exhaust system. This flap-type bypass valve is usually integrated into the turbine housing.

The wastegate is actuated by the boost-pressure control valve (6). This valve is connected pneumatically to the pulse valve (1) through a control line (2). The pulse valve changes the boost pressure upon being triggered by an electrical signal from the engine ECU. This electrical signal is a function of the current boost pressure, information on which is provided by the Boost-Pressure Sensor (BPS).

If the boost pressure is too low, the pulse valve is triggered so that a somewhat lower pressure prevails in the control line. The boost-pressure control valve then closes the wastegate and the proportion of the

exhaust-gas mass flow used to power the turbine is increased.

If, on the other hand, the boost pressure is excessive, the pulse valve is triggered so that a somewhat higher pressure is built up in the control line. The boost-pressure control valve then opens the wastegate and the proportion of the exhaust-gas mass flow used to power the turbine is reduced.

VTG turbocharger

Variable turbines (VTG: Variable Turbine Geometry) offer another way of limiting the exhaust-gas mass flow at higher engine speeds (Fig. 4). The VTG turbocharger is state-of-the-art technology in diesel engines. It has not been able to establish itself as the preferred choice for gasoline engines because of the high thermal stresses caused by the higher exhaust-gas temperatures.

By varying the geometry, the adjustable guide vanes (3) adapt the flow cross-section, and with it the gas pressure at the turbine, to the required boost pressure. At low speeds, they open up a small cross-section so that the exhaust-gas mass flow in the turbine reaches a high speed and in doing so also brings the exhaust-gas turbine up to high speed (Fig. 4a).

At high engine speeds, the adjustable guide vanes open up a larger cross-section so that more exhaust gas can enter without accelerating the exhaust-gas turbine to excessive speeds (Fig. 4b). This limits the boost pressure.

The guide-vane angle is adjusted very simply by turning an adjusting ring (2). This sets the guide vanes to the desired angle by operating them either directly using adjusting levers (4) attached to the vanes or indirectly by means of adjuster cams. The adjusting ring is rotated pneumatically via a barometric adjustment cell (5) using vacuum pressure, or electrically. The engine-management system actuates the adjusting mechanism. Thus, the turbocharger pressure can be adjusted to the optimum setting in response to the engine operating state.

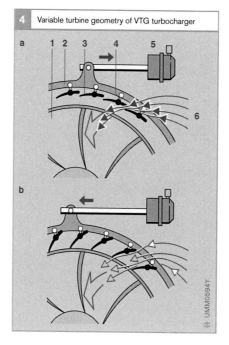

4   Variable turbine geometry of VTG turbocharger

Fig. 4

a   Guide-vane setting for high charge-air pressure

b   Guide-vane setting for low charge-air pressure

1   Turbine
2   Adjusting ring
3   Guide vanes
4   Adjusting lever
5   Barometric cell
6   Exhaust-gas flow

◀–   High flow velocity
◁–   Low flow velocity

UMM0594Y

Advantages and disadvantages of exhaust-gas turbocharging

Compared with a naturally aspirated engine with the same output power, the major advantages are to be found in the turbocharged engine's lower weight and smaller size (downsizing). The turbocharged engine's torque characteristic is better throughout the usable speed range (Fig. 5, curve 4 compared to curve 3). All in all, at a given speed, this results in a higher output (A→B).

Due to its more favorable torque characteristic at full load, the turbocharged engine generates the required power as shown in Fig. 5 (B or C) at lower engine speeds than the naturally aspirated engine. At part load the throttle valve must be opened further. In this way, the operating point is moved to a range with smaller frictional and throttling losses (C→B). This results in lower fuel-consumption figures even though turbocharged engines in fact feature less favorable thermodynamic efficiency figures due to their lower compression ratio.

The low torque that is available at very low engine speeds is a disadvantage of the turbocharger. In such speed ranges, there is not enough energy in the exhaust gas to drive the exhaust-gas turbine. In transient operation, even in the medium-speed range, the torque curve is less favorable than that of the naturally aspirated engine (curve 5). This is due to the delay in building up the exhaust-gas mass flow. On acceleration from slow speeds, therefore, the "turbo lag" effect occurs.

Turbo lag can be minimized by making full use of dynamic charge. This supports the turbocharger's running-up characteristic.

Intercooling

Air is heated as it is compressed in the turbocharger. Since hot air is less dense than cold air, a higher air temperature has a negative effect on cylinder charge. The compressed, warmed air must therefore be cooled off again by the intercooler. Charge-air cooling, or intercooling, therefore increases the cylinder charge compared with turbocharged engines without intercooling. A further increase in torque and power output is thus possible.

The drop in the combustion-air temperature also leads to a reduction in the temperature of the cylinder charge compressed during the compression cycle. This has a number of advantages:
- Reduced tendency to knock
- Improved thermal efficiency resulting in lower fuel-consumption figures
- Reduced thermal loading of the pistons
- Lower NO_X emissions
- Higher power yield

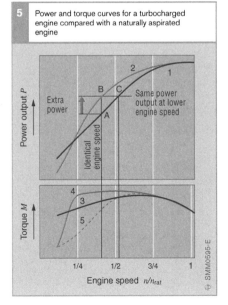

5 Power and torque curves for a turbocharged engine compared with a naturally aspirated engine

Fig. 5
1, 3 Naturally aspirated engine in steady-state operation
2, 4 Supercharged engine in steady-state operation
5 Torque curve of the supercharged engine in transient (dynamic) operation

Controlled charge flow

The airflow behavior in both intake manifold and cylinder is an essential factor in
forming an ideal air/fuel mixture. A vigorous charge-flow movement ensures that the
air-fuel mixture is well blended to achieve
excellent, low-pollutant combustion.

Charge-flow control valve

In systems with gasoline direct injection, a
controlled or switched charge-flow control
valve (i.e., continuous adjustment or two
settings) is used to generate a vigorous
charge flow. In the intake-valve area, the
intake manifold is typically split into two
channels, one of which can be closed by
means of a flap, the charge-flow control
valve (Fig. 1). This valve, in combination
with the geometry of the intake area, causes
the mixture to adopt a tumbling or swirling
motion in the combustion chamber (Fig. 2).

The charge-flow control valve allows the intensity of the charge flow to be controlled.
In stratified-charge mode (wall-directed
combustion), the flow moved in this way
makes sure that the mixture is conveyed to
the spark plug. It supports mixture formation at the same time.

In homogeneous-charge mode, the
charge-flow control valve is normally closed
at low torques and rotational speeds. The
charge-flow control valve must be opened at
high torques and speeds. Otherwise, it is not
possible to introduce the air required for
high power into the combustion chamber
because the charge-flow control valve closes
part of the flow cross-section. Ideal mixture
formation is achieved, even without increased charge flow, simply by advancing
fuel injection into the combustion chamber
(as early as the induction stroke) at a high
temperature.

Technical implementation of a charge-air
control valve is difficult in manifold-injection systems. This is because it is necessary
to stop fuel from accumulating ahead of the
valve when it is closed and from passing into
the combustion chamber when it is opened.

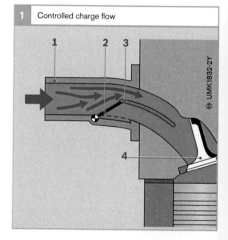

1   Controlled charge flow

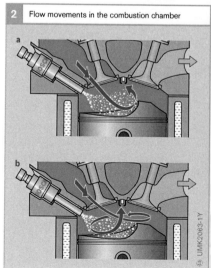

2   Flow movements in the combustion chamber

Fig. 1

1   Intake manifold
2   Charge-flow control
    valve
3   Separating ridge
4   Intake valve

Fig. 2

a   Tumble
b   Swirl

Exhaust-gas recirculation (EGR)

The mass of the residual gas remaining in the cylinder, and with it the inert-gas content of the cylinder charge, can be influenced by varying the valve timing. In this case, one refers to "internal" EGR. The inert-gas content can be influenced far more by applying "external" EGR with which part of the exhaust gas which has already left the cylinder is directed back into the intake manifold through a special line (Fig. 1, Pos. 3). EGR leads to a reduction of the NO_X emissions and to a slightly lower fuel-consumption figure.

Limiting the NO_X emissions
Since they are highly dependent upon temperature, EGR is highly effective in reducing NO_X emissions. When peak combustion temperature is lowered by introducing burnt exhaust gas to the air/fuel mixture, NO_X emissions drop accordingly.

Lowering fuel consumption
When EGR is applied, the overall cylinder charge increases while the charge of fresh air remains constant. This means that the throttle valve (2) must reduce the engine throttling if a given torque is to be achieved. Fuel consumption drops as a result.

EGR: Method of operation
Depending upon the engine's operating point, the engine ECU (4) triggers the EGR valve (5) and defines its opened cross-section. Part of the exhaust gas (6) is diverted via this opened cross-section (3) and mixed with the incoming fresh air (1). This defines the exhaust-gas content of the cylinder charge.

EGR with gasoline direct injection
EGR is also used on gasoline direct-injection engines to reduce NO_X emissions and fuel consumption. In fact, it is absolutely essential since with it NO_X emissions can be lowered to such an extent in lean-burn operations that other emissions-reduction measures can be reduced accordingly (for instance, rich homogeneous operation for NO_X "Removal" from the NO_X accumulator-type catalytic converter). EGR also has a favorable effect on fuel consumption.

There must be a pressure gradient between the intake manifold and the exhaust-gas tract in order that exhaust gas can be drawn in via the EGR valve. At part load though, direct-injection engines are operated practically unthrottled. Furthermore a considerable amount of oxygen is drawn into the intake manifold via EGR during lean-burn operation.

Non-throttled operation and the introduction of oxygen into the intake manifold via the EGR therefore necessitate a control strategy which coordinates throttle valve and EGR valve. This results in severe demands being made on the EGR system with regard to precision and reliability, and it must be robust enough to withstand the deposits which accumulate in the exhaust-gas components as a result of the low exhaust-gas temperatures.

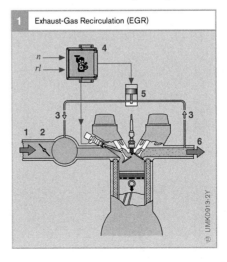

1 Exhaust-Gas Recirculation (EGR)

Fig. 1
1   Fresh-air intake
2   Throttle valve
3   Recirculated exhaust gas
4   Engine ECU
5   EGR valve
6   Exhaust gas

n Engine rpm
rl Relative air charge

Gasoline injection systems over the years

The primary purpose of the fuel-injection system is to provide the engine with an air/fuel mixture which is best suited to the prevailing operating state. Over the years, these systems have been continually improved, a significant feature of this improvement being the constant increase in the amount of electronic components used.

While the development objective in the 1970s lay primarily in increasing power and comfort, the emphasis switched from the 1980s onwards to reducing emissions. A further requirement, which began to be taken increasingly seriously, was the reduction of fuel consumption and with it also the reduction of CO_2 emissions.

Overview

Development of gasoline injection systems

An important milestone in the development history of control systems for gasoline engines was the introduction of electronically controlled fuel-injection systems. Where previously mechanically controlled fuel-injection systems had been used, Bosch with D-Jetronic introduced in 1967 for the first time an electronic system in which fuel was injected via electromagnetically actuated fuel injectors intermittently onto the intake valve of each cylinder (multi-point injection).

However, wide-range use of fuel-injection systems was only possible with lower-cost designs. Mechanical K-Jetronic and Mono-Jetronic with only one single, centrally situated electromagnetic fuel injector (single-point injection) enabled fuel-injection technology to stretch also to mid-size and small cars.

The carburetor was rendered completely obsolete on account of the advantages of gasoline injection with regard to fuel consumption, power output, engine performance and emission behavior. In particular, the reduction of pollutant emissions could only be achieved thanks to the advances made in fuel-injection technology in conjunction with exhaust-gas treatment (three-way catalytic converter). The emission limits laid down by legislators for hydrocarbons (HC), carbon monoxide (CO) and nitrogen oxide (NO_X) called for fuel-injection systems which were able to adjust the mixture composition within very narrow limits.

Table 1 shows the development of Bosch fuel-injection systems. Today, only the Motronic engine-management system with multi-point injection is still used. Only with this type of fuel injection in conjunction with complex engine management is it possible to comply with today's stringent emission limits.

1	Development of fuel-injection systems	
Year	System	Features
1967	D-Jetronic	– Analog-technology multi-point injection system – Intermittent fuel injection – Intake-manifold-controlled
1973	K-Jetronic	– Mechanical-hydraulic multi-point injection system – Continuous fuel injection
1973	L-Jetronic	– Electronic multi-point injection system (initially analog, later digital technology) – Intermittent fuel injection – Air-flow sensing
1981	LH-Jetronic	– Electronic multi-point injection system – Intermittent fuel injection – Air-mass sensing
1982	KE-Jetronic	– K-Jetronic with electronically controlled additional functions
1987	Mono-Jetronic	– Single-point injection system – Intermittent fuel injection – Air-flow calculation via throttle-valve angle and engine speed

Table 1

The story of fuel injection extends back to cover a period of almost one hundred years. The Gasmotorenfabik Deutz was manufacturing plunger pumps for injecting fuel in a limited production series as early as 1898.

A short time later the uses of the venturi-effect for carburetor design were discovered, and fuel-injection systems based on the technology of the time ceased to be competitive.

Bosch started research on gasoline-injection pumps in 1912. The first aircraft engine featuring Bosch fuel injection, a 1200-hp unit, entered series production in 1937; problems with carburetor icing and fire hazards had lent special impetus to fuel-injection development work for the aeronautics field. This development marks the beginning of the era of fuel injection at Bosch, but there was still a long path to travel on the way to fuel injection for passenger cars.

1952 saw a Bosch direct-injection unit being featured as standard equipment on a small car for the first time. A unit was then installed in the 300 SL, the legendary production sports car from Daimler-Benz. In the years that followed, development on mechanical injection pumps continued, and ...

In 1967 fuel injection took another giant step forward: The first electronic injection system: the intake-pressure-controlled D-Jetronic!

In 1973 the air-flow-controlled L-Jetronic appeared on the market, at the same time as the K-Jetronic, which featured mechanical-hydraulic control and was also an air-flow-controlled system.

In 1976, the K-Jetronic was the first automotive system to incorporate a Lambda closed-loop control.

1979 marked the introduction of a new system: Motronic, featuring digital processing for numerous engine functions. This system combined L-Jetronic with electronic program-map control for the ignition. The first automotive microprocessor!

In 1982, the K-Jetronic model became available in an expanded configuration, the KE-Jetronic, including an electronic closed-loop control circuit and a Lambda oxygen sensor.

These were joined by Bosch Mono-Jetronic in 1987: This particularly cost-efficient single-point injection unit made it feasible to equip small vehicles with Jetronic, and once and for all made the carburetor absolutely superfluous.

▶ Bosch gasoline fuel injection from the year 1954

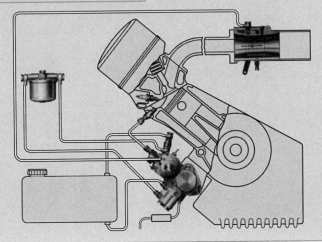

UMK1355Y

Beginnings of mixture formation

At a time when the first atmospheric engines were being designed by various inventors, a common problem was forming an ignitable mixture. The question of whether or not such an internal-combustion engine would be able to function at all. The solution was very much dependent on the ignition mechanism.

The basics of the carburetor were developed as early as the 18th Century. Inventors' efforts at the time were directed at vaporizing liquid fuels in such a way that they

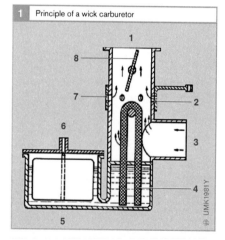

1   Principle of a wick carburetor

Fig. 1
1   Air/fuel mixture to engine
2   Annular slide valve
3   Air inlet
4   Wick
5   Float chamber with float
6   Fuel inlet
7   Auxiliary air
8   Throttle valve

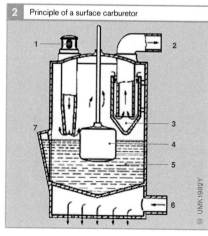

2   Principle of a surface carburetor

Fig. 2
1   Air inlet
2   Air/fuel mixture to engine
3   Fuel separator
4   Float
5   Fuel
6   Engine exhaust gases
7   Fuel filler neck

could be used to operate a lighting or heating device.

In 1795 Robert Street was the first to suggest to vaporize turpentine or creosote in an atmospheric engine. Around 1825 Samuel Morey and Eskine Hazard developed a two-cylinder engine for which they also designed the first carburetor. It was granted British patent no. 5402. Up to that time, mixture-formation systems were essentially fueled with turpentine or kerosene.

However, this situation changed in 1833, when Eilhardt Mitscherlich, professor of chemistry at the University of Berlin, managed to split benzoic acid by thermal cracking. The result was the so-called "Faraday's olefiant gas", which he called "benzene", the precursor of today's gasoline.

William Barnett designed the first carburetor for gasoline. He was awarded patent no. 7615 in 1838.

During this time, these designs were either wick carburetors (Fig. 1) or surface carburetors (Fig. 2). The first carburetor to be used in a vehicle was a wick carburetor. The wick drew up the fuel, similar to the oil-lamp principle. The wick was then exposed to an air stream in the engine, causing air and fuel to blend. By contrast, in the surface carburetor, in which the fuel was heated by the engine's exhaust gases, the result was a layer of vapor just above the fuel surface. This was then blended to form the required air-fuel mixture by introducing an air stream.

In Berlin in 1882 Siegfried Marcus applied for a patent for the brush carburetor (Fig. 3) that he had invented. This mixture generator used the interaction of a rapidly rotating cylindrical brush (3) driven by a drive pulley (1) and a fuel stripper (2) to form a mist of atomized fuel in the brush chamber (4). The fuel mist was then drawn into the engine via the inlet (5). The brush carburetor maintained its dominance for about 11 years.

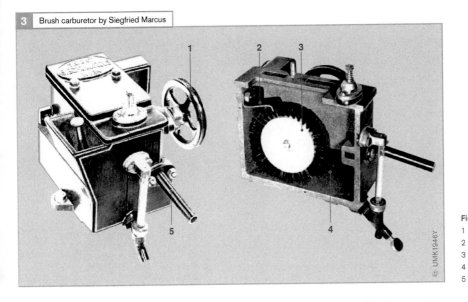

3   Brush carburetor by Siegfried Marcus

Fig. 3
1   Drive pulley
2   Fuel stripper
3   Rotating brush
4   Brush chamber
5   Intake fitting

In 1885 Nikolaus August Otto succeeded in his struggle to master the engine powered by hydrocarbon fuels (alcohol/gasoline); he had been working toward this goal since 1860. The first gasoline-engine, working on the four-stroke principle and equipped with a surface carburetor and an electrical ignition device of Otto's own construction, garnered the highest praise and profound recognition at the World Fair at Antwerp. This design was later built and sold in great numbers by the firm of Otto & Langen in Deutz over many years (Fig. 4).

Fig. 4
A   Carburetor
B   Engine with ignition system

1   Air inlet
2   Air tube
3   Gravel canister (flame shield)
4   Water funnel
5   Fuel filler neck
6   Float
7   Fuel reservoir
8   Exhaust-gas inlet
9   Shutoff valve
10   Heating pad
11   Cooling-water jacket
12   Water tubing
13   Coolant inlet
14   Gas inlet
15   Ignition device
16   Gas shutoff valve
17   Air inlet
18   Air shutoff valve

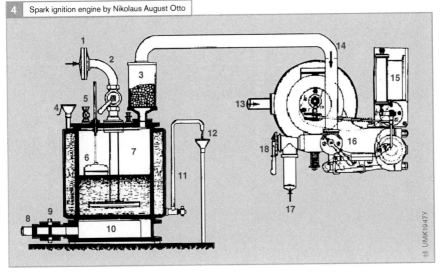

4   Spark ignition engine by Nikolaus August Otto

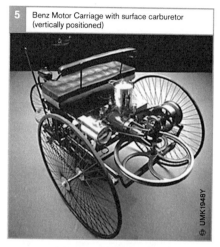

5 Benz Motor Carriage with surface carburetor (vertically positioned)

In the same year Carl Benz installed a surface carburetor of his own design in his first "Patent Motor Carriage" (Fig. 5). A short while later, he improved the original design of his carburetor by adding a valve float, as he put it, "to always maintain the fuel level automatically at the same height".

In 1893 Wilhelm Maybach introduced his jet-nozzle carburetor (Fig. 7). In this device, the fuel was sprayed from a fuel nozzle onto a baffle surface, which caused the fuel to distribute in a cone-shaped pattern (Fig. 8).

6 Surface carburetor, 1885 (cutaway model)

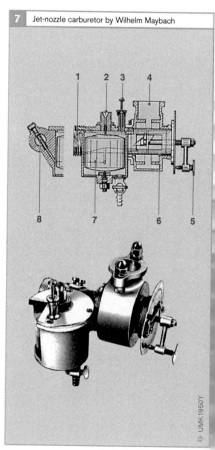

7 Jet-nozzle carburetor by Wilhelm Maybach

Fig. 7
1 Air inlet
2 Fuel inlet
3 Spring-loaded swab
4 Air/fuel mixture outlet
5 Rotating-slide stop
6 Rotating-slide for mixture control
7 Float
8 Jet nozzle

1906/07 saw the introduction of the Claudel carburetor and the carburetor designs of François Bavery, both of which brought

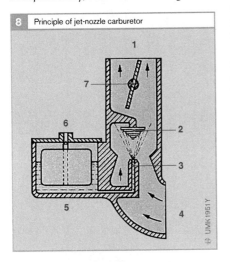

8 | Principle of jet-nozzle carburetor

9 | ZENITH carburetor, type 22, 1910

10 | SOLEX carburetor, type DHR, 1912

11 | PALLAS carburetor, type I, 1914

further advancement to carburetor design. In these carburetors, which were to become famous under the ZENITH brand name, the lean-fuel auxiliary or compensation jet delivers a virtually unchanged air/fuel mixture despite increasing air velocity (Fig. 9).

The same period also saw applications for the carburetor patents of Mennesson and Goudard. Their designs became world-famous under the SOLEX brand (Fig. 10).

The years that followed produced a proliferation of carburetor designs. In this context, some of their names, e.g. SUM, CUDELL, FAVORIT, ESCOMA, and GRAETZIN, deserve special mention. After the Haak carburetor was patented in 1906 and manufactured by the firm of PALLAS, Scüttler and Deutrich developed the PALLAS carburetor in 1912. It had a ring float and combination jet (Fig. 11).

In 1914 the Royal Prussian War Ministry sponsored a competition for benzol (benzene) carburetors. Even at that time the test specifications included the condition that the exhaust gases should be as clean as possible. Of the competing products bearing 14 different brand names, it was a ZENITH carburetor that won 1st prize. All the carburetors were examined at the test facility of the Technical University at Charlottenburg, and subjected to a demanding 800-km winter trial by the German army administration in automobiles of identical horsepower.

Fig. 8
1   Air/fuel mixture to engine
2   Baffle surface
3   Fuel jet (jet nozzle)
4   Air inlet
5   Float chamber with float
6   Fuel inlet
7   Throttle valve

The period that followed marked the beginning of various attempts at detail work and specialization. A variety of model configurations and auxiliary devices was developed, e.g. rotating slides and preliminary throttles serving as start-assist measures, diaphragm systems replacing the float in aircraft carburetors, and pump systems providing acceleration aids. The diversity of these modifications was so extensive that any descriptive attempt would exceed the scope of this chapter.

In the 1920s, to obtain greater engine power, single and twin carburetors (carburetors featuring two throttle valves) were installed in the form of multiple carburetor systems (several single or twin carburetors with synchonized controls). In the decades that followed, the great variety of carburetor variants made by various manufacturers increased even further.

In parallel with the ongoing carburetor development for aircraft engines, the 1930s saw the development of the first gasoline-injection systems with direct injection (example in Fig. 12). This engine required two 12-cylinder in-line fuel-injection pumps, each of which

was mounted on the crankcase between the cylinder banks (not visible in Fig. 12). A pump of this type is shown in Fig. 13. It has a total length of about 70 cm (27.5 in.).

In the late 1930s direct-injection systems were used in conjunction with the 9-cylinder radial BMW engines (Fig. 14) in the legendary three-engine Junkers Ju 52 aircraft. Especially noteworthy is the "boxer" (reciprocal) configuration of the

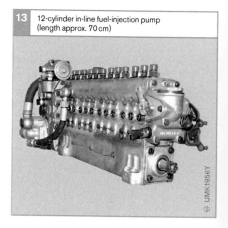

13 12-cylinder in-line fuel-injection pump (length approx. 70 cm)

UMK1956Y

12 Aircraft engine by Daimler-Benz, with 24 cylinders in in-line-X configuration, type DB 604

Fig. 12
This aircraft engine was produced by the Daimler-Benz AG between 1939 and 1942.
There were models ranging from 48.5-liter displacement and 2350 HP (1741 kW) to 50.0-liter displacement and 3500 HP (2593 kW), all featuring gasoline direct injection by Bosch. The engine had a total length of 2.15 m.
(Photo: DaimlerChrysler Classic, Corporate Archives)

UMK1955Y

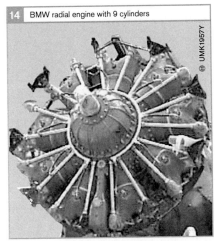

14   BMW radial engine with 9 cylinders

15   Bosch fuel-injection pump in "boxer" (reciprocal) variant (length approx. 35 cm)

16   Gutbrod Superior 600 convertible (1950–1954; 1952 and later with direct injection)

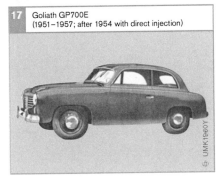

17   Goliath GP700E (1951–1957; after 1954 with direct injection)

18   Two-cylinder fuel-injection pump (length approx. 15 cm)

mechanical fuel-injection pump (Fig. 15) by Bosch.

In the 1950s this type of fuel-injection system working with direct fuel injection also made its debut in passenger cars. One example was the "Gutbrod Superior" of 1952 (Fig. 16), and the Goliath GP700E introduced in 1954 (Fig. 17). These two vehicles were compact cars powered by two-cylinder, two-stroke engines with a displacement of less than 1000 cc. Their fuel-injection pumps were also compact in size (Fig. 18).

A component diagram of this two-cylinder fuel-injection system, which entered the annals of automotive history as the first gasoline direct-injection system in passenger cars, appears in Figure 19 (next page).

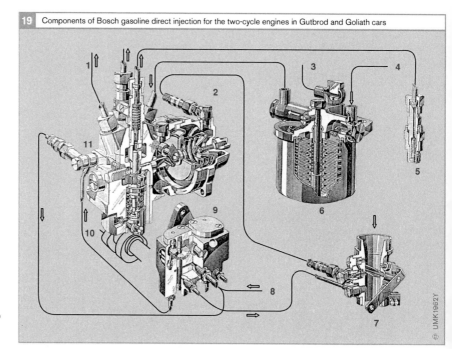

19   Components of Bosch gasoline direct injection for the two-cycle engines in Gutbrod and Goliath cars

Fig. 19

1   Venting tube
2   Mixture-control-unit
    diaphragm block
3   Venting tube
4   From fuel tank
5   Fuel injector
6   Fuel filter
7   Mixture-control-unit
    flap supports
8   From oil reservoir
9   Lubricating-oil pump
10   Fuel-injection pump
11   Overflow valve

However, the Mercedes-Benz 300 SL sports car (Fig. 20) featured a Bosch-built gasoline direct-injection system. It was presented to the public on February 6, 1954 at the International Motor Sports Show in New York. Installed at a 50-degree slant, the car's 6-cylinder in-line engine (M198/11) had a displacement of 2996 cc, and delivered 215 HP (159 kW).

A fundamentally different facet of air/fuel mixture formation for gasoline engines appeared during the latter part of WWII and for a while thereafter: the wood-gas generator.

The wood gas emitted by the glowing charcoal was used to form an ignitable air/fuel mixture (Fig. 21). However, it was hard to overlook the physical size of these wood gasifier systems (Fig. 22).

20   Mercedes-Benz 300 SL (1954)

Fig. 20
Photo:
DaimlerChrysler Classic,
Corporate Archives

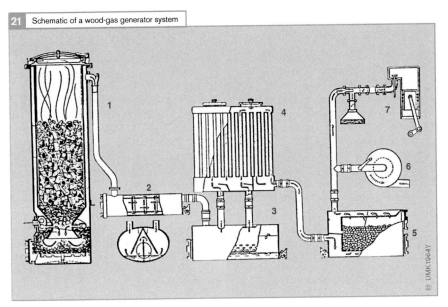

21   Schematic of a wood-gas generator system

Fig. 21
1   Gas generator
2   Baffle plate cleaner
3   Settling tank
4   Gas cooler
5   Secondary cleaner
6   Bellows blower
7   Regulator group

Due to the gradual tightening-up of emission standards, the automotive industry has increasingly shunned the carburetor. In the early 1990s, however, there was a successful design by Bosch and Pierburg that equipped a modified conventional carburetor with modern-day actuators: the Ecotronic carburetor (Fig. 23). This carburetor made it possible to comply with the prevailing emission standards at the time, while at the same time ensuring economical fuel consumption.

22   Wood-gas generator system on a 1936 Adler Diplomat

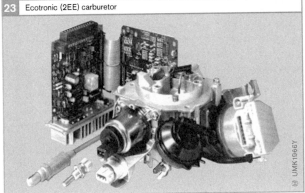

23   Ecotronic (2EE) carburetor

At the conclusion of this brief review of the history of air/fuel mixture formation, it should be stated that the various carburetor types continued to be used well into the 1990s as standard equipment in passenger cars. In compact cars in particular, the carburetor enjoyed sustained popularity due to its lower cost.

Evolution of gasoline injection systems

Since 1885, when the first manifold-injection system was used on a fixed industrial engine, many changes have occurred in the field of gasoline-injection systems. Later attempts included a floatless carburetor with attached fuel-injection device installed in aircraft engines in 1925, and an electrically triggered fuel-injection device in a racing motorcycle in 1930. That was before Bosch finally developed a mechanically driven gasoline-injection pump for the Gutbrod Superior 600 and Goliath GP 700 vehicles in 1951. These were the first passenger cars with gasoline injection. The system was designed as a direct-injection system. Even the legendary Mercedes 300 SL was equipped with gasoline direct injection featuring a mechanical inline pump.

After passing though several development stages of manifold-injection systems (described below), the current trend is again headed for direct-injection systems.

D-Jetronic

System overview

The Bosch-engineered D-Jetronic comprises a gasoline-injection system that is essentially controlled by intake-manifold pressure and engine speed. Hence the designation D-Jetronic (D stands for "drucksensorgesteuert" – German for pressure-sensor-controlled).

The electronic control unit (Fig. 1, Pos. 1) receives signals for intake-manifold pressure, intake-air temperature, cooling-water (coolant) and/or cylinder-head temperature, throttle-valve position and movement, and starting, engine speed, and start of injection. The ECU processes these data, and sends electrical pulses to the fuel injectors (2). The ECU is interconnected with the electrical components via a multiple connector and wiring harness.

The fuel injectors spray the fuel into the intake manifolds of the cylinders. The pressure sensor (3) sends engine-load data to the ECU. The temperature sensors communicate the temperatures of air (13) and coolant (4) to the ECU. The thermo-time switch (5)

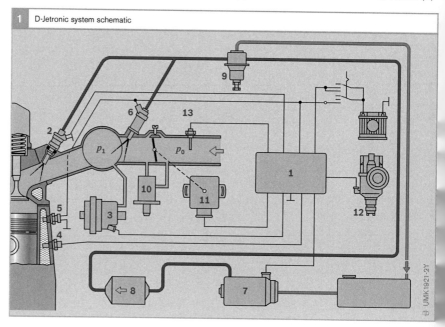

1   D-Jetronic system schematic

Fig. 1

1 ECU
2 Fuel injector
3 Pressure sensor
4 Coolant-temperature sensor
5 Thermo-time switch
6 Electric start valve
7 Electric fuel pump
8 Fuel filter
9 Fuel-pressure regulator
10 Auxiliary-air device
11 Throttle-valve switch
12 Injection trigger
13 Air-temperature sensor

p_0 Atmospheric pressure
p_1 Pressure in the intake manifold

switches the electric start valve (6), which injects additional fuel into the intake manifold during low-temperature starts. The electric fuel pump (7) continuously delivers fuel to the fuel injectors. The fuel filter (8) is integrated in the fuel line to remove contaminants. The fuel-pressure regulator (9) maintains a constant fuel pressure in the fuel lines. The temperature-dependent function of the auxiliary-air device (10) provides additional air during engine warm-up. The throttle-valve switch (11) sends engine idle and full-load states to the ECU.

Method of operation
Pressure measurement
Upstream of the throttle valve inside the intake manifold, the pressure is equal to the ambient atmospheric pressure. The pressure downstream of the throttle valve is lower, and changes with the throttle-valve position. This reduced intake-manifold pressure serves as the measured quantity for engine load, which is the most important information. Engine load is derived from the measure of the volume of air drawn into the engine. The information pertaining to the pressure in the intake manifold is determined by the pressure sensor.

The pressure sensor (Fig. 2) contains two aneroid capsules which shift the armature of a coil. The measuring system is pneumatically connected via a line to the intake manifold. As the load increases, i.e., as pressure in the intake manifold increases, the aneroid capsules are compressed and the armature is drawn deeper into the coil. This changes its inductance. The device is therefore a measuring transducer that converts a pneumatic pulse into an electrical signal. The induction-type pulse generator in the pressure sensor is connected to an electronic timer in the ECU. It determines the duration of the electrical pulses required to trigger the fuel injectors. In this way, the intake-manifold pressure is directly converted to an injection duration.

Fuel injection
Specific contacts in the ignition distributor (injection trigger, Fig. 1, Pos. 12) determine – in accordance with camshaft adjustment – the beginning of the pulse for opening the fuel injectors. Injection duration is mainly dependent on engine load and engine speed. Pressure sensor and injection trigger supply the required signals to the ECU. This enables the fuel injectors to inject fuel in metered quantities by means of electrical pulses.

Adaptation to operating conditions
Adaptations under different operating conditions are required to ensure good engine performance:
- Full load: The fuel quantity is determined for maximum power
- Acceleration enrichment: Additional injection pulses are applied during acceleration
- Altitude compensation: It is possible by taking into account the pressure differential between intake manifold and free atmosphere to obtain good adaptation of fuel injection to different altitudes
- Intake-air temperature: The temperature-dependent density differences of the air can be taken into account by recording the outside temperature

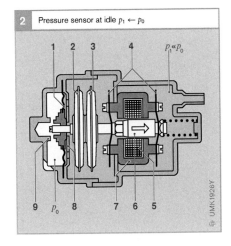

2 Pressure sensor at idle $p_1 \leftarrow p_0$

$p_1 \ll p_0$

Fig. 2
Basic function:
Aneroid capsules
2 and 3 expanded
1   Diaphragm
2   Aneroid capsule
3   Aneroid capsule
4   Leaf spring
5   Coil
6   Armature
7   Core
8   Part-load stop
9   Full-load stop

p_0 Atmospheric
        pressure
p_1 Pressure in the
        intake manifold

K-Jetronic

System overview

K-Jetronic is a mechanically-hydraulically controlled fuel-injection system which needs no form of drive and which meters the fuel as a function of the intake air quantity and injects it continuously onto the engine intake valves. Hence the system designation K-Jetronic (Kontinuierlich = German for continuous).

Specific operating conditions of the engine require corrective intervention in mixture formation and this is carried out by K-Jetronic in order to optimize starting and driving performance, power output and exhaust-gas composition.

The K-Jetronic fuel-injection system covers the following functional areas:

- Fuel supply
- Air-flow measurement and
- Fuel metering

Mode of operation

An electrically driven roller-cell pump (Fig. 3, Pos. 2) pumps the fuel from the fuel tank (1) at a pressure of over 5 bar to a fuel accumulator (3) and through a filter (4) to the fuel distributor (9). The pressure regulator integrated in the fuel distributor holds the delivery pressure in the fuel system (system pressure) at about 5 bar.

From the fuel distributor, the fuel flows to the fuel injectors (6). The fuel injectors inject the fuel continuously into the engine's intake ports engine. When the intake valve is opened, the air drawn in by the engine carries the waiting "cloud" of fuel with it into the cylinder. An ignitable air-fuel mixture is formed during the induction stroke due to the swirl effect.

The amount of air, corresponding to the position of the throttle valve (16), drawn in by the engine serves as the criterion for metering of the fuel to the individual cylinders. The amount of air drawn in by the engine is measured by the air-flow sensor (10), which, in turn, controls the fuel distributor.

Injection occurs continuously, i.e., without regard to the position of the intake valve. When the intake valve is closed, the mixture is stored.

Fig. 3

1 Fuel tank
2 Electric fuel pump
3 Fuel accumulator
4 Fuel filter
5 Warm-up regulator
6 Fuel injector
7 Intake manifold
8 Cold-start valve
9 Fuel distributor
10 Air-flow sensor
11 Timing valve
12 Lambda sensor
13 Thermo-time switch
14 Ignition distributor
15 Auxiliary-air device
16 Throttle-valve switch
17 Primary-pressure regulator
18 ECU (for version with lambda closed-loop control)
19 Ignition/starting switch
20 Battery

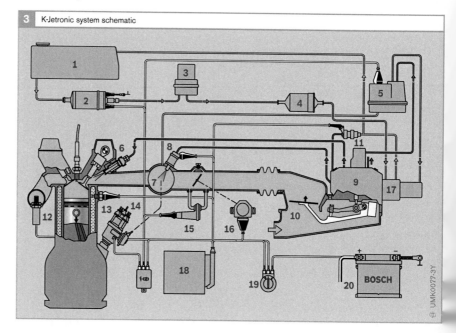

3   K-Jetronic system schematic

Mixture enrichment is controlled in order to adapt to various operating conditions such as start, warm-up, idle and full load. In addition, supplementary functions such as overrun fuel cutoff, engine-speed limitation and lambda closed-loop control are possible.

Mixture-control unit

The task of the fuel-management system is to meter a quantity of fuel corresponding to the intake air quantity. Basically, fuel metering is carried out by the mixture-control unit. This comprises the air-flow sensor and the fuel distributor.

Air-flow sensor

The quantity of air drawn in by the engine is a precise measure of its power. The air-flow sensor (Fig. 4) installed ahead of the throttle valve operates according to the suspended-solid-particle principle and measures the quantity of air drawn in by the engine. The intake air quantity serves as the main control variable for determining the basic injection quantity.

The air-flow sensor consists of an air funnel (1), in which a moving sensor plate (2) is located. The air flowing through the air funnel deflects the sensor plate by a specific distance from its rest position. A lever system (12) transmits the movements of the sensor plate to a control plunger (8), which determines the basic injection quantity required for basic functions. A counterweight compensates for the weight of the sensor plate and lever system (this is carried out by an extension spring on the downdraft air-flow sensor). A leaf spring (13) ensures the correct zero position in the switched-off phase.

Fuel distributor

Depending upon the position of the plate in the air-flow sensor, the fuel distributor (10) meters the basic injection quantity to the individual engine cylinders.

Depending upon its position in the barrel with metering slits (9), the control plunger opens or closes the slits to a greater or lesser extent. The fuel flows through the open section of the slits to the differential-pressure valves and then to the fuel injectors.

If sensor-plate travel is only small, then the control plunger is lifted only slightly and, as a result, only a small section of the slit is opened for the passage of fuel. In the event of larger sensor-plate travel, the control plunger opens a larger section of the slits and more fuel can flow. There is a linear relationship between sensor-plate travel and the slit section in the barrel which is opened for fuel flow.

Differential-pressure valves

Differential-pressure valves (Fig. 5, next page) in the fuel distributor result in a specific pressure drop at the metering slits. If the sensor-plate travel is to result in a change of basic injection quantity in the same proportion, then a constant drop in pressure must be guaranteed at the metering slits, regardless of the amount of fuel flowing through them. The differential-pressure valves maintain the differential pressure between the upper chamber (5) and the lower chamber (8) at a constant level, regardless of fuel throughflow.

Flat-seat valves are used as differential-pressure valves. They are fitted in the fuel distributor and one such valve is allocated to each metering slit. A diaphragm separates the upper and lower chambers of the valve.

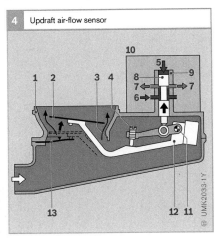

4   Updraft air-flow sensor

Fig. 4

1 Air funnel
2 Sensor plate
3 Relief cross-section
4 Mixture adjusting
   screw
5 Control pressure
6 Fuel inlet
7 Metered quantity
   of fuel
8 Control plunger
9 Barrel with metering
   slits
10 Fuel distributor
11 Pivot
12 Lever
13 Leaf spring

The control pressure is tapped from the primary system pressure through a throttle bore. This throttle bore serves to isolate the control-pressure circuit and the primary-pressure circuit from one another. A connection line joins the fuel distributor and the warm-up regulator (control-pressure regulator). When the cold engine is started, the control pressure is about 0.5 bar. As the engine warms up, the warm-up regulator increases the control pressure to about 3.7 bar. The control pressure acts through a damp-

ing orifice on the control plunger and thereby develops the force which opposes the force of the air in the air-flow sensor. In doing so, the orifice dampens a possible oscillation of the sensor plate which could result due to pulsating air-intake flow.

The level of the control pressure influences fuel distribution. If the control pressure is low, the air drawn in by the engine can deflect the sensor plate further. This results in the control plunger opening the metering slits (11) further and the engine being allocated more fuel. On the other hand, if the control pressure is high, the air drawn in by the engine cannot deflect the sensor plate so far and, as a result, the engine receives less fuel.

Fuel injectors

The fuel injectors (Fig. 6) open at a given pressure and inject the fuel metered to them into the intake manifolds and onto the cylinder intake valves. They open automatically as soon as the opening pressure exceeds, for example, 3.5 bar. The valve needle (3) vibrates at high frequency (chatter) as the fuel is injected. When the engine is switched off, the injectors close tightly when the pressure in the fuel-supply system drops below their opening pressure. This means that no more fuel can enter the intake manifolds once the engine has stopped.

Adaptation to operating states

In addition to the basic functions described up to now, the mixture has to be adapted in particular operating states. These adaptations (corrections) are necessary in order to optimize the power delivered, and to improve exhaust-gas composition, starting behavior and performance.

● Basic mixture adaptation: Basic adaptation of the air/fuel mixture to the idle, part-load and full-load operating conditions is effected by shaping the air funnel appropriately in the air-flow sensor. This adaptation is achieved by designing the air funnel so that it becomes wider in stages.

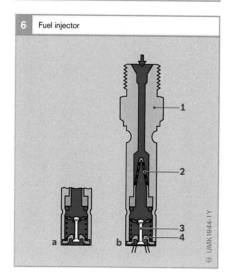

5 Differential-pressure valve with system and control pressures

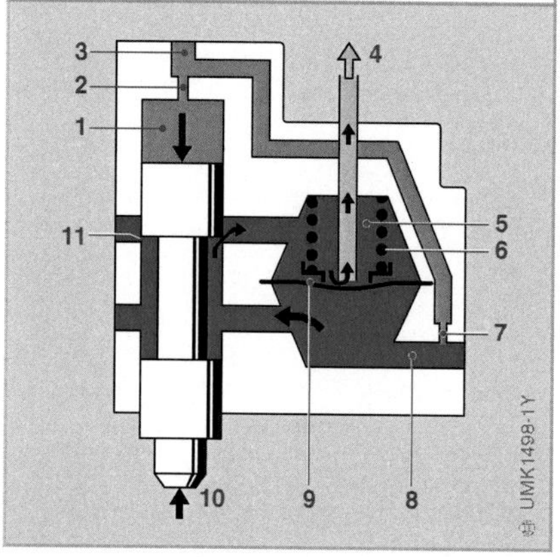

6 Fuel injector

Fig. 5
1   Control-pressure effect (hydraulic force)
2   Damping orifice
3   Line to warm-up regulator
4   To intake valve
5   Pressure in upper chamber of differential-pressure valve (0.1 bar < primary pressure)
6   Control spring
7   Isolating throttle bore
8   Pressure in lower chamber = primary pressure (delivery pressure)
9   Diaphragm
10 Effect of air pressure via sensor-plate lever
11 Metering slits

Fig. 6
a   In rest position
b   In actuated position

1   Valve housing
2   Filter
3   Valve needle
4   Valve seat

- Cold-start enrichment: As a function of engine temperature, the cold-start valve injects an additional quantity of fuel for a limited period during starting in order to compensate for the fuel lost through condensation from the inducted mixture. The injection period of the cold-start valve is limited by a thermo-time switch depending on the engine temperature. This consists of an electrically heated bimetal strip which, depending on its temperature, opens or closes a contact.
- Warm-up enrichment: Mixture control for warm-up operation is effected via the control pressure by the warm-up regulator (control-pressure regulator). When the engine is cold, the warm-up regulator reduces the control pressure to a degree dependent on engine temperature and thus causes the metering slits to open further.
- Idle stabilization: As it warms up, the engine receives more air via the auxiliary-air device. Due to the fact that this auxiliary air is measured by the air-flow sensor and taken into account for fuel metering, the engine is provided with more air-fuel mixture. This results in idle stabilization when the engine is cold.
- Full-load enrichment: Engines operated in the part-load range with a very lean mixture require enrichment during full-load operation in addition to mixture correction resulting from the shape of the air funnel. This extra enrichment is performed by a specially designed warm-up regulator, which regulates the control pressure as a function of the intake-manifold pressure.
- Transition response during acceleration: If, at constant engine speed, the throttle valve is quickly opened, the amount of air which enters the combustion chambers plus the amount of air which is needed to bring the manifold pressure up to the new level flows through the air-flow sensor. This causes the sensor plate to briefly "overswing" past the fully opened throttle point. This "overswing" results in more fuel being metered to the engine (acceleration enrichment) and ensures good transition response.

- Overrun fuel cutoff: A solenoid valve actuated by a speed relay opens the air bypass to the sensor plate in overrun mode at a specific engine speed. The sensor plate then reverts to the zero position and interrupts fuel metering.

Lambda closed-loop control
The lambda closed-loop control system (Fig. 7) required for operation of a three-way catalytic converter necessitates the use of an electronic control unit (2). The important input variable for this ECU is the signal supplied by the lambda oxygen sensor (1).

In order to adapt the injected fuel quantity to the required air-fuel ratio at $\lambda = 1$, the pressure in the lower chambers (5) of the fuel distributor (4) is varied. If, for instance, the pressure in the lower chambers is reduced, the differential pressure at the metering slits (6) increases, whereby the injected fuel quantity is increased. In order to permit the pressure in the lower chambers to be varied, these chambers, by comparison

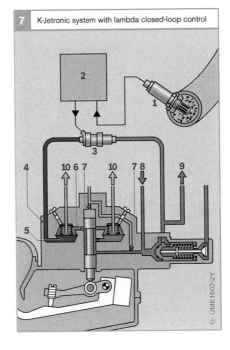

7   K-Jetronic system with lambda closed-loop control

Fig. 7
1 Lambda sensor
2 ECU
3 Timing valve
   (variable restrictor)
4 Fuel distributor
5 Lower chambers of
   differential-pressure
   valves
6 Metering slits
7 Isolating throttle
   bore (fixed restrictor)
8 Fuel inlet
9 Fuel return
10 To fuel injector

with the standard K-Jetronic fuel distributor, are isolated from the primary pressure via a fixed restrictor (7).

A further restrictor (3) connects the lower chambers and the fuel return. This restrictor is variable: If it is open, the pressure in the lower chambers can drop. If it is closed, the primary pressure builds up in the lower chambers. If this restrictor is opened and closed in a fast rhythmic succession, the pressure in the lower chambers can be varied dependent on the ratio of closing time to opening time. An electromagnetic valve, the timing valve, is used as the variable restrictor. It is controlled by electrical pulses from the lambda controller.

KE-Jetronic

The system design of KE-Jetronic is essentially identical to that of K-Jetronic. The distinguishing difference is the electronic mixture control, which in this system is performed by means of an electrohydraulic pressure actuator (Fig. 8). This pressure actuator is mounted on the fuel distributor. The actuator is a differential-pressure controller which functions according to the nozzle/baffle-plate principle, and its pressure drop is controlled by the current input from the ECU.

L-Jetronic

System overview

L-Jetronic (Fig. 9) is an electronically controlled fuel-injection system which injects fuel intermittently into the intake manifolds. It does not require any form of drive.

The electric fuel pump (2) supplies the fuel to the engine and generates the pressure necessary for injection. Fuel injectors (5) inject the fuel into the individual intake manifolds. An electronic control unit (4) controls the fuel injectors. The ECU evaluates the signals delivered by the sensors and generates the appropriate control pulses for the fuel injectors. The amount of fuel to be injected is defined by the opening time of the fuel injectors.

Acquisition of operating data

Sensors detect the operating mode of the engine and communicate this in the form of electrical signals to the ECU. The main measured variables are the engine speed and the amount of air drawn in by the engine. These variables determine the basic injection period. The engine speed is either detected by the breaker point in the ignition distributor (in breaker-triggered ignition systems) or communicated by terminal 1 of the ignition coil to the ECU (in breakerless ignition systems).

The sensor plate in the air-flow sensor measures the total amount of air inducted by the engine. The measurement allows for all changes which may take place in the engine during the service life of the vehicle, e.g.

- Wear
- Deposits in the combustion chamber, and
- Changes in valve adjustment

Fuel metering

As the central unit of the system, the ECU evaluates the data delivered by the sensors pertaining to the engine operating state. The ECU uses these data to generate control pulses for fuel metering by the fuel injectors, whereby the quantity to be injected is determined by the length of time

8   Electrohydraulic pressure actuator

UMK0159-2Y

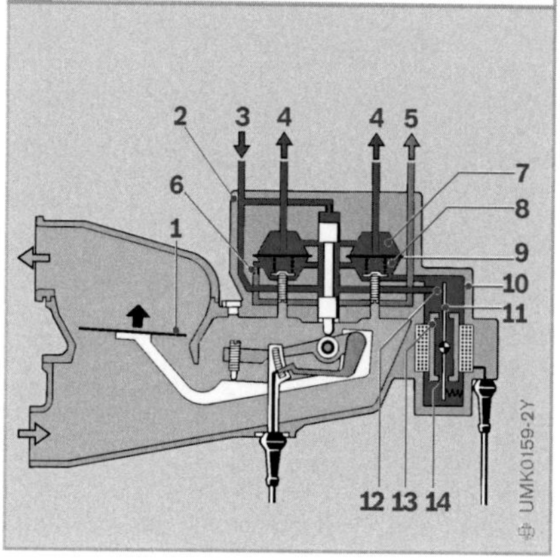

Fig. 8

1 Sensor plate
2 Fuel distributor
3 Fuel inlet (primary pressure)
4 Fuel to fuel injectors
5 Fuel-return line to pressure regulator
6 Fixed restrictor
7 Upper chamber
8 Lower chamber
9 Diaphragm
10 Pressure actuator
11 Baffle plate
12 Nozzle
13 Magnetic pole
14 Air gap

the injection valves are opened (intermittent injection).

The air/fuel ratio can be maintained at $\lambda = 1$ by means of lambda closed-loop control. The ECU compares the lambda-sensor signal with an ideal value (setpoint) and thereby activates a two-state controller.

The L-Jetronic output stage activates three or four fuel injectors in parallel. ECUs for six- and eight-cylinder engines have two output stages for three or four injectors each. Both output stages operate in unison. The injection cycle is selected so that for each revolution of the camshaft half the amount of fuel required by each working cylinder is injected twice.

Adaptation to operating states

In addition to the basic functions, the mixture has to be adapted during particular operating states. These adaptations (corrections) are necessary in order to improve the power delivered by the engine, the exhaust-gas composition, starting behavior, and performance. With additional sensors for the engine temperature and the throttle-valve

position (load signal), the L-Jetronic ECU can perform these adaptation tasks. The characteristic curve of the air-flow sensor determines the fuel-requirement curve, specific to the particular engine, for all operating ranges.

The following adaptations are possible with L-Jetronic:

- Cold-start enrichment: As a function of engine temperature, an additional quantity of fuel is injected for a limited period during starting. Enrichment is achieved by extending the injection period or by injecting an additional quantity of fuel via the cold-start valve. The injection period of the cold-start valve is limited by a thermo-time switch depending on the engine temperature.

- Post-start and warm-up enrichment: Engine cold starting is followed by its warming-up phase. During this phase, the engine needs warm-up enrichment since some of the fuel condenses on the still cold cylinder walls. In addition, without supplementary fuel enrichment during the warm-up period, a major drop in

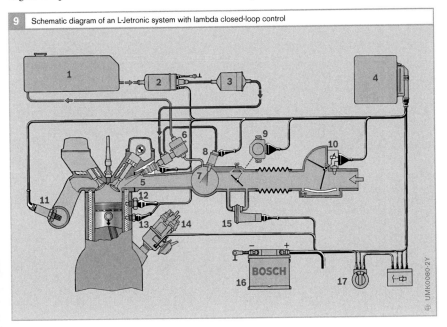

9   Schematic diagram of an L-Jetronic system with lambda closed-loop control

Fig. 9
1   Fuel tank
2   Electric fuel pump
3   Fuel filter
4   ECU
5   Fuel injector
6   Fuel rail and
    pressure regulator
7   Intake manifold
8   Cold-start valve
9   Throttle-valve switch
10   Air-flow sensor
11   Lambda sensor
12   Thermo-time switch
13   Engine-temperature
    sensor
14   Ignition distributor
15   Auxiliary-air device
16   Battery
17   Ignition/starting
    switch

engine speed would be noticed after the additional fuel from the cold-start valve has been cut off. When post-start enrichment has finished, the engine needs only a slight mixture enrichment, this being controlled by the engine temperature.

- Acceleration enrichment: If the throttle valve is quickly opened (acceleration), the amount of air which enters the combustion chambers plus the amount of air which is needed to bring the manifold pressure up to the new level flows through the air-flow sensor. This causes the sensor plate to "overswing" past the wide-open-throttle point. This "overswing" results in more fuel being metered to the engine and ensures good transition response. Since this acceleration enrichment is not sufficient during the warming-up phase, the ECU also evaluates an electrical signal representing the speed at which the sensor plate is deflected in this operating state.

- Idle-speed control: The air-flow sensor contains an adjustable bypass via which a small quantity of air can bypass the sensor plate. The idle-mixture-adjusting screw permits a basic setting of the air/fuel ratio by varying the bypass cross-section. An auxiliary-air device, which is connected as a bypass to the throttle valve, directs auxiliary air to the engine, depending on the engine temperature, in order to achieve smooth idling when the engine is cold.

- Air-temperature adaptation: The air mass crucial to combustion is dependent on the temperature of the inducted air. The intake port of the air-flow sensor incorporates a temperature sensor so that this influence can be taken into account.

L3-Jetronic

Specific systems have been developed from L-Jetronic. The most recent stage of development is L3-Jetronic, which differs from L-Jetronic in respect of the following details:
- The ECU is attached to the air-flow sensor and therefore no longer requires space in the passenger compartment
- The use of digital technology permits new functions with improved adaptation capabilities to be implemented as compared with the previous analog technology used

L3-Jetronic is available both with and without lambda closed-loop control. Both versions have what is called a "limp-home" function, which enables the driver to drive the vehicle to the nearest garage/workshop if the microcomputer fails.

By contrast with L-Jetronic, the ECU of this system adapts the air/fuel ratio by means of a load/engine-speed map. On the basis of the input signals from the sensors, the ECU computes the injection duration as a measure of the amount of fuel to be injected. It permits the required functions to be influenced.

L3-Jetronic performs corrective interventions in mixture formation with components such as the throttle-valve switch, auxiliary-air device, engine-temperature sensor, and lambda closed-loop control.

LH-Jetronic

LH-Jetronic is closely related to L-Jetronic. The difference lies in air-mass metering. The result of measurement is thus independent of the air density, which is itself dependent on temperature and pressure.

Air-mass meters

The hot-wire air-mass meter (HLM) and the hot-film air-mass meter (HFM) are thermal load sensors. They are installed between the air filter and the throttle valve and register the air-mass flow drawn in by the engine (kg/h). Both sensors operate according to the same principle.

Hot-wire air-mass meter

In the case of the hot-wire air-mass meter (Fig. 10), the electrically heated element is in the form of a 70 µm thick platinum wire. The intake-air temperature is registered by a temperature sensor. The hot wire and the intake-air temperature sensor are part of a bridge circuit and function as temperature-dependent resistors. A voltage signal which is proportional to the air-mass flow is transmitted to the ECU.

Hot-film air-mass meter

In the case of the hot-film air-mass meter (Fig. 11), the electrically heated element is in the form of a platinum film resistor (heater element). The temperature of the heater element is registered by a temperature-dependent resistor (flow sensor). The voltage across the heater element is a measure for the air-mass flow. It is converted by the hot-film air-mass meter's electronic circuitry into a voltage which is suitable for the ECU.

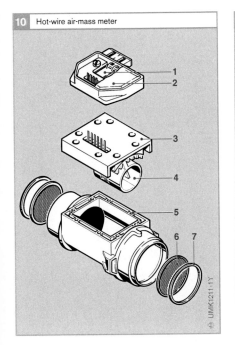

10 Hot-wire air-mass meter

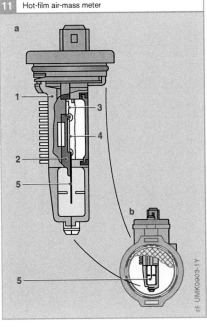

11 Hot-film air-mass meter

Fig. 10
1 Hybrid circuit
2 Cover
3 Metal insert
4 Inner tube with hot wire
5 Housing
6 Protective screen
7 Retaining ring

Fig. 11
a Hot-film sensor
b Plug-in tube with built-in hot-film sensor

1 Heat sink
2 Intermediate module
3 Power module
4 Hybrid circuit
5 Sensor element (heater element)

Mono-Jetronic

System overview

Mono-Jetronic (Fig. 12) is an electronically controlled low-pressure central injection system for four-cylinder engines with a centrally situated electromagnetic fuel injector (single-point injection) – in contrast to one fuel injector for each cylinder in the multipoint injection systems K-, KE-, L-, L3- and LH-Jetronic. The heart of Mono-Jetronic is the central injection unit with an electromagnetic fuel injector for intermittent fuel injection above the throttle valve.

The intake manifold distributes the fuel to the individual cylinders. A variety of different sensors are used to determine all the engine's operating parameters which are required for optimum mixture adaptation.

Input variables are, for example:
- Throttle-valve angle
- Engine speed
- Engine and intake-air temperatures
- Throttle-valve positions (idle/full load)
- Residual-oxygen content in the exhaust gas
- Automatic-transmission setting (depending on the vehicle equipment specification)
- Air-conditioner settings, and
- A/C-compressor switch setting.

Input circuits in the ECU condition these data for the microprocessor. The microprocessor processes the operating data, identifies the engine's operating state from these data, and calculates actuating signals depending on the data. Output stages amplify the signals and actuate the fuel injector, throttle-valve actuator and canister-purge valve (evaporative-emissions control system).

12 Mono-Jetronic system overview

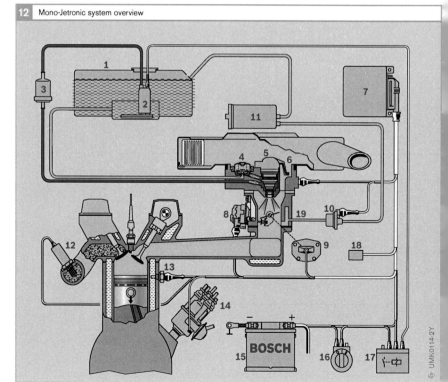

Fig. 12
1   Fuel tank
2   Electric fuel pump
3   Fuel filter
4   Pressure regulator
5   Electromagnetic
    fuel injector
6   Air-temperature
    sensor
7   ECU
8   Throttle-valve
    actuator
9   Throttle valve with
    throttle-valve
    potentiometer
10 Canister-purge valve
11 Carbon canister
12 Lambda sensor
13 Engine-temperature
    sensor
14 Ignition distributor
15 Battery
16 Ignition/starting
    switch
17 Relay
18 Diagnosis
    connection
19 Central injection
    unit

A typical Mono-Jetronic design is broken down into the following functional areas:
- Fuel supply
- Acquisition of operating data, and
- Processing of operating data

Mono-Jetronic's essential function is to control the fuel-injection process. Mono-Jetronic also incorporates a number of supplementary closed-loop and open-loop control functions with which it monitors components that influence exhaust-gas composition. These include:
- Idle-speed control
- Lambda closed-loop control, and
- Open-loop control of the evaporative-emissions control system

Central injection unit

The central injection unit (Fig. 13) is bolted directly to the intake manifold. It supplies the engine with finely atomized fuel, and is the heart of the Mono-Jetronic system. Its design is dictated by the fact that, in contrast to multi-point injection systems (e.g., L-Jetronic), gasoline injection takes place at a central point and the quantity of air inducted by the engine is determined indi-

rectly by a combination of the two variables throttle-valve angle α and engine speed n.

The lower section of the central injection unit comprises the throttle valve (3) together with the throttle-valve potentiometer. The upper section accommodates the fuel system with the fuel injector (1), the pressure regulator (4) and the fuel passages (5, 6). In addition, the air-temperature sensor (2) is located on the upper-section cap.

The fuel flows to the injector via the lower passage (6). The upper passage (5) is connected to the lower chamber of the pressure regulator, from which point excess fuel enters the fuel-return line via the plate valve.

A shoulder on the fuel-injector strainer limits the open cross-section between the inlet and return passages to a defined dimension in such a way that the excess, uninjected fuel is split into two partial flows. One flow passes through the fuel injector, while the other flows around the fuel injector for cooling purposes.

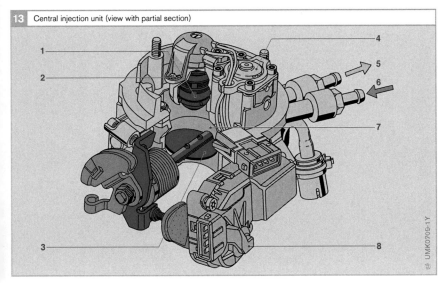

13 Central injection unit (view with partial section)

UMK0709-1Y

Fig. 13
1 Fuel injector
2 Air-temperature sensor
3 Throttle valve
4 Fuel-pressure regulator
5 Fuel return
6 Fuel inlet
7 Throttle-valve potentiometer (on throttle-valve-shaft extension, not shown)
8 Throttle-valve actuator

Fuel injector

One of Mono-Jetronic's most important functions is to distribute the air/fuel mixture uniformly to all the cylinders. Apart from intake-manifold design, distribution depends mainly on the fuel injector's location and position, and on the quality of its air-fuel mixture preparation.

The fuel injector is installed in the housing of the central injection unit's upper section, which is centered in the intake-air flow by a bracket. This installation position above the throttle valve ensures that the injected fuel is mixed thoroughly with the air that flows past. To this end, the fuel is finely atomized and injected in a cone-shaped jet between the throttle valve and the throttle-valve housing.

The fuel injector (Fig. 14) comprises the valve housing and the valve group. The valve housing contains the solenoid winding (4) and the electrical connection (1). The valve group includes the valve body, which holds the valve needle (6) and its solenoid armature (5). When no voltage is applied to the solenoid winding, a helical spring assisted by the primary system pressure forces the valve needle onto its seat. When the winding is energized, the needle lifts about 0.06 mm (depending on valve design) from its seat

so that fuel can exit through an annular gap. The front end of the valve needle incorporates a pintle (7), which projects out of the valve-body bore; the shape of this pintle atomizes the fuel.

The size of the gap between the pintle and the valve body determines the static injector quantity, i.e., the maximum fuel throughflow when the injector is permanently open. The dynamic quantity injected during intermittent operation is also dependent on the valve spring, the mass of the valve needle, the magnetic circuit, and the ECU output stage. Because the fuel pressure is constant, the amount of fuel actually injected by the fuel injector depends solely on the valve's opening time (injection duration).

The fuel injector's pickup and dropout times vary according to battery voltage. The battery voltage is therefore measured continuously by the ECU and used to correct the injection duration.

Throttle-valve actuator

Through its actuator shaft (Fig. 15, Pos. 4), the throttle-valve actuator (Fig. 15) can adjust the throttle-valve lever and thereby influence the amount of air made available to the engine. In this way, the idle speed can be regulated when the accelerator pedal is not pressed (idle).

Fig. 14
1 Electrical connection
2 Fuel return
3 Fuel inlet
4 Solenoid winding
5 Solenoid armature
6 Valve needle
7 Pintle

Fig. 15
1 Housing with
   electric motor
2 Worm
3 Worm gear
4 Actuator shaft
5 Idle contact
6 Rubber bellows

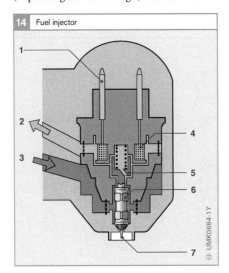

14 Fuel injector

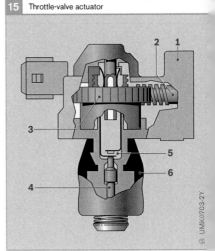

15 Throttle-valve actuator

The throttle-valve actuator is powered by a DC motor (1), which drives the actuator shaft through a worm (2) and a worm gear (3). Depending upon the motor's direction of rotation (which in turn depends upon the polarity applied to it), the actuator shaft either extends and opens the throttle valve or retracts and reduces the throttle-valve angle.

The actuator shaft incorporates a switching contact, which closes when the shaft abuts against the throttle-valve lever and thus signals the idle operating state to the ECU.

Fuel-pressure regulator

The fuel-pressure regulator keeps constant at 100 kPa the difference between fuel and ambient pressures at the fuel-injector metering point. It is integrated in the hydraulic section of the central injection unit. A rubber-fabric diaphragm divides the fuel-pressure regulator into a fuel-pressurized lower chamber (Fig. 16, Pos. 6) and an upper chamber (5), in which a preloaded helical spring (4) is supported on the diaphragm. A movable valve plate, which is connected to the diaphragm through the valve holder, is pressed onto the valve seat by spring force (flat-seat valve). When the force resulting from the fuel pressure and the diaphragm

surface exceeds the opposing spring force, the valve plate is lifted off its seat slightly. This allows fuel to flow back through the opened cross-section to the fuel tank. In this state of equilibrium, the differential pressure between the upper and lower chambers of the pressure regulator is 100 kPa. Venting ports maintain the spring chamber's ambient pressure at levels corresponding to those at the injector nozzle. The valve-plate lift varies depending on the delivery quantity and the actual fuel quantity required.

Because of the constant fuel pressure compared with the ambient pressure at the point of injection, the injected fuel quantity is determined solely by the length of time (injection duration) the injector remains open for each triggering pulse.

Acquisition of operating data

Sensors monitor all essential operating data to furnish instantaneous information on current engine operating conditions. This information is transmitted to the ECU in the form of electrical signals, which are then converted into digital form and processed for use in actuating the various actuators or final controlling elements.

Throttle-valve angle

The air charge, which is crucial to calculating the injection duration, is determined indirectly by combining the two variables throttle-valve angle α and engine speed n (α/n-system). Here, the throttle-valve potentiometer in the central injection unit records the throttle-valve angle α.

The engine map range in which the air charge varies the most as a function of α is encountered with small throttle-valve angles α and at low engine speeds n, i.e., at idle and at low part load. It follows from this that a high angle resolution is required in these operating states.

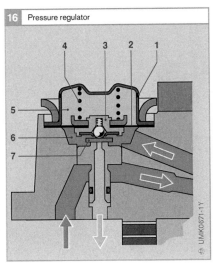

16   Pressure regulator

UMK0671-1Y

The required high level of signal resolution is achieved by distributing the throttle-valve angle in the throttle-valve potentiometer for the range between idle and full load between two resistance paths (0°...24° and 18°...90°). In the ECU, the angle signals α are read in separately, each via its own analog-digital converter channel.

Engine speed
The engine-speed information required for α/n control is obtained by monitoring the periodicity of the ignition signal. The signals provided by the ignition system are processed in the ECU. At the same time, these signals are also used for triggering the injection pulses, whereby each ignition pulse triggers an injection pulse.

Further operating states
In addition, Mono-Jetronic records the following information:
- Engine temperature in order to be able to take into account the increased fuel demand when the engine is cold.
- Intake-air temperature for compensating the temperature-dependent air density.
- Idle position for activating overrun fuel cutoff; this information is supplied by the idle contact on the throttle-valve actuator.
- Full load for activating full-load enrichment; this information is derived from the throttle-valve signal.
- Battery voltage in order to be able to compensate the voltage-dependent pickup time of the electromagnetic fuel injector and the voltage-dependent delivery rate of the electric fuel pump.
- Switching signals of the air conditioner and automatic transmission in order to be able to adapt the idle speed to the increased power demand.

Processing of operating data
The ECU generates from the data supplied by the sensors the triggering signals for the fuel injector, the throttle-valve actuator and the canister-purge valve.

Lambda program map
In order to ensure a desired air/fuel ratio, it is necessary to select the injection duration so that it is proportional to the recorded air charge. In other words: The injection duration can be directly allocated to α and n. This allocation is effected by means of a lambda program map (Fig. 17) with the input variables α and n. The influence of the air density is fully compensated here. The intake-air temperature is measured as the air enters the central injection unit and is taken into account in the ECU with a correction factor.

Mixture adaptation
When the engine is started cold, effective fuel vaporization is inhibited by the following factors:
- Cold intake air
- Cold manifold walls
- High manifold pressure
- Low air-flow velocity in the intake manifold, and
- Cold combustion chambers and cylinder walls

These conditions call for mixture adaptation in the starting phase and in the post-start and warming-up phases.

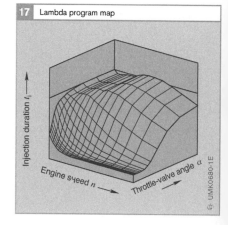

17 Lambda program map

Injection duration t_i

Engine speed n

Throttle-valve angle α

UMK0680-1E

Fig. 17
Injection duration as a function of engine speed n and throttle-valve angle α

Likewise, various corrections are made to the mixture composition when the engine is hot:
- Intake-air-dependent mixture correction
- Transient compensation in the case of load changes which are triggered by throttle-valve movements: The injected fuel quantity must be increased in the event of rapid acceleration/opening up of the throttle (buildup of wall-applied fuel film in the intake manifold) and conversely reduced in the event of rapid deceleration/closing of the throttle (reduction of wall-applied fuel film again).
- Lambda closed-loop control
- Mixture adaptation which takes into account influences on account of tolerances or changes made to the engine and fuel-injection components over the course of time.

Further Mono-Jetronic functions are:
- Idle-speed control, which guarantees a constant engine speed at idle over the entire service life of the vehicle. The engine speed is maintained under all conditions (e.g. vehicle electrical system subjected to load, switched-on air conditioner).
- Altitude compensation, which compensates for lower intake-air density at high altitudes.
- Full-load enrichment so that maximum engine power can be delivered when the accelerator pedal is pressed to the floor.
- Engine-speed limitation in order to prevent the engine from incurring damage caused by excessive engine speeds.
- Overrun fuel cutoff so that exhaust-gas emissions are reduced when driving with the throttle valve closed (vehicle on overrun).

Lambda closed-loop control
Mono-Jetronic is equipped with a lambda closed-loop control function, designed to maintain the air/fuel ratio for the three-way catalytic converter at $\lambda = 1$. Adaptive mixture corrections are also performed, i.e., the self-learning system adapts itself to the changing conditions.

With this adaptive mixture control and the additionally superimposed lambda control loop, indirect recording of the inducted air mass by means of α/n control facilitates constancy of mixture. Air-mass metering no longer needs to be performed here.

Fuel supply

The function of the fuel-supply system is to deliver fuel at a defined pressure to the fuel injectors. The fuel injectors inject the fuel into the intake manifold (manifold injection) or directly into the combustion chamber (gasoline direct injection). In the case of manifold injection, an electric fuel pump delivers the fuel from the tank to the fuel injectors. In the case of gasoline direct injection, the fuel is likewise delivered from the tank by means of an electric fuel pump; then it is compressed to a higher pressure by a high-pressure pump and supplied to the high-pressure injectors.

Fuel delivery with manifold injection

An electric fuel pump delivers the fuel and generates the injection pressure, which for manifold injection is typically about 0.3...0.4 MPa (3...4 bar). The built-up fuel pressure to a large extent prevents vapor bubbles from forming in the fuel system. A non-return valve integrated in the pump stops fuel from flowing back through the pump to the fuel tank and thereby maintains the system pressure for a certain amount of time, even after the electric fuel pump has been switched off. This prevents the formation of vapor bubbles in the fuel system when the fuel heats up after the engine has been switched off.

System with fuel return
The fuel is drawn from the fuel tank (Fig. 1, Pos. 1) and passes through the fuel filter into a high-pressure line, from where it flows to the engine-mounted fuel rail (7). The rail supplies the fuel to the fuel injectors (6). A mechanical pressure regulator (5) mounted on the rail keeps the differential pressure between the fuel injectors and the intake manifold constant, regardless of the absolute intake-manifold pressure, i.e., the engine load.

The fuel not needed by the engine flows through the rail via a return line (8) connected to the pressure regulator back to the fuel tank. The excess fuel heated in the engine compartment causes the fuel temperature in the tank to rise. Fuel vapors are formed in the tank as a function of fuel temperature. Ensuring adherence to environmental-protection regulations, the vapors are routed through a tank-ventilation system for intermediate storage in a carbon canister until they can be returned through the intake manifold for combustion in the engine (evaporative-emissions control system).

Fig. 1
1   Fuel tank
2   Electric fuel pump
3   Fuel filter
4   High-pressure line
5   Pressure regulator
6   Fuel injectors
7   Fuel rail
8   Return line

Fig. 2
1   Suction-jet pump for tank filling
2   Electric fuel pump with fuel filter
3   Fuel-pressure regulator
4   High-pressure line
5   Fuel rail
6   Fuel injectors

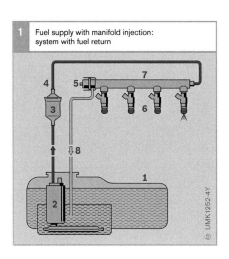

1 Fuel supply with manifold injection: system with fuel return

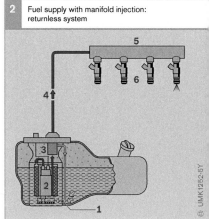

2 Fuel supply with manifold injection: returnless system

Returnless system

In a returnless fuel-supply system (Fig. 2), the pressure regulator (3) is located in the fuel tank or in its immediate vicinity. A return line from the engine to the fuel tank is therefore rendered superfluous.

Since the pressure regulator on account of its installation location has no reference to the intake-manifold pressure, the relative injection pressure here is not dependent on the engine load. This is taken into account in the calculation of the injection duration in the engine ECU.

Only the amount of fuel which is to be injected is delivered to the rail (5). The excess flow volume delivered by the electric fuel pump (2) returns directly to the fuel tank without taking the circuitous route through the engine compartment. In this way, fuel heating in the fuel tank and thus also evaporative emissions are significantly lower than in systems with fuel return.

Because of these advantages, it is returnless systems which are predominantly used today.

Demand-controlled system

In a demand-controlled system (Fig. 3), the fuel-supply pump delivers only that amount of fuel that is currently used by the engine and that is required to set up the desired pressure. Pressure control is effected by means of a closed control loop in the engine ECU, whereby the current fuel pressure is recorded by a low-pressure sensor. A mechanical pressure regulator is rendered superfluous. To adjust the delivery volume of the fuel-supply pump, its operating voltage is altered by means of a clock module that is triggered by the engine ECU.

The system is equipped with a pressure-relief valve (3) to prevent the buildup of excessive pressure even during overrun fuel cutoff or after the engine has been switched off.

As a result of demand control, no excess fuel is compressed and thus the capacity of the electric fuel pump minimized. Compared with systems with maximum-delivery electric fuel pumps, this lowers fuel consumption and can also reduce still further the fuel temperatures in the tank.

Further advantages of a demand-controlled system are derived from the variably adjustable fuel pressure. On the one hand, the pressure can be increased during hot starting to prevent the formation of vapor bubbles. On the other hand, it is possible above all in turbocharging applications to extend the metering range of the fuel injectors by effecting a pressure increase at full load and a pressure decrease at very low loads.

Furthermore, the measured fuel pressure provides for improved diagnostic options for the fuel system compared with previous systems. In addition, the fact that the current fuel pressure is taken into account in the calculation of the injection duration results in higher-precision fuel metering.

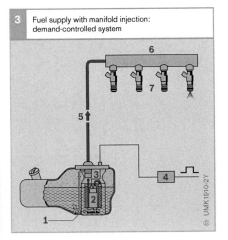

3 Fuel supply with manifold injection: demand-controlled system

Fig. 3
1   Suction-jet pump for tank filling
2   Electric fuel pump with fuel filter
3   Pressure-relief valve and pressure sensor
4   Clock module for controlling electric fuel pump
5   High-pressure line
6   Fuel rail
7   Fuel injectors

Fuel delivery with gasoline direct injection

Compared with injecting fuel into the intake manifold, there is only a limited time window available for injecting fuel directly into the combustion chamber. Increased importance is also attached to mixture preparation. For this reason, fuel must be injected at significantly higher pressure with direct injection than with manifold injection.

The fuel system is divided into:
● Low-pressure circuit, and
● High-pressure circuit

Low-pressure circuit

Low-pressure circuits for gasoline direct injection essentially use the fuel systems and components known in manifold-injection systems. Due to the fact that currently used high-pressure pumps require increased predelivery pressure (admission pressure) in order to prevent vapor-bubble formation during hot starts and high-temperature operation, it is advantageous to use systems with variable low pressure. Demand-controlled low-pressure systems are particularly well suited here in that the optimum admission pressure in each case can be set for every engine operating state. However, other systems are used. They may be returnless systems with selectable admission pressure (controlled by means of a shutoff valve) or systems featuring a constant, high admission pressure.

High-pressure circuit

The high-pressure circuit consists of
● High-pressure pump
● High-pressure fuel rail
● High-pressure sensor

and, depending on the system,
● Pressure-control valve, or
● Pressure-limiting valve

Where both continuous-delivery and demand-controlled high-pressure systems are used in 1st-generation gasoline direct injection, 2nd-generation systems are demand-controlled.

Depending on the operating point, a system pressure of between 5 and 12 MPa and in 2nd-generation systems of up to 20 MPa is set by means of high-pressure control in the engine ECU. The high-pressure injectors injecting the fuel directly into the engine's combustion chamber are mounted on the fuel rail.

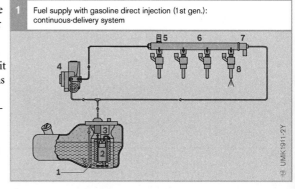

1 Fuel supply with gasoline direct injection (1st gen.): continuous-delivery system

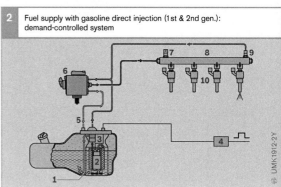

2 Fuel supply with gasoline direct injection (1st & 2nd gen.): demand-controlled system

Continuous-delivery system

A high-pressure pump (Fig. 1, Pos. 4) driven by the engine camshaft, normally a three-barrel radial-piston pump, forces fuel into the rail against the system pressure. The pump's delivery quantity is not adjustable. The excess fuel not required for fuel injection or to maintain the pressure is depressurized by the pressure-control valve (7) and returned to the low-pressure circuit. For this purpose, the pressure-control valve is actuated by the engine ECU in such a way as to obtain the injection pressure required at a given operating point. The pressure-control valve doubles up as a mechanical pressure-limiting valve.

In continuous-delivery systems, most of the operating points cause significantly more fuel to be compressed to high system pressure than is needed by the engine. This involves an unnecessary expenditure of energy and with it increased fuel consumption; furthermore, the excess fuel depressurized by the pressure-control valve contributes to increasing the temperature in the fuel system. To avoid this problem, demand-controlled high-pressure systems are now preferred.

Demand-controlled system

In a demand-controlled system, the high-pressure pump – usually a single-barrel radial-piston pump – delivers to the fuel rail only that amount of fuel which is actually needed for injection and to maintain the pressure (Fig. 2). The pump (6) is driven by the engine camshaft. The delivery quantity is adjusted by a fuel-supply control valve: The engine ECU actuates the pump's fuel-supply control valve in such a way as to obtain in the rail the necessary system pressure for a given operating point.

For safety reasons, the high-pressure circuit features an integrated mechanical pressure-limiting valve; this valve is mounted on the fuel rail (8) in the case of 1st-generation gasoline direct injection, and integrated directly in the high-pressure pump in the case of the 2nd generation. Should the pressure

exceed the permissible level, fuel is returned via the pressure-limiting valve to the low-pressure circuit.

Evaporative-emissions control system

Vehicles with gasoline engines are equipped with an evaporative-emissions control system to prevent fuel that evaporates from the fuel tank from escaping to atmosphere. The maximum permissible limits for evaporative hydrocarbon emissions are laid down in emission-control legislation.

Design and method of operation

Fuel vapor is routed via a vent line (Fig. 3, Pos. 2) from the fuel tank (1) to the carbon canister (3). The activated carbon absorbs the fuel contained in the fuel vapor and allows the air to escape to atmosphere through the fresh-air inlet opening (4). In order to ensure that the carbon canister is always able to absorb freshly evaporating fuel, the activated carbon must be regenerated at regular intervals. The carbon canister is connected to the intake manifold (8) via a canister-purge valve (5) for this purpose.

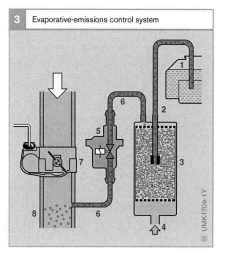

3 Evaporative-emissions control system

Fig. 3
1 Fuel tank
2 Fuel-tank vent line
3 Carbon canister
4 Fresh air
5 Canister-purge valve
6 Line to intake manifold
7 Throttle valve
8 Intake manifold

To regenerate, the canister-purge valve is actuated by the engine-management system and opens the line connecting the canister to the intake manifold. Fresh air (4) is drawn in through the activated carbon as a result of the vacuum pressure in the intake manifold. The fresh air takes up the absorbed fuel from the carbon canister and carries it to the intake manifold. From there, it passes with the air inducted by the engine into the combustion chamber. The injected fuel quantity is simultaneously reduced so that the correct fuel quantity is available. The fuel quantity drawn in through the carbon canister is calculated by means of the measured excess-air factor λ and regulated to a setpoint value.

The purge-gas quantity, i.e. the air/fuel mixture that flows in through the canister-purge valve, is limited due to possible fluctuations of the fuel concentration; this is because the greater the proportion of fuel supplied through the valve, the quicker and more intensively the system will have to correct the injected fuel quantity. This correction is effected by means of lambda closed-loop control, whereby fluctuations of concentration are necessarily compensated for with a time delay. In order not to impair exhaust-gas values and driveability, it is essential for fluctuations of the lambda value to be kept to a minimum by limiting the purge-gas quantity.

Gasoline direction injection: special features
The effect of purging is limited in systems with gasoline direct injection in stratified-charge mode since the extensive dethrottling gives rise to a low intake-manifold vacuum. This results in reduced purge-gas flow compared to homogeneous operation. For instance, if the purge-gas flow is inadequate for coping with high levels of gasoline evaporation, the engine must be operated in homogeneous mode until the high concentrations of gasoline in the purge-gas flow have dropped far enough.

Fuel-vapor generation
Increased evaporation of fuel from the fuel tank is caused when
● The fuel in the fuel tank is heated up on account of increased ambient temperature, by adjoining hot components (e.g., exhaust system) or by the return of heated fuel to the fuel tank
● The ambient pressure drops, e.g. when driving up a hill in mountain environments

Electric fuel pump

Function
The electric fuel pump must in all operating states deliver enough fuel to the engine at a high enough pressure to permit efficient fuel injection. The most important performance demands made on the pump are:
● Delivery quantity between 60 and 250 l/h at nominal voltage
● Pressure in the fuel system between 300 and 650 kPa (3.0...6.5 bar)
● Buildup of system pressure from 50...60 % of nominal voltage; the decisive factor here is operation during cold starting

Apart from this, the electric fuel pump is increasingly being used as the pre-supply pump for modern direct-injection systems used on both gasoline and diesel engines. In the case of gasoline direct injection, sometimes pressures of up to 700 kPa must be provided during hot-delivery operation.

Design
The electric fuel pump comprises:
● Fitting cover (Fig. 1, A) with electrical connections, non-return valve (preventing fuel from escaping from the fuel system) and hydraulic outlet. The fitting cover usually also contains the carbon brushes for operating the commutator drive motor and interference-suppression elements (inductance coils and, if necessary, capacitors).

- Electric motor (B) with armature and permanent magnet (a copper commutator is standard, carbon commutators are used for special applications and diesel systems).
- Pump section (C), designed as a positive-displacement or flow-type pump.

Types

Positive-displacement pump

In a positive-displacement pump, volumes of liquid are basically drawn in and transported in a closed chamber (apart from leaks) by rotation of the pump element to the high-pressure side. A *roller-cell pump* (Fig. 2a), an *internal-gear pump* (Fig. 2b) or a *screw-spindle pump* may be used for the electric fuel pump.

Positive-displacement pumps are advantageous at high system pressures (450 kPa and above) and have a good low-voltage characteristic, i.e., they have a relatively "flat" delivery-rate characteristic over the operating voltage. Efficiency can be as high as 25%. Pressure pulsations, which are unavoidable, can cause audible noise depending on the particular design details and installation conditions.

Whereas in electronic gasoline-injection systems the positive-displacement pump has to a large extent been superseded by the flow-type pump for the classical electric-fuel-pump requirements, it has gained a new field of application as the pre-supply pump on direct-injection systems (gasoline and diesel) with their significantly greater pressure requirements and viscosity range.

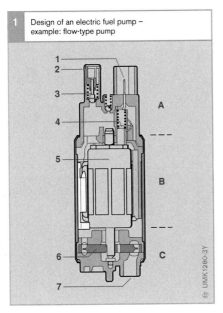

1   Design of an electric fuel pump – example: flow-type pump

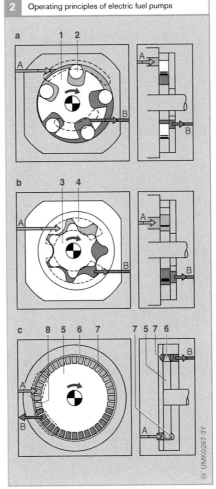

2   Operating principles of electric fuel pumps

Fig. 1

1   Electrical connection
2   Hydraulic connection (fuel outlet)
3   Non-return valve
4   Carbon brushes
5   Motor armature with permanent magnet
6   Flow-type-pump impeller
7   Hydraulic connection (fuel inlet)

Fig. 2

a   Roller-cell pump
b   Internal-gear pump
c   Peripheral pump

A   Intake port
B   Outlet

1   Slotted rotor (eccentric)
2   Roller
3   Inner drive wheel
4   Rotor
5   Impeller
6   Impeller blades
7   Passage (peripheral)
8   "Stopper"

Flow-type pump

The flow-type pump has become the accepted solution for gasoline applications up to 500 kPa. An impeller with numerous blades (Fig. 2c, Pos. 6) around its periphery rotates in a chamber consisting of two fixed housing sections. These housing sections feature a passage (7) in each case in the area of the impeller blades. These passages begin at the height of the intake port (A) and end at the point where the fuel exits the pump section at system pressure (B). For the purpose of improving the hot-delivery characteristics, a small degassing bore is provided at a given angle and distance from the intake port, which (at the expense of a very slight internal leakage) facilitates the exit of any gas bubbles which may be in the fuel.

Pressure builds up along the passage as a result of the exchange of pulses between the impeller blades and the liquid particles. This leads to spiral-shaped rotation of the liquid volume trapped in the impeller and in the passages.

Flow-type pumps feature a low noise level since pressure buildup takes place continuously and is practically pulsation-free. They are much simpler in terms of design and construction than positive-displacement pumps. Single-stage pumps can generate system pressures of up to 500 kPa. The efficiency of these pumps can be as high as 22 %.

Outlook

Some modern vehicles are already supplied with fuel by demand-controlled fuel-supply systems. In these systems, an electronic module drives the pump as a function of the required pressure, which is monitored by a fuel-pressure sensor. The advantages of such systems are:

- Low current consumption
- Reduced heat entry through the electric motor
- Reduced pump noise, and
- Possibility of setting variable pressures in the fuel system

In future systems, pure pump control will be extended to included further functions. Examples include:

- Tank-leakage diagnosis and evaluation of the fuel-level-sensor signal
- Actuation of valves, e.g., for fuel-vapor management

In order to comply with the increasing demands with regard to pressure and service life and the differing fuel grades around the world, non-contact motors with electronic commutation will play a more important role in the future.

Fuel-supply modules

Whereas in the early stages of electronic gasoline injection the electric fuel pump was mounted exclusively outside the fuel tank (in-line), it is common practice today to install the electric fuel pump inside the tank itself. In this case, the electric fuel pump forms an integral part of a fuel-supply module which may comprise further elements:

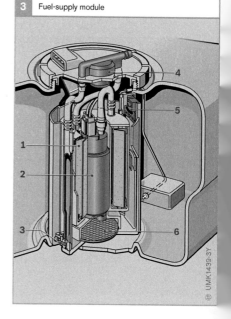

3   Fuel-supply module

Fig. 3
1   Fuel filter
2   Electric fuel pump
3   Jet pump
    (closed-loop-
    controlled)
4   Fuel-pressure
    regulator
5   Fuel-level sensor
6   Prefilter

- A bowl as fuel reservoir for cornering (usually actively filled by a suction-jet pump or passively by a flap system, switchover valve or similar)
- A fuel-level sensor
- A pressure regulator in returnless systems (RLFS)
- A suction filter for protecting the pump
- A pressure-side fuel fine filter, which does not need to be changed over the entire service life of the vehicle
- Electrical and hydraulic connections
- Furthermore, tank-pressure sensors (for tank-leakage diagnosis), fuel-pressure sensors (for demand-controlled systems) and valves can be integrated.

Gasoline filter

The function of the gasoline filter is to absorb and permanently accumulate dirt particles from the fuel so as to protect the fuel-injection system against wear caused by particle erosion.

Design
Fuel filters for gasoline engines are located on the pressure side after the fuel-supply pump. In-tank filters are the preferred choice in newer vehicles, i.e., the filter is integrated in the fuel tank. In this case, it must always be designed as a lifetime filter, which does not need to be changed over the full service life of the vehicle. Furthermore, in-line filters, which are installed in the fuel line, continue to be used. These can be designed as replacement parts or lifetime parts.

The filter housing is manufactured from steel, aluminum or plastic. It is connected to the fuel feed line by a thread, tube or quick-action connection. The housing contains the filter element, which filters the dirt particles out of the fuel. The filter element is integrated in the fuel circuit in such a way that fuel passes through the entire surface of the filter medium as much as possible at the same flow velocity.

Filter medium
Special resin-impregnated microfiber papers which are also bonded for higher-duty applications to a synthetic-fiber (meltblown) layer are used as the filter medium. This bond must ensure high mechanical, thermal and chemical stability. The paper porosity and the pore distribution of the filter paper determine the filtration efficiency and throughflow resistance of the filter.

Filters for gasoline engines are either spiral vee-form or radial vee-form in design. In a spiral vee-form filter (Fig. 1), an embossed filter paper is wrapped round a support tube. The unfiltered fuel flows through the filter in the longitudinal direction.

In a radial vee-form filter (Fig. 2), the filter paper is folded and inserted into the housing in the shape of a star. Plastic, resin or metal end rings and, if necessary, an inner protective jacket provide stability. The unfiltered fuel flows through the filter from the outside inwards, during which the dirt particles are separated from the filter medium.

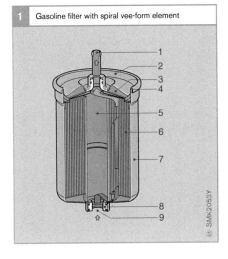

1 Gasoline filter with spiral vee-form element

SMK2053Y

Fig. 1
1 Fuel outlet
2 Filter cover
3 Support plate
4 Double flange
5 Support tube
6 Filter medium
7 Filter housing
8 Screw-on fitting
9 Filter inlet

Filtration effects

Solid dirt particles are separated both by means of the straining effect and by means of impact, diffusion and barrier effects. The straining effect is based on the fact that larger particles on account of their dimensions cannot pass through the filter's pores.

Smaller particles, on the other hand, adhere to filter-medium fibers when they strike these fibers. Three different mechanisms are distinguished here: In the case of the barrier effect, the particles are flushed around the fibers with the fuel flow, but touch the edges of these fibers and are retained on these edges by intermolecular forces. Heavier particles, because of their mass inertia, do not follow the fuel flow around the filter fibers; instead, they strike the fibers frontally (impact effect). In the case of the diffusion effect, very small particles, on account of their proper motion (Brownian molecular motion), touch filter fibers by chance, at which point they adhere to the fibers.

The filtration efficiency of the individual effects is dependent on the size, the material and the rate of flow of the particles.

Requirements

The required filter fineness is dependent on the fuel-injection system. For manifold-injection systems, the filter element has a mean pore size of approximately $10\,\mu m$. Gasoline direct injection requires finer filtration. The mean pore size is in the range of $5\,\mu m$. Particles which are more than $5\,\mu m$ in size must be separated at a rate of 85 %. In addition, a filter for gasoline direct injection, when new, must satisfy the following residual-dirt requirement: Metal, mineral and plastic particles and glass fibers with diameters of more than $200\,\mu m$ must be reliably filtered out of the fuel.

Filter efficiency depends on the through-flow direction. When replacing in-line filters, it is imperative that the flow direction specified by the arrow be observed.

The interval for changing conventional in-line filters is, depending on filter volume and fuel contamination, normally between 30,000 km and 90,000 km. In-tank filters generally have change intervals of at least 160,000 km. There are in-tank and in-line filters available for use with gasoline direct-injection systems which feature service lives in excess of 250,000 km.

Fig. 2
1   Fuel outlet
2   Filter cover
3   Sealing ring
4   Internally welded
    edge
5   Support ring
6   Filter medium
7   Filter housing
8   Filter inlet

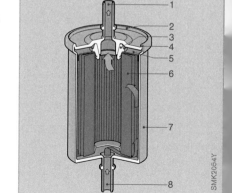

2   Gasoline filter with radial vee-form element

High-pressure pumps for gasoline direct injection

Function
The function of the high-pressure pump (German: Hochdruckpumpe, hence HDP) is to compress a sufficient quantity of the fuel delivered by the electric fuel pump at an admission pressure of 0.3...0.5 MPa (3...5 bar) to the level required for high-pressure injection of 5...12 MPa (1st-generation direct injection) or 5...20 MPa (2nd-generation direct injection).

Different high-pressure pumps are used in the various direct-injection systems.

Types
HDP1 (1st-generation direct injection, continuous-delivery)
Design and method of operation
The HDP1 is a radial-piston pump with three delivery barrels situated at circumfer-ential offsets of 120°. Figure 1 shows the longitudinal and cross-sections of the HDP1.

Driven by the engine camshaft, the drive shaft (13) rotates with the eccentric cam (1). The eccentric cam converts the rotational motion via the cam ring (10) and the slipper (2) in a vertical motion of the pump pistons (4). The drive runs in gasoline for cooling and lubrication purposes.

The fuel delivered by the electric fuel pump passes enters the HDP1 through the fuel inlet (9). The pump pistons contain transverse and longitudinal ports, through which the fuel enters the displacement chambers of the three delivery barrels. As the pump piston travels from top to bottom dead center, the fuel is drawn in through the inlet valve (7). In the delivery stroke, the drawn-in fuel is compressed as the pump piston travels from bottom to top dead center and delivered through the outlet valve into the high-pressure area.

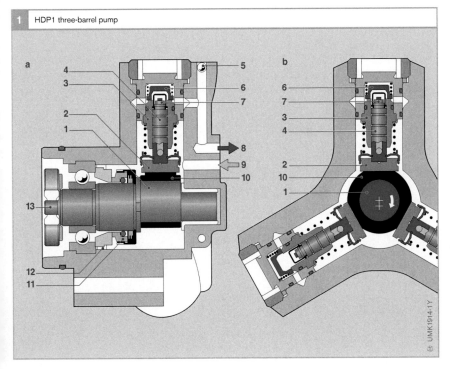

1   HDP1 three-barrel pump

Fig. 1
a   Longitudinal section
b   Cross-section

1   Eccentric cam
2   Slipper
3   Pump barrel
4   Pump piston
     (hollow piston,
     fuel inlet)
5   Sealing ball
6   Outlet valve
7   Inlet valve
8   High-pressure
     connection to
     fuel rail
9   Fuel inlet
     (low pressure)
10   Cam ring
11   Axial face seal
12   Static seal
13   Drive shaft

UMK1914-1Y

The HDP1 is a continuous-delivery fuel-supply pump, its delivery quantity being proportional to rotational speed. The three barrels deliver fuel at circumferential offsets of 120° in order to ensure overlapping and therefore continuous delivery. This gives rise to minimal pressure pulsations only. This means that, when compared with demand-controlled systems with single-barrel pumps, less demands have to be placed on the pump connections and piping. Furthermore, there is no need for a low-pressure attenuator. The system can therefore also be integrated relatively easily in already existing engine platforms from manifold-injection development.

To ensure that the system pressure can be varied at sufficient speed even in the event of maximum engine fuel demand, the maximum delivery quantity of the HDP is configured for maximum demand. Factors influencing delivery performance (e.g., hot gasoline, pump ageing, dynamics) are taken into account. When operating at constant rail pressure or at part load, the pressure-control valve depressurizes the excess delivered fuel quantity to admission pressure level, and the fuel is returned to the suction side of the HDP. The pressure level in the high-pressure circuit is regulated and adjusted by means of the engine ECU, which specifically actuates the pressure-control valve.

If one or more of the delivery barrels should fail, emergency operation is possible with the intact barrels or by means of the electric fuel pump with admission pressure.

Technical features
- Continuous-delivery three-barrel pump
- Pressure range up to 12 MPa (120 bar)
- Delivery rate 0.4...0.5 cm³/rev_{cam} (camshaft revolution)
- Speed up to 7000 rpm (engine speed)
- Weight approx. 1000 g
- Dimensions: dia. ≈ 125 mm, l ≈ 65 mm
- Drive via camshaft
- Drive runs in gasoline
- Suitable for engines up to V_H = 2.2 l, P_{max} = 125 kW

HDP2 (1st-generation direct injection, demand-controlled)

Design and method of operation
The HDP2 (Fig. 3) is a cam-driven single-barrel pump that runs in oil with an integrated fuel-supply control valve (10) and a pressure attenuator (11) on the low-pressure side. The fuel-supply control valve facilitates control intervention on the high-pressure side. The HDP2 is driven via the engine's intake or exhaust camshaft such that it is ideally mounted as a plug-in pump directly on the cylinder head.

The rotational motion of the camshaft is – depending on the engine's fuel demand – is transmitted via two or three cams to the pump piston (Fig. 2). A barrel tappet provides the connection between the camshaft and the delivery barrel for all cam variants.

The fuel delivered by the electric fuel pump is drawn into the delivery chamber via the inlet valve (Fig. 3, Pos. 5). In the suction stroke, the fuel-supply control valve is not actuated (unenergized) and the fuel is drawn in through the inlet valve (spring-loaded non-return valve). In the delivery stroke, the fuel-supply control valve is closed from bottom dead center; the fuel is compressed and delivered to the high-pressure circuit. Actuation of the fuel-supply control valve is deactivated when the fuel quantity required in the relevant load state is reached.

2   Drive of single-plunger/barrel pump

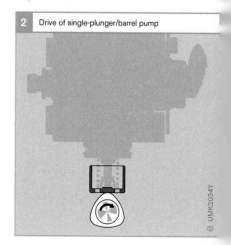

UMK2034-Y

The fuel that is not required is returned at admission pressure to the low-pressure circuit.

The volumetric efficiency is derived from the ratio of actually delivered fuel quantity to theoretically possible quantity. This is dependent on the piston diameter and stroke. The volumetric efficiency is not constant over the full rotational speed (Fig. 4). It is dependent on the following factors:

● In the lower speed range: piston and other leakages
● In the upper speed range: inertia and opening pressure of the inlet valve
● In the total speed range: dead volume of the delivery chamber and temperature dependence of fuel compressibility

The HDP2.1 is an aluminum variant of the HDP2. A stainless-steel variant, the HDP2.5, is a further development in terms of resistance to media (e.g., fuels containing ethanol).

Pressure attenuator DD

The DD pressure attenuator (German: Druckdämpfer) (Fig. 3, Pos. 11) is integrated in the HDP2. Its function is to limit the pressure pulsations that occur in the lower-pressure circuit to ±1 bar. This is achieved by the pressure attenuator taking up the diverted fuel quantity and releasing it again in the subsequent suction stroke. In addition, it supports the filling process of the

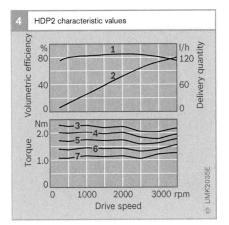

4   HDP2 characteristic values

Fig. 4
1   Volumetric efficiency
2   Delivery quantity

Torque at
3   4 MPa
4   6 MPa
5   8 MPa
6   10 MPa
7   12 MPa

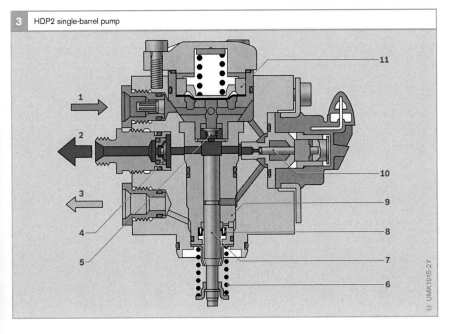

3   HDP2 single-barrel pump

Fig. 3
1   Fuel inlet
    (low pressure)
2   High-pressure
    connection to
    fuel rail
3   Leakage return
4   Outlet valve
5   Inlet valve
6   Plunger spring
7   Pump plunger
8   Plunger seal
9   Pump barrel
10 Fuel-supply
    control valve
11 Pressure attenuator

■ High-pressure area
■ Low-pressure area
■ Zero-pressure area
   (return)

pump interior during the intake process. This ensures that an additional vacuum is not generated by the inertia of the fuel during intake.

In the pressure attenuator, an elastomer diaphragm separates the fuel-filled admission-pressure chamber from a spring chamber. The pressure range of the pressure attenuator is adjusted in accordance with the vehicle manufacturer's instructions (depending on the maximum occurring pump or fuel temperature) via the inserted spring (Fig. 5). As fuel temperature increases, a higher pressure is needed to prevent vapor-bubble formation (Fig. 6).

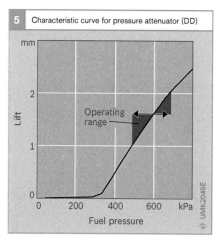

5 | Characteristic curve for pressure attenuator (DD)

Operating range

MSV1 fuel-supply control valve
Demand control of the HDP2 high-pressure pump is effected with the MSV1-type fuel-supply control valve (German: **M**engen-**steuerv**entil, hence MSV). Only the required fuel quantity is delivered at high pressure to the fuel rail. The MSV is therefore referred to as a metering unit.

The MSV (Fig. 7) is an electrically switched solenoid valve which is open at zero current. When the solenoid coil (3) is energized, the valve needle (4) is drawn into its seat so that the required delivery pressure can be built up in the delivery stroke. The delivery period of the pump is determined as a function of load by the actuation period of the MSV. It is also dependent on the rail pressure, which is recorded by a pressure sensor and regulated to a specified value.

During the suction stroke, fuel flows from the low-pressure are through the inlet valve into the delivery chamber. The delivery stroke begins after the pump piston has reached bottom dead center. The fuel is compressed and delivered via the outlet valve to the high-pressure area as soon as the fuel pressure in the pump exceeds the pressure obtained in the rail. The opening pressure of the outlet valve is negligible by comparison with the rail pressure.

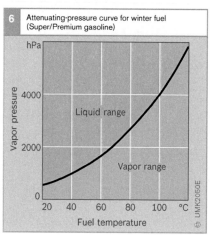

6 | Attenuating-pressure curve for winter fuel (Super/Premium gasoline)

Liquid range

Vapor range

Fig. 7
1 Electrical connection
2 Solenoid armature
3 Solenoid coil
4 Valve needle
5 Valve body

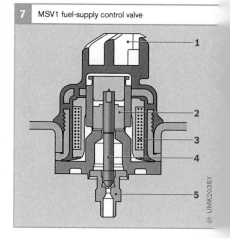

7 | MSV1 fuel-supply control valve

When the delivery period has ended, the MSV is deactivated, the valve opens and the compressed fuel confined in the delivery chamber is delivered to the low-pressure area (Fig. 8).

Technical features of HDP2
- Demand-controlled single-barrel pump
- Pressure range up to 12 MPa (120 bar)
- Delivery rate:
  0.5 cm³/rev_{cam} for two cams,
  0.75 cm³/rev_{cam} for three cams
- Speed up to 7000 rpm (engine speed)
- Weight: 1000 g (aluminum version HDP2.1) or 2500 g (stainless-steel version HDP2.5)
- Dimensions h = 85 mm, b = 110 mm, s = 80 mm
- Integrated pressure attenuator
- Integrated fuel-supply control valve with demand control by variation of the end of delivery
- Drive via intake or exhaust camshaft, power transmitted via barrel tappet
- Drive runs in oil
- Mounted on the cylinder head or adapter housing

- Several high-pressure pumps can be used, depending on the fuel demand (e.g., for 8-cylinder engines)

HDP5 (2nd-generation direct injection)
Design and method of operation
The HDP5 (Fig. 10) is a cam-driven single-barrel pump that runs in oil with an integrated fuel-supply control valve (metering unit), a pressure-limiting valve on the high-pressure side and an integrated pressure attenuator (on the low-pressure side). Like the HDP2, it is mounted as a plug-in pump on the cylinder head.

The point of connection between the camshaft and the delivery barrel is provided in the case of two cams by a barrel tappet and in the case of three and four cams by a roller tappet (Fig. 9). This ensures that the cam-lifting curve is transmitted to the delivery plunger/piston. Here, the requirements with regard to criteria such as lubrication, Hertzian stress and mass inertia are greater than for the HDP2. During the cam lift, the roller tappet travels with it roller down the cam profile. This results in the vertical motion, or stroke, of the delivery plunger.

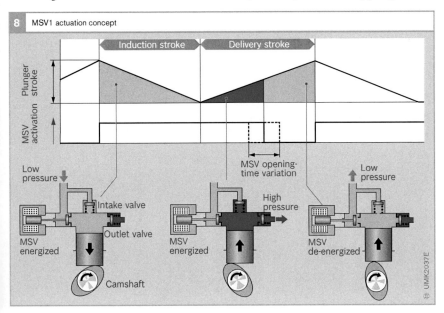

8 MSV1 actuation concept

Induction stroke   Delivery stroke

Plunger stroke

MSV activation

Low pressure

Intake valve

Outlet valve

MSV energized

MSV opening-time variation

High pressure

MSV energized

Low pressure

MSV de-energized

Camshaft

UMK2037E

In the delivery stroke, the roller tappet absorbs the applied forces, such as pressure, mass, spring and contact forces. It is rotationally secured in the process.

With four cams, time synchronization of delivery and injection is possible in a 4-cylinder engine, i.e., each injection is also

accompanied by a delivery. In this way, it is possible on the one hand to reduce excitation of the high-pressure circuit and on the other hand to reduce the rail volume.

MMD pressure attenuator
The variable pressure attenuator (0.05...0.6 MPa) of the HDP5 attenuates the pressure pulsations excited by the high-pressure pump in the low-pressure circuit and also guarantees good filling at high speeds. The pressure attenuator takes up the fuel quantity diverted at the relevant operating point via the deformation of its gas-filled diaphragms and releases it again in the suction stroke to fill the delivery chamber. Operation with variable admission pressure – i.e., the use of demand-controlled low-pressure systems – is possible here.

MSV5 fuel-supply control valve
Demand control of the HDP5 high-pressure pump is effected with the MSV5-type fuel-supply control valve. The fuel delivered by

9   Drive of HDP5

UMK2038Y

10   HDP5 single-barrel pump

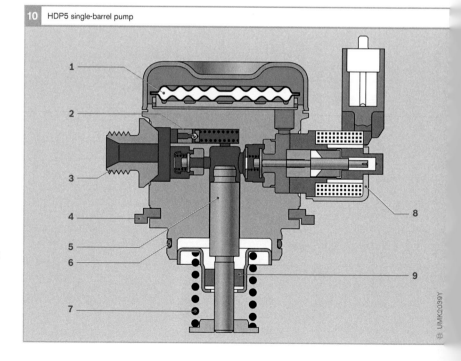

UMK2039Y

Fig. 10
1   Variable pressure attenuator MMD
2   Pressure-limiting valve
3   High-pressure port
4   Mounting flange
5   Delivery plunger
6   O-ring
7   Plunger spring
8   MSV5 fuel-supply control valve
9   Plunger seal

the electric fuel pump is drawn into the
delivery chamber via the inlet valve of the
open fuel-supply control valve. In the subsequent delivery stroke, the MSV remains
open after bottom dead center so that fuel
that is not needed at the relevant load point
is returned at admission pressure to the low-
pressure circuit. After the MSV is actuated,
the inlet valve closes, the fuel is compressed
by the pump plunger and delivered to the
high-pressure circuit (Fig. 11). The engine-
management system calculates the time
from which the MSV is actuated as a function of delivery quantity and rail pressure.
In contrast to the MSV of the HDP2, the
start of delivery is varied for demand
control.

Technical features of HDP5
- Demand-controlled single-barrel pump
- Pressure range up to 20 MPa (200 bar)
- Delivery rate:
  0.5 cm^3/rev_{cam} for two cams,
  0.75 cm^3/rev_{cam} and 0.9 cm^3/rev_{cam}
  (2 variants) for three cams,
  1.0 cm^3/rev_{cam} for four cams
- Speed up to 8600 rpm (engine speed)

- Weight: approx. 780 g
- Dimensions h = 50 mm, b = 90 mm,
  s = 50 mm
- Integrated pressure-limiting valve.
- Passive pressure-reducing function
  (optional), i.e., slow pressure reduction
  in the high-pressure area via a bypass in
  the outlet valve to the low-pressure area
- Integrated pressure attenuator for variable
  admission pressure
- Integrated fuel-supply control valve, open
  at zero current, with demand control by
  variation of the start of delivery
- Drive via intake or exhaust camshaft,
  power transmitted via barrel or roller
  tappet
- ZEVAP (**Z**ero **Evap**oration) capability, i.e.,
  no fuel is evaporated from the valve
- Drive runs in oil
- Mounted on the cylinder head or adapter
  housing
- Good media compatibility thanks to
  stainless-steel housing
- Several high-pressure pumps can be used,
  depending on the fuel demand (e.g., for
  8-cylinder engines)

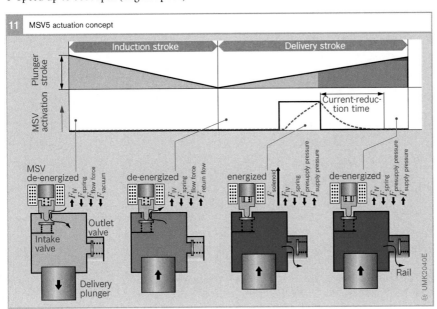

11 MSV5 actuation concept

Fuel rail

Manifold injection

The function of the fuel rail is to store the fuel required for injection and to ensure a uniform distribution of fuel to all the fuel injectors. The fuel injectors are mounted directly on the fuel rail. In addition to the injectors, the fuel rail usually accommodates the fuel-pressure regulator and possibly even a pressure attenuator.

Local pressure fluctuations caused by resonance when the injectors open and close is prevented by careful selection of the fuel-rail dimensions. As a result, irregularities in injected fuel quantity which can arise as a function of load and engine speed are avoided.

Depending upon the particular requirements of the vehicle in question, stainless-steel or plastic fuel rails are used. The fuel rail can incorporate a diagnosis valve for garage/workshop testing purposes.

Gasoline direct injection

The function of the KSZ-HD fuel rail (Fig. 12) is to store and distribute the required fuel quantity for the respective operating point. The fuel is stored by way of its volume and compressibility. In this way,

the volume is application-dependent and must be adapted to the relevant engine demand and pressure range.

The volume ensures attenuation in the high-pressure range, i.e., pressure fluctuations in the rail are compensated.

The add-on components for the direct-injection system are mounted on the rail: the high-pressure fuel injectors (German: Hochdruckeinspritzventile, hence HDEV), the pressure sensor for regulating the high pressure and – for the 1st-generation direct-injection system – the pressure-control or pressure-limiting valve.

The fuel rail for 1st-generation direct-injection systems are designed for a pressure range of 0.4...12 MPa (plus 0.5 MPa opening pressure of the pressure-limiting valve). For the 2nd generation, the pressure range stretches up to 25 MPa (plus 1.2 MPa opening pressure of the pressure-limiting valve). The burst pressure is higher.

The KSZ-HD fuel rail is manufactured for direct-injection systems with HDP1 and HDP2 from shell cast aluminum. When used with the HDP5, the fuel rail is made from hard-soldered stainless steel.

Fig. 12

1   Fuel rail
2   Intermediate fitting for HDEV
3   Support ring
4   O-ring
5   Pressure sensor
6   Pressure-control valve
7   Connection tube
8   Screw-on fitting
9   O-ring

12 Fuel rail (example of 1st-generation direct-injection system)

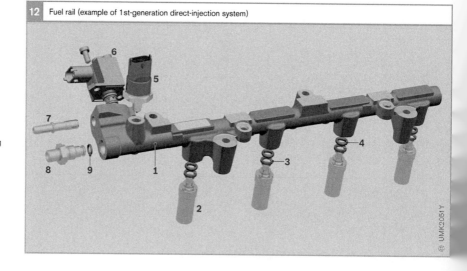

Pressure-control valve

Function

A pressure-control valve (German: <u>D</u>ruck<u>s</u>teuer<u>v</u>entil, hence DSV) is required for 1st-generation direct-injection systems with the HDP1 high-pressure pump. This valve is mounted on the fuel rail between the high-pressure and low-pressure areas. The function of the DSV is to set the desired pressure

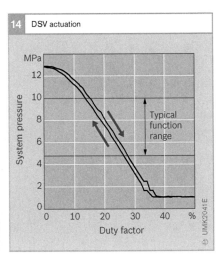

13   Pressure-control valve (DSV)

in the fuel rail. This is achieved by altering the flow cross-section in the valve.

Design and method of operation

The pressure-control valve (Fig. 13) is a proportional control valve which is closed at zero current and actuated by means of a pulse-width-modulated signal. During operation, the energizing of the solenoid coil (3) sets a magnetic force which relieves the load on the spring, lifts the valve ball (8) off the valve seat (9), and thereby alters the flow cross-section. The DSV sets the desired rail pressure as a function of the pulse duty factor (Fig. 14). The excess fuel delivered by the HDP1 is diverted into the low-pressure circuit.

A pressure-limiting function is integrated by way of the valve spring to protect the components against unacceptably high rail pressures, for example, in the event of actuation failure.

Pressure-limiting valve

Function

The pressure-limiting valve (German: <u>D</u>ruck<u>b</u>egrenzungs<u>v</u>entil, hence DBV) is used in 1st-generation direct-injection systems with the demand-controlled HDP2 high-pressure pump. It prevents the fuel pressure from reaching unacceptably high levels when the fuel-supply control valve is non-operational (HDP2 delivers continuously). The DBV limits the fuel pressure in the high-pressure system to a value below the burst pressure. On the other hand, the DBV ensures the operation of the high-pressure fuel injectors through the flat pressure curve over the volumetric flow even in normal operation at operating points without fuel-supply control intervention (overrun and shutdown). To be exact, no fuel is injected at these operating points such that the stored quantity is heated up by the heat from the engine. Each degree of temperature rise results in a pressure increase of roughly 1 MPa. If the pressure is too high, the fuel

Fig. 13
1 Electrical connection
2 Valve spring
3 Solenoid coil
4 Solenoid armature
5 Valve needle
6 Sealing rings
   (O-rings)
7 Outlet passage
8 Valve ball
9 Valve seat
10 Inlet with inlet
   strainer

injector is no longer able to open against the high fuel pressure. This means that reuse after overrun fuel cutoff or short-term restarting of the engine after a hot shutdown would not be possible without pressure limitation.

Design and method of operation

The DBV is mounted on the fuel rail in direct-injection systems with a HDP2 (1st-generation direct injection) and separates the high-pressure area from the low-pressure area. At low fuel pressure, the valve spring (Fig. 15, Pos. 1) presses the valve ball (4) into its seat and seals the high-pressure area from the low-pressure area. In the event that the pressure increases to levels above the opening pressure, the ball is lifted off its seat to allow the compressed fuel to flow off into the low-pressure area. The pressure is relieved via the DBV.

The DBV is not designed for continuous operation at full delivery flow and must be replaced together with the high-pressure pump in the event of a fault.

The pressure-limiting valve is integrated in the HDP5 in 2nd-generation direct-injection systems.

Fuel-pressure regulator

Function

The DR2 fuel-pressure regulator is used in manifold-injection systems. The amount of fuel injected by the injector (injected fuel quantity) depends upon the injection period and the pressure differential between the fuel pressure in the fuel rail and the counter-pressure in the manifold. On fuel systems with fuel return, the influence of pressure is compensated for by a pressure regulator which maintains the differential between fuel pressure and manifold pressure at a constant level. This pressure regulator permits just enough fuel to return to the fuel tank so that the pressure drop across the injectors remains constant. In order to ensure that the fuel rail is efficiently flushed, the fuel-pressure regulator is normally located at the end of the rail.

On returnless fuel systems, the pressure regulator is part of the in-tank unit installed in the fuel tank. The fuel-rail pressure is maintained at a constant level with reference to the ambient pressure. This means that the pressure differential between fuel-rail pressure and manifold pressure is not constant and must be taken into account when the injection duration is calculated.

Fig. 15
1   Valve spring
2   Sealing rings
    (O-rings)
3   Fuel outlet
4   Valve ball
5   Filter strainer
6   High-pressure port

Fig. 16
1   Intake-manifold
    connection
2   Spring
3   Valve holder
4   Diaphragm
5   Valve
6   Fuel inlet
7   Fuel return

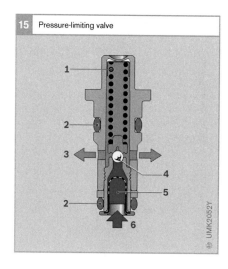

15   Pressure-limiting valve

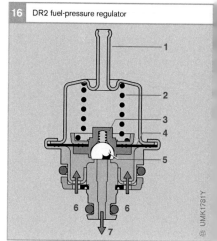

16   DR2 fuel-pressure regulator

Design and method of operation

The fuel-pressure regulator (Fig. 16) is of the diaphragm-controlled overflow type.
A rubber-fabric diaphragm (4) divides the pressure regulator into a fuel chamber and a spring chamber. Through a valve holder (3) integrated in the diaphragm, the spring (2) forces a movable valve plate against the valve seat so that the valve closes. As soon as the pressure applied to the diaphragm by the fuel exceeds the spring force, the valve opens again and permits just enough fuel to flow back to the fuel tank that equilibrium of forces is achieved again at the diaphragm.

On multipoint fuel-injection systems, in order that the manifold vacuum can be applied to the spring chamber, this is connected pneumatically to the intake manifold at a point downstream of the throttle plate. There is therefore the same pressure ratio at the diaphragm as at the injectors. This means that the pressure drop across the injectors is solely a function of spring force and diaphragm surface area, and therefore remains constant.

Fuel-pressure damper

The repeated opening and closing of the injectors, together with the periodic supply of fuel when electric positive-displacement fuel pumps are used, leads to fuel-pressure oscillations. These can cause pressure resonances which adversely affect fuel-metering accuracy. It is even possible that under certain circumstances, noise can be caused by these oscillations being transferred to the fuel tank and the vehicle bodywork through the mounting elements of the fuel rail, fuel lines, and fuel pump.

These problems are alleviated by the use of special-design mounting elements and fuel-pressure dampers.

The fuel-pressure damper is similar in design to the fuel-pressure regulator. Here too, a spring-loaded diaphragm separates the fuel chamber from the air chamber. The spring force is selected such that the diaphragm lifts from its seat as soon as the fuel pressure reaches its working range. This means that the fuel chamber is variable and not only absorbs fuel when pressure peaks occur, but also releases fuel when the pressure drops. In order to always operate in the most favorable range when the absolute fuel pressure fluctuates due to conditions at the manifold, the spring chamber can be provided with an intake-manifold connection.

Similar to the fuel-pressure regulator, the fuel-pressure damper can also be attached to the fuel rail or installed in the fuel line. In the case of gasoline direct injection, it can also be attached to the high-pressure pump.

Manifold injection

In gasoline engines with manifold injection, formation of the air/fuel mixture begins outside the combustion chamber in the intake manifold. Since they were introduced to the market, these engines and their control systems have been vastly improved. Their superior fuel-metering characteristics have enabled them to almost completely supersede the carburetor engine, which also operates with external air/fuel-mixture formation.

Overview

High standards of smooth running and emission behavior are made on motor vehicles. This leads to strict requirements with respect to the formation of the air/fuel mixture. In addition to precise metering of the injected fuel mass – matched to the air mass inducted by the engine – perfectly timed injection and targeting of the spray are crucial. Because of the continual tightening of emission-control legislation, these requirements

are increasingly taking center stage. Accordingly, this situation necessitates constant development of the fuel-injection systems.

In the field of manifold injection, the electronically controlled multi-point fuel-injection system represents state-of-the-art technology. This system injects the fuel intermittently, and individually, for each cylinder directly onto its intake valves. The electronic control facility is integrated in the ECU of the engine-management system. Bosch's version of this system is called Motronic.

Mechanical, continuous-injection multi-point injection systems and single-point injection systems no longer play a significant role in new developments. With single-point injection, the fuel is injected intermittently but only via a single fuel injector ahead of the throttle valve into the intake manifold.

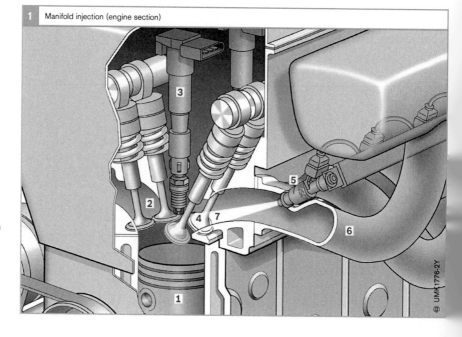

1   Manifold injection (engine section)

Fig. 1

1   Cylinder with piston
2   Exhaust valves
3   Ignition coil with
    spark plug
4   Intake valves
5   Fuel injector
6   Intake manifold
7   Intake passage

Method of operation

Creation of the air/fuel mixture

In gasoline injection systems with manifold injection, the fuel is introduced into the intake manifold or into the intake passage. For this purpose, the electric fuel pump delivers fuel to the fuel injectors. The fuel is applied to the injectors at system pressure. In multipoint injection systems, each cylinder is allocated its own injector (Fig. 1, Pos. 5), which injects fuel intermittently into the intake manifold (6) or into the intake passage (7) ahead of the intake valves (4).

Mixture formation begins outside the combustion chamber in the intake passage. After injection, the air/fuel mixture created flows in the subsequent induction stroke through the opened intake valves into the cylinder (1). In this process, the air mass is metered by the throttle valve (Fig. 2, Pos. 2). Depending on the engine type, each cylinder utilizes one up to a maximum of three intake valves.

Fuel metering by the fuel injectors is configured in such a way that fuel demand is covered for all engine operating states. This means that, on the one hand, sufficient fuel must be introduced in the available time at high engine speeds and loads. On the other hand, it is also necessary to ensure a sufficient minimal injected fuel quantity during idle operation so as to facilitate stoichiometric engine operation ($\lambda = 1$).

Measurement of the air mass
In order that the air/fuel mixture can be precisely adjusted, it is imperative that the mass of the air which is used for combustion can be measured exactly. The air-mass meter (1), which is situated upstream of the throttle valve, measures the air-mass flow entering the intake manifold and sends a corresponding electrical signal to the engine ECU.

As an alternative, there are also systems on the market which use a pressure sensor to measure the intake-manifold pressure. Together with the throttle-valve setting and the engine speed, these data are then used to calculate the intake-air mass.

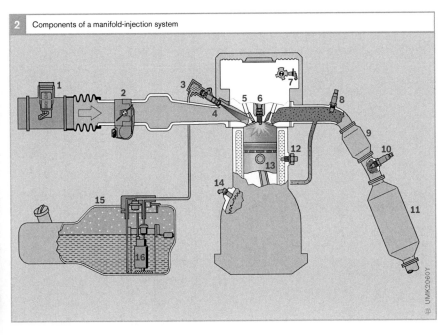

2   Components of a manifold-injection system

Fig. 2
1 Hot-film air-mass meter
2 Throttle device
3 Fuel rail
4 Fuel injector
5 Intake valve
6 Spark plug
7 Camshaft phase sensor
8 Lambda sensor upstream of primary catalytic converter
9 Primary catalytic converter (three-way catalytic converter)
10 Lambda sensor downstream of primary catalytic converter
11 Main catalytic converter (three-way)
12 Engine-temperature sensor
13 Cylinder with piston
14 Speed sensor
15 Fuel tank
16 Electric fuel pump

The Motronic ECU then utilizes the data on intake-air mass and the engine's current operating state to calculate the required fuel mass.

Injection duration

The injection period, or duration, which is required to introduce the calculated fuel mass is derived depending on the narrowest cross-section in the fuel injector, its opening and closing behavior, and the pressure differential between the intake-manifold pressure and the fuel pressure.

Reduction of emissions

In recent years, advances in engine technology have led to improved combustion processes, producing lower untreated emissions. Electronic engine-management systems make it possible to inject precisely the required amount of fuel in accordance with the inducted air mass, to control precisely the moment of ignition, and to optimize actuation of all the available components (e.g., electronic throttle device DV-E) depending on the various engine operating points. Along with enhancements in engine power and performance, these advances also lead to substantial improvements in the quality of the exhaust gas.

It is possible in combination with the exhaust-gas-treatment system to adhere to the statutory exhaust-emission limits. Provided the air/fuel mixture is stoichiometric ($\lambda = 1$), the pollutants generated during the combustion process can to a great extent be reduced using the three-way catalytic converter. At the majority of their operating points, manifold-injection engines are therefore operated with this air/fuel-mixture composition.

On-engine measures

On-engine measures can be used to reduce untreated emissions. The following measures are commonly encountered today:
● Optimized combustion-chamber geometry
● Multi-valve technology

● Variable valve timing
● Central spark-plug position
● Increased compression
● Exhaust-gas recirculation

The task of reducing emissions is faced with the additional challenge posed by the operating range of engine cold starting.

Cold starting

When the ignition key is turned, the starter rotates and turns the engine over at starter speed. The signals from the speed and phase sensors (Fig. 2, Pos. 14 and 7) are collected. The Motronic ECU uses these signals to determine the piston positions of the individual cylinders. The injected fuel quantities are calculated in accordance with the program maps stored in the ECU and transmitted via the fuel injectors. The ignition is then activated. Engine speed is increased with the first combustion.

Cold starting is characterized by different phases (Fig. 3):
● Starting phase
● Post-start phase
● Warming-up
● Catalytic-converter heating

Starting phase
The range from the first combustion to the first time the catalytic-converter-heating speed is exceeded is called the starting phase. An increased fuel quantity is required for engine starting (roughly 3 to 4 times the full-load delivery at approx. 20 °C).

Post-start phase
In the subsequent post-start phase, the charge and the injected fuel quantity are successively reduced as a function of the engine temperature and the time already elapsed since the end the starting phase.

Warming-up phase
The post-start phase is followed by the warming-up phase. An increased torque demand is required on account of the still low engine temperature, i.e., there is still an

increased fuel demand compared with the demand when the engine is hot. In contrast to the post-start phase, this additional demand is only dependent on the engine temperature and is required up to a specific temperature threshold.

Catalytic-converter-heating phase
The catalytic-converter-heating phase refers to the cold-starting range in which additional measures are used to achieve a quicker heating-up of the catalytic converter.

The limits of the different phases are fluid. Depending on the respective engine system, the warming-up phase can also extend beyond the catalytic-converter-heating phase.

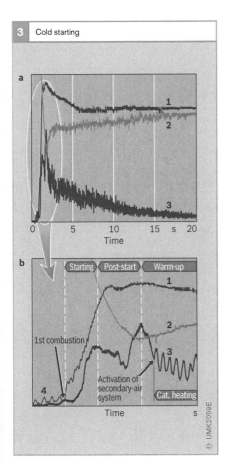

3   Cold starting

a

1
2

3

0   5   10   15   s   20
Time

b

Starting   Post-start   Warm-up

1

2

1st combustion

3

4

Activation of
secondary-air
system   Cat. heating

Time   s

UMK2059E

Cold-starting problems
Fuel which precipitates during starting when the engine is cold on the cold cylinder wall does not evaporate immediately and therefore does not take part in the following combustion. It enters the exhaust system during the exhaust stroke and therefore does not contribute to reducing torque. In order to ensure stable engine revving-up, it is therefore necessary to provide an increased fuel quantity in the starting and post-start phases.

Untreated HC and CO emissions rise dramatically when these unburnt fuel constituents are discharged. Another complication is the fact that the catalytic converter must attain a minimum temperature of roughly 300 °C before it can start to convert pollutant emissions. To encourage the catalytic converter to reach its operating temperature quickly, there are measures available which facilitate a quick heating-up of the catalytic converter as well as additional systems for thermal aftertreatment of the exhaust gas which are activated in the catalytic-converter-heating phase.

Measures for heating up the catalytic converter
The following measures can be used to heat up the catalytic converter quickly during cold starting:
- High exhaust-gas temperatures induced by late ignition timing and high mass gas flow
- Catalytic converters mounted close to the engine
- Increased exhaust-gas temperature by means of thermal aftertreatment

Thermal aftertreatment
The unburnt hydrocarbons are reduced in the exhaust system by thermal afterburning at high temperatures. In the case of a rich air/fuel mixture, an injection of air (secondary-air injection) is required. In the case of a lean air/fuel mixture, afterburning takes place with the residual oxygen in the exhaust gas.

Fig. 3
Qualitative progression
a   Up to approx.
    20 secs.
b   Extract after
    engine start

1   Engine speed
2   Lambda
3   Untreated
    exhaust gas
    (HC concentration)

Secondary-air injection

In the case of secondary-air injection, additional air is introduced into the exhaust system after the starting sequence in the warming-up phase ($\lambda_{engine} < 1$). This results in an exothermal reaction with the unburnt hydrocarbons, which reduce the high HC and CO concentrations in the exhaust gas. In addition, this oxidation process releases heat to the exhaust gas, which in turn quickly heats up the catalytic converter as it flows through.

Instants of injection

In addition to the correct injection duration, a further parameter which is important for optimization of the fuel-consumption and exhaust-gas figures is the instant of injection referred to the crankshaft angle.

In considering the instant of injection for each individual cylinder, a distinction is made between pre-intake and intake-synchronous injection. Injection is said to be pre-intake when the end of injection for the cylinder in question occurs before the intake valve and a majority of the fuel spray strikes the base of the intake passage and the intake valves. In contrast, intake-synchronous injection takes place when the intake valves are open.

If, on the other hand, the instant of injection of all the cylinders is considered in relation to each other, a distinction is made between the follow instants of injection (Fig. 4):
- Simultaneous fuel injection
- Group fuel injection
- Sequential fuel injection
- Cylinder-individual fuel injection

4   Types of manifold injection

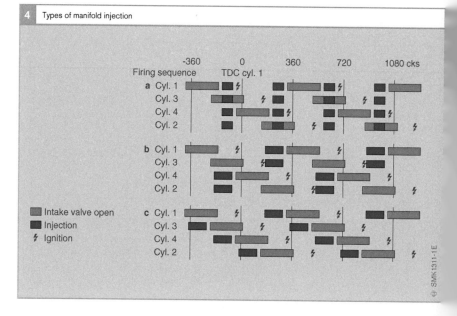

Fig. 4
a   Simultaneous fuel injection
b   Group fuel injection
c   **Sequential Fuel Injection (SEFI)** and Cylinder-Individual Fuel Injection (CIFI)

SMK1311-1E

Here, the possible variations are dependent on the instant of injection actually used. Today, serial-individual fuel injection is used almost exclusively. Simultaneous or group fuel injection is still used only occasionally during cold starting in the first combustion processes.

Simultaneous fuel injection
All injectors open and close together in this form of fuel injection. This means that the time which is available for fuel evaporation is different for each cylinder. In order nevertheless to obtain efficient air/fuel-mixture formation, the fuel quantity needed for combustion is injected in two portions: half in one revolution of the crankshaft and the remainder in the next.

In the case of this instant of injection, pre-intake injection is not possible for all the cylinders. It is sometimes necessary to inject fuel into the open intake valve in that the start of injection is already prespecified.

Group fuel injection
Here, the fuel injectors are combined to form two groups. For one revolution of the crankshaft, one injector group injects the total fuel quantity required for its cylinders, and for the next revolution the second group injects. This arrangement enables the injection timing to be selected as a function of the engine operating point, and avoids the undesired injection of fuel into open intake ports in wide program-map ranges. Here too, the time available for the evaporation of fuel is different for each cylinder.

Sequential fuel injection (SEFI)
In the case of SEFI, the fuel is injected individually for each cylinder. The fuel injectors are actuated in succession in firing sequence. The injection duration and the instant of injection referred to the top dead center of the respective cylinder are the same for all cylinders. Injection timing for each cylinder is thus identical. Start of injection is freely programmable and can be adapted to the engine's operating state.

Cylinder-individual fuel injection (CIFI)
CIFI offers the greatest degrees of freedom. Compared to serial-individual fuel injection, CIFI has the advantage that the duration of injection can be individually varied for each cylinder. This permits compensation of irregularities, for instance with respect to cylinder charge.

Stoichiometric operation of each cylinder requires a cylinder-specific lambda recording here. This necessitates an optimization of the manifold geometry so as to avoid as far as possible a mixing of the exhaust gas in the individual cylinders.

CIFI is only used in engines in which there are large cylinder-specific differences in the intake geometry.

Mixture formation

Mixture formation begins with the injection of fuel into the intake manifold and extends through the intake phase into the compression phase of the cylinder in question. It is subject to many requirements, such as, for example:
- Provision of an ignitable mixture at the spark plug at the moment of ignition
- Good homogenization of the mixture in the cylinder
- Good dynamic behavior in non-stationary operation
- Low HC emissions during cold starting

Mixture formation in manifold-injection systems is complex. It ranges from the characteristic of the primary fuel spray through transportation of the spray in the intake manifold, introduction of the spray into the combustion chamber down to homogenization of the mixture at the moment of ignition. At the end of the day, optimal adaptation of these processes will result in good mixture preparation. Mixture preparation differs to some extent for cold and hot engine operation and is decisively influenced by:

- Engine temperature
- Primary-droplet spray
- Instant of injection
- Spray targeting, and
- Air flow

The objective is to have a homogeneous mixture of fuel vapor and air in the combustion chamber at the moment of ignition of the respective cylinder.

Primary-droplet spray

The fuel spray directly after it emerges from the fuel injector is referred to as the primary-droplet spray. Small primary droplets broadly encourage the fuel to evaporate. However, it is important to bear in mind here that only a very small proportion of the introduced fuel evaporates in the intake manifold when the engine is cold on account of the low temperature. The majority of the fuel exists as wall-applied film and is entrained in the air flow during the intake phase. The actual mixture preparation takes place in the cylinder.

On the other hand, when the engine is hot, a large proportion of the injected fuel spray and some of the existing wall-applied film already evaporates in the intake manifold.

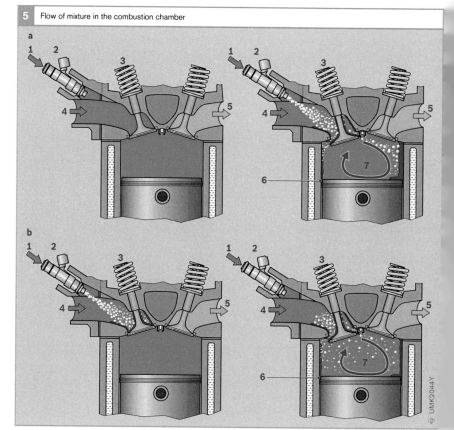

5   Flow of mixture in the combustion chamber

Fig. 5

a Intake-synchronous
   injection
1. No injection with
   intake valve closed
2. Injection into opened
   intake valve
b Pre-intake injection
1. Injection onto closed
   intake valve
2. Induction of pre-
   intake air/fuel mixture

1 Fuel inlet
2 Fuel injector
3 Intake valve
4 Air flowing in
5 Exhaust
6 Top land

Instant of injection

The instant of injection has a significant influence on mixture formation and untreated HC emissions, above all when the engine is cold.

Intake-synchronous injection

In the case of intake-synchronous injection, some of the fuel is transported by the air flow to the opposite cylinder wall in the direction of the exhaust valves (Fig. 5a). This fuel film (wall-applied film) does not evaporate on the cold cylinder walls, does not take part in the combustion process, and therefore passes unburnt into the exhaust port. This results in increased untreated emissions.

Nowadays, intake-synchronous injection is used only rarely in cold starting. It is used to increase power when the engine is running at normal (hot) operating temperature. Because evaporation of the fuel takes place to a large extent in the combustion chamber in the case of intake-synchronous injection, the fresh-air charge can be increased. The reason for this is that the liquid fuel droplets in the intake manifold have a smaller volume than vapor. Furthermore, the process of the fuel evaporating in the combustion chamber cools the cylinder charge, which has a positive effect on the engine's knock tendency.

Pre-intake injection

It is possible with pre-intake injection (Fig. 5b) to achieve a significant reduction in pollutant emissions during cold starting. The injected fuel is forced towards the center of the combustion chamber and the undesired formation of wall-applied film on the exhaust-side cylinder wall is avoided.

Spray targeting

In addition to pre-intake injection, it is possible in combination with optimal spray targeting (Fig. 6) to further reduce HC emissions during cold starting. When the spray is targeted at the base of the intake port, the drawn-in spray is transported with increased intensity towards the center of the combustion chamber. This further reduces the depositing of fuel on the exhaust-side cylinder wall, which in turn manifests itself in lower HC emissions in the starting phase.

On the other hand, the depositing of fuel on the base of the intake port results in an increased formation of wall-applied film in the intake manifold. Here, the application expenditure for non-stationary operation (load changes) is somewhat higher. On engines with manifold injection, it is necessary in the event of load changes to take into account the accumulated mass of wall-applied

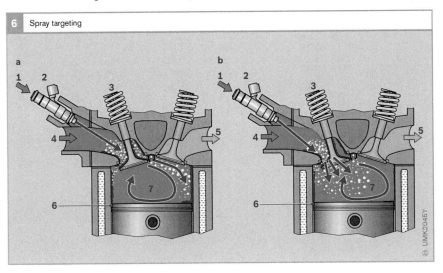

6 Spray targeting

Fig. 6
a Central spray targeting onto intake valves
b Optimal spray targeting

1 Fuel inlet
2 Fuel injector
3 Intake valve
4 Air flowing in
5 Exhaust
6 Top land

film in the intake manifold. More wall-applied film builds up in the event of a sharp load increase. The engine would lean briefly if the accumulated mass of wall-applied film and its delayed introduction into the combustion chamber were not taken into account in the calculation of the required injected fuel quantity. For this purpose, the Motronic ECU incorporates wall-applied-film compensation functions, for which data must be obtained during application with regard to the respective engine geometry and spray targeting so as to guarantee to a large extent $\lambda = 1$ operation even in the non-stationary operating state.

Air flow

The air flow is decisively influenced by the engine speed, by the geometric layout of the intake port, and by the opening times and the lift curve of the intake valves. Sometimes charge-flow control valves (tumble) are used in order additionally to bring an influence to bear on the flow direction as a function of the engine operating point. The objective is to introduce the necessary air in the available time into the combustion chamber and to achieve good homogenization of the air/fuel mixture in the combustion chamber up to the moment of ignition.

A strong internal cylinder flow encourages good homogenization and facilitates an increase in EGR compatibility, by means of which it is possible to achieve fuel-consumption and NO_X reductions. However, a strong internal cylinder flow reduces the charge at full load, which results in a lowering of the maximum torque and maximum power.

Secondary mixture preparation

The air flow also aids fuel preparation (secondary mixture preparation). If there is a differential pressure between the intake manifold and the combustion chamber at the moment when the intake valves open (IO), the flow created influences the preparation and transportation of the fuel. If the intake-manifold pressure is significantly greater than the combustion-chamber pressure when the intake valve opens, the air/fuel mixture and the wall-applied film in the valve gap are drawn into the combustion chamber at greater speed.

If the intake-manifold pressure is smaller than the combustion-chamber pressure when the intake valve opens, then hot exhaust gas from the previous combustion flows back into the intake manifold. Here, on the one hand, the flow encourages the formation of wall-applied film and fuel droplets, while, on the other hand, the hot exhaust gas also supports evaporation of the fuel. This process is particularly important in cold starting in the warming-up and catalytic-converter-heating phases.

Ignition of homogeneous air/fuel mixtures

At the moment of ignition, the air/fuel mixture in the combustion chamber is ideally fully homogenized. Combustion of the mixture is initiated in a gasoline engine by an electrical ignition spark. Here, very hot spark plasma (ionized gas) is generated (Fig. 7) when high voltage is applied across the spark-plug electrodes. The surrounding air/fuel-vapor mixture is heated very intensively in the area of the plasma, the ignition volume, to start an accelerated chain reaction. As a result, the temperature rises very rapidly and a flame front is created.

A safe ignition process is dependent among other things on
- The introduced ignition energy
- The spark duration
- The local flow conditions in the combustion chamber close to the spark-plug electrodes
- The air/fuel ratio λ, and
- The spark-plug geometry and configuration in the combustion chamber

The flame front spreads in the event of non-knocking combustion at subsonic speed in the combustion chamber. A burnt zone is created after the flame front. The homogeneous air/fuel mixture burns above the soot-emission limit with a characteristically bluish flame.

7 Spark plasma

The speed of the spreading flame front increases as the turbulence of the combustion-chamber flow rises. The degree of flow turbulence and thus the rate of combustion also rise as engine speed increases. The result of this is that the duration of combustion assumes the same crankshaft-angle range virtually independently of the engine speed.

The flame goes out close to the combustion-chamber wall due to high heat losses, or "flame quenching". This results in the emission of unburnt fuel.

Electromagnetic fuel injectors

Function

Electrically actuated fuel injectors spray fuel into the intake manifold at system pressure. They allow fuel to be metered in the precise quantity required by the engine. They are actuated by output stages which are integrated in the engine ECU with the signal calculated by the engine-management system.

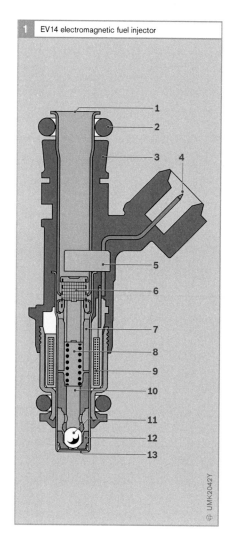

1   EV14 electromagnetic fuel injector

Fig. 1
1   Hydraulic port
2   O-ring
3   Valve housing
4   Electrical connection
5   Plastic clip with injected pins
6   Filter strainer
7   Internal pole
8   Valve spring
9   Solenoid coil
10   Valve needle with armature
11   Valve ball
12   Valve seat
13   Injection-orifice plate

Design and method of operation

Essentially, electromagnetic injectors (Fig. 1) are comprised of the following components:
- Valve housing (3) with electrical connection (4) and hydraulic port (1)
- Solenoid coil (9)
- Moving valve needle (10) with solenoid armature and valve ball (11)
- Valve seat (12) with injection-orifice plate (13), and
- Valve spring (8)

In order to ensure trouble-free operation, stainless steel is used for the parts of the injector which come into contact with fuel. The injector is protected against contamination by a filter strainer (6) at the fuel inlet.

Connections

On the injectors presently in use, fuel supply to the injector is in the axial direction, i.e., from top to bottom ("top feed"). The fuel line is secured to the hydraulic port by means of a clamp. Retaining clips ensure reliable alignment and fastening. The sealing ring (O-ring) (2) on the hydraulic port (1) seals off the injector at the fuel rail.

The injector is electrically connected to the engine ECU.

Injector operation

When the solenoid coil is de-energized, the valve needle and valve ball are pressed against the cone-shaped valve seat by the spring and the force exerted by the fuel pressure. The fuel-supply system is thus sealed off from the intake manifold. When the solenoid coil is energized, this generates a magnetic field which attracts valve-needle armature solenoid armature. The valve ball lifts off the valve seat and the fuel is injected. When the excitation current is switched off, the valve needle closes again due to spring force.

Fuel outlet

The fuel is atomized by means of an injection-orifice plate in which there are a number of holes. These holes (injection orifices) are stamped out of the plate and ensure that the injected fuel quantity remains highly constant. The injection-orifice plate is insensitive to fuel deposits. The spray pattern of the fuel leaving the injector is produced by the number of orifices and their configuration.

The injector is efficiently sealing at the valve seat by the cone/ball sealing principle. The injector is inserted into the opening provided for it in the intake manifold. The lower sealing ring provides the seal between the injector and the intake manifold.

Essentially, the injected fuel quantity per unit of time is determined by
- The system pressure in the fuel-supply system
- The back pressure in the intake manifold, and
- The geometry of the fuel-exit area

Electrical activation

An output module in the Motronic ECU actuates the injector with a switching signal (Fig. 2a). The current in the solenoid coil rises (b) and causes the valve needle (c) to lift. The maximum valve lift is achieved after the time t_{pk} (pickup time) has elapsed. Fuel is sprayed as soon as the valve ball lifts off its seat. The total quantity of fuel injected during an injection pulse is shown in Fig. 2d.

Current flow ceases when activation is switched off. Mass inertia causes the valve to close, but only slowly. The valve is fully closed again after the time t_{dr} (dropout time) has elapsed.

When the valve is fully open, the injected fuel quantity is proportional to time. The non-linearity during the valve pickup and dropout phases must be compensated for throughout the period that the injector is activated (injection duration). The speed at which the valve needle lifts off its seat is also dependent on the battery voltage.

Battery-voltage-dependent injection-duration extension (Fig. 3) corrects these influences.

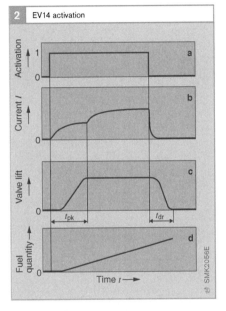

2 EV14 activation

Activation | Current I | Valve lift | Fuel quantity

t_{pk} t_{dr}

Time $t \longrightarrow$

SMK2056E

Fig. 2
a   Activation signal
b   Current curve
c   Valve lift
d   Injected fuel quantity

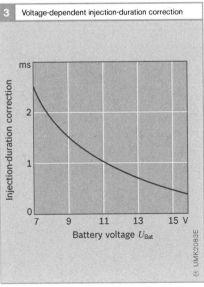

3 Voltage-dependent injection-duration correction

Injection-duration correction

ms

2

1

0

7   9   11   13   15 V

Battery voltage U_{Bat}

UMK2083E

Designs

Over the course of time, the injectors have been continually developed to adapt them to the ever-increasing demands regarding engineering, quality, reliability, and weight. This has led to a variety of different injector designs.

EV14 injector

The EV14 injector is the standard injection valve for today's modern fuel-injection systems (Fig. 4). It is characterized by its small external dimensions and its low weight. This injector therefore already provides one of the prerequisites for the design of compact intake modules.

In addition, the EV14 demonstrates excellent hot-fuel performance, i.e., there is very little tendency for vapor-bubble formation when the fuel is hot. This facilitates the use of returnless fuel-supply systems because in such systems the fuel temperature in the injector is higher than in systems featuring fuel return.

Thanks to wear-resistant surfaces, the EV14 also demonstrates high endurance stability and a long service life. Thanks to their highly efficient sealing, these injectors fulfill all future requirements regarding "zero evaporation", i.e., no fuel vapor escapes from them.

For better fuel atomization, the injection-orifice plates with four orifices usually used are replaced by multi-orifice plates with up to 12 injection orifices. This reduces the droplet size by up to 35%.

There are a wide variety of injectors available for different areas of application. These feature different lengths, flow classes, and electrical properties. The EV14 is also suitable for use with fuels which have an ethanol content of up to 85%. The EV14 is currently undergoing tests to ascertain its suitability for use with E100 (pure ethanol).

The EV14 is available in 3 different lengths: compact, standard and long (Fig. 4). This makes it possible to adapt individually to the engine's intake-manifold geometry.

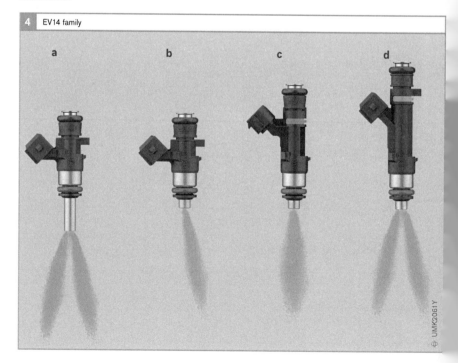

4 EV14 family

a b c d

Fig. 4
a Compact with advanced spray point
b Compact
c Standard
d Long

Spray formation

An injector's spray formation, i.e., its spray shape, spray angle, and fuel-droplet size, influences the formation of the air/fuel mixture. It is therefore an important function of the fuel injector.

The EV14's atomizing capability has been further improved compared with the EV6 thanks to the use of new processes for manufacturing multihole injection-orifice plates and to modified flow guidance. In this way, highly homogeneous fuel sprays with droplet sizes reduced by up to 50 % when compared with four-hole plates are created.

Individual intake-manifold and cylinder-head geometries make it necessary to have different types of spray formation. Different spray-formation variants are available in order to satisfy with these requirements. The spray shapes depicted in Figure 5 can be created both as a four-hole variant and as a multihole variant with reduced droplet size.

Tapered spray

Individual jets of fuel are discharged through the holes in the injection-orifice plate. The tapered spray cone results from the combination of these fuel jets.

Tapered-spray injectors are typically used in engines which feature only one intake valve per cylinder. However, the tapered spray is also suitable for two intake valves.

Dual spray

Dual-spray formation is often used in engines which feature two intake valves per cylinder. Engines with three intake valves per cylinder must be equipped with dual-spray injectors.

The holes in the injection-orifice plate are arranged in such a way that two fuel sprays – which can be formed from a number of individual sprays (two tapered sprays) – leave the injector and impact against the respective intake valve or against the web between the intake valves.

Gamma angle

Referred to the injector's principle axis, the fuel spray in this case (single spray and dual spray) is at an angle, the offset spray angle.

Injectors with this spray shape are mostly used when installation conditions are difficult.

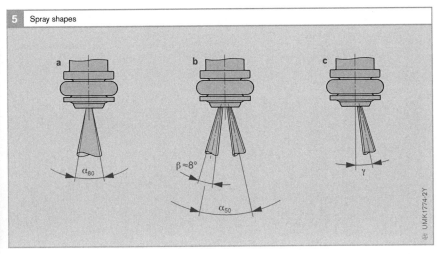

5 Spray shapes

Fig. 5
a   Tapered spray
b   Dual spray
c   Gamma angle

α_{80}: 80 % of fuel situated within angle
α_{50}: 50 % of fuel situated within angle
: 70 % of fuel in individual spray situated within angle
: Offset spray angle

Gasoline direct injection

Gasoline direct-injection engines generate the air/fuel mixture in the combustion chamber. During the induction stroke, only the combustion air flows through the open intake valve. The fuel is injected directly into the combustion chamber by special fuel injectors.

Overview

The demand for higher-power spark-ignition engines, combined with the requirement for reduced fuel consumption, were behind the "rediscovery" of gasoline direct injection. The principle is not a new one. As far back as 1937, an engine with mechanical gasoline direct injection took to the air in an airplane. In 1951 the "Gutbrod" was the first passenger car with a series-production mechanical gasoline direct-injection engine, and in 1954 the "Mercedes 300 SL" with a four-stroke engine and direct injection followed.

At that time, designing and building a direct-injection engine was a very complicated business. Moreover, this technology made extreme demands on the materials used. The engine's service life was a further problem. These facts all contributed to it taking so long for gasoline direct injection to achieve its breakthrough.

Method of operation

Gasoline direct-injection systems are characterized by injecting the fuel directly into the combustion chamber at high pressure (Fig. 1). As in a diesel engine, air/fuel-mixture formation takes place inside the combustion chamber (internal mixture formation).

High-pressure generation
The electric fuel pump (Fig. 2, Pos. 19) delivers fuel to the high-pressure pump (4) at a presupply pressure of 3...5 bar. The latter pump generates the system pressure depending on the engine operating point (requested torque and engine speed). The highly pressurized fuel flows into and is stored in the fuel rail (Fig. 1, Pos. 6).

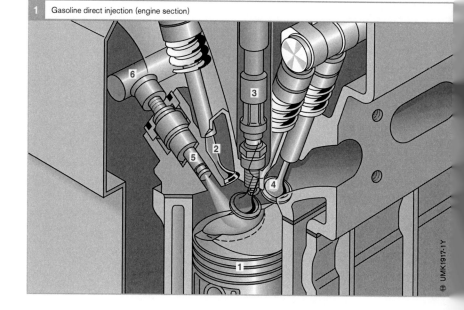

1   Gasoline direct injection (engine section)

Fig. 1
1   Piston
2   Intake valve
3   Ignition coil and
     spark plug
4   Exhaust valve
5   High-pressure
     fuel injector
6   Fuel rail

The fuel pressure is measured with the high-pressure sensor and adjusted via the pressure-control valve (in the HDP1) or the fuel-supply control valve integrated in the HDP2/HDP5 to values ranging between 50 and 200 bar.

The high-pressure fuel injectors (5) are mounted on the fuel rail, also known as the "common rail". These injectors are actuated by the engine ECU and spray the fuel into the cylinder combustion chambers.

Combustion process

In the case of gasoline direct injection, the combustion process is defined as the way in which mixture formation and energy conversion take place in the combustion chamber. The mechanisms are determined by the geometries of the combustion chamber and the intake manifold, and the injection point and the moment of ignition. Depending on the combustion process concerned, flows of air are generated in the combustion chamber. The relationship between injected fuel and air flow is extremely important, above all in

relation to those combustion processes which work with charge stratification (stratified concepts). In order to obtain the required charge stratification, the injector fuel injects the fuel into the air flow in such a manner that it evaporates in a defined area. The air flow then transports the mixture cloud in the direction of the spark plug so that it arrives there at the moment of ignition.

A combustion process is often made up of several different operating modes between which the process switches as a function of the engine operating point. Basically, the combustion processes are divided into two categories: stratified-charge and homogeneous combustion processes.

Homogeneous combustion process
In the case of the homogeneous combustion process, usually a generally stoichiometric mixture is formed in the combustion chamber in the engine map (Fig. 3a), i.e. an air ratio of $\lambda = 1$ always exists. In this way, the expensive exhaust-gas treatment of NO_X emissions which is required with lean mixtures is avoided. Homogeneous concepts are therefore set out to be emission-reducing concepts.

Fig. 2
1 Hot-film air-mass meter
2 Throttle device (ETC)
3 Intake-manifold pressure sensor
4 High-pressure pump
5 Charge-flow control valve
6 Fuel rail with high-pressure injector
7 Camshaft adjuster
8 Ignition coil with spark plug
9 Camshaft phase sensor
10 Lambda sensor
11 Primary catalytic converter
12 Lambda sensor
13 Exhaust-gas temperature sensor
14 NO_X accumulator-type catalytic converter
15 Lambda sensor
16 Knock sensor
17 Engine-temperature sensor
18 Speed sensor
19 Fuel-supply module with electric fuel pump

2   Components of gasoline direct-injection system

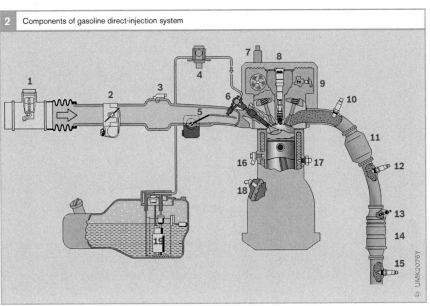

The homogeneous combustion process is always run in homogeneous mode; there may, however, also be a few special operating modes which can be used differently from engine to engine for specific application purposes (see section entitled "Operating modes").

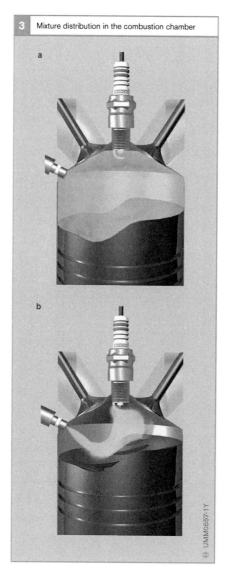

3   Mixture distribution in the combustion chamber

a

b

Stratified-charge combustion process

In the case of the stratified-charge combustion process, the fuel is injected in a specific map range (small load, low engine speed) first during the compression stroke into the combustion chamber, and transported as a stratified-charge cloud to the spark plug (Fig. 3b, shown here in the wall-guided combustion process). The cloud here is ideally surrounded by pure fresh air. In this way, an ignitable mixture is only present in the local cloud. An air ratio of greater than 1 exists generally in the combustion chamber. This enables the engine to be operated unthrottled in greater ranges, which results in increased efficiency on account of the reduced pumping losses. Stratified-charge combustion processes are therefore run predominantly as a fuel-consumption concept.

Two different stratified-charge concepts may be used: the wall/air-guided process and the spray-guided process.

Wall/air-guided combustion process
In wall/air-guided stratified-charge mode, the injector is usually situated between the intake valves. In this combustion process, the fuel is usually injected at a pressure of 50 to 150 bar. The mixture is transported via the piston recess, which either interacts directly with the fuel (wall-guided) or guides the air flow in the combustion chamber in such a way that the fuel is directed on an air cushion to the spark plug (air-guided). Real stratified-charge combustion processes with side injector installation usually combine both processes, depending on the installation angle of the injectors and the injected fuel quantity. At idle (low injected fuel quantity), a wall-guided combustion process barely strikes the piston recess; at higher loads (high injected fuel quantity), a certain quantity of fuel strikes the piston directly, even in the case of the air-guided combustion process.

The air flow can be configured as a swirl or tumble flow.

Swirl air flow

The air drawn in through the open intake valve generates a turbulent flow (rotational air movement) along the cylinder wall (Fig. 4a). This process is also known as the "swirl combustion process".

Tumble air flow

This process produces a tumbling air flow, which in its movement from top to bottom is deflected by a pronounced piston recess so that it then moves upwards in the direction of the spark plug (Fig. 4b).

Spray-guided combustion process

In the case of the spray-guided combustion process, the injector is situated centrally at the top in the roof of the combustion chamber. The spark plug is installed next to the injector (Fig. 4c). The advantage of this arrangement is the possibility of the fuel spray being directly guided to the spark plug without having to take a circuitous route through the piston or air flows. The disadvantage, however, is the short amount of time available for mixture preparation. Spray-guided stratified-charge combustion processes therefore require a fuel pressure increased to approximately 200 bar.

In order to be able to ignite the mixture at the correct moment in time, the spray-guided combustion process requires that the spark plug and fuel injector be exactly positioned, and that the spray be precisely targeted. With this process, the spark plug is subjected to considerable thermal stressing since under certain circumstances the hot spark plug can be directly impacted by the relatively cold jet of injected fuel.

When the combustion process is properly configured, the spray-guided combustion process demonstrates greater efficiency than the other stratified-charge combustion processes to such an extent that it can achieve even greater consumption savings in comparison with stratified-charge mode with the wall/air-guided combustion process.

Outside the stratified-charge-mode range, the engine is also run homogeneous mode in the case of the stratified-charge combustion process.

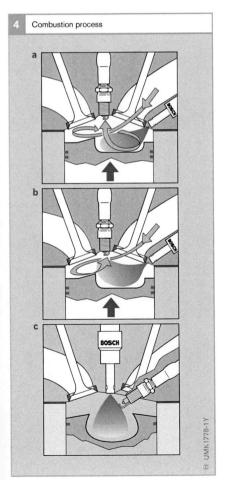

4 Combustion process

Fig. 4

a   Wall-guided swirl air flow

b   Wall-guided tumble air flow

c   Spray-guided combustion process

Operating modes

The different operating modes which are used in gasoline direct injection are described in the following. The appropriate operating mode is set by the engine-management system, depending on the engine operating point (Fig. 1).

Homogeneous

In homogeneous mode, the injected fuel quantity is metered precisely to the fresh air in the stoichiometric ratio of 14.7:1. Here, the fuel is injected during the induction stroke so that there is sufficient time remaining to homogenize the entire mixture. For the purpose of protecting the catalytic converter or increasing power at full load, the engine is also operated with a slight fuel excess in parts of the operating map ($\lambda < 1$).

Since the whole of the combustion chamber is utilized, the homogeneous mode is required when high levels of torque are demanded. Because of the stoichiometric air/fuel mixture, emissions of untreated pollutants are low in this operating mode; these pollutants can also be fully converted by the three-way catalytic converter.

In homogeneous operation, combustion to a great extent corresponds to the combustion for manifold injection.

Stratified charge

In stratified-charge mode, the fuel is first injected during the compression stroke. Here, the fuel is prepared only with part of the air. A stratified-charge cloud which is ideally surrounded by pure fresh air is created. The start of injection is very important in stratified-charge mode. The stratified-charge cloud must be not only sufficiently homogenized at the moment of ignition but also positioned at the spark plug.

Because a stoichiometric mixture is only present locally in stratified-charge mode, the mixture is on average lean on account of the surrounding fresh air. This setup requires more expensive exhaust-gas treatment since the three-way catalytic converter is unable to reduce NO_X emissions in lean-burn operation.

Stratified-charge mode can only be run within certain limits because at higher loads

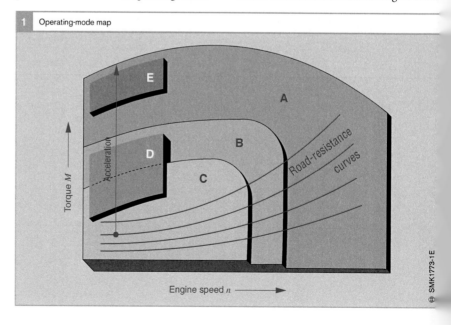

Fig. 1
A   Homogeneous
mode at $\lambda = 1$,
this operating mode
is possible in all
operating ranges
B   Lean-burn or
homogeneous mode
$\lambda = 1$ with EGR;
this operating mode
is possible in area C
and area D
C   Stratified-charge
mode with EGR

Operating modes
with dual injection:
C   Stratified-charge/
cat.-heating mode,
same area as
stratified-charge
mode with EGR
D   Homogeneous
stratified-charge
mode
E   Homogeneous
knock-protection
mode

1   Operating-mode map

Torque M

Acceleration

Engine speed n

Road-resistance curves

SMK1773-1E

soot and/or NO_X emissions increase dramatically and the fuel-consumption advantages over homogeneous mode are lost. At lower loads, stratified-charge mode is limited by low exhaust-gas enthalpy, i.e., the exhaust-gas temperatures become so low that the catalytic converter cannot be kept at operating temperature by the exhaust gas alone. The speed range is limited to approximately 3000 rpm in stratified-charge mode because at speeds above this threshold the time available is no longer sufficient to homogenize the stratified-charge cloud.

The stratified-charge cloud becomes lean in the peripheral zone adjoining the surrounding air. For this reason, there is an increase in untreated NO_X emissions in this zone during combustion. A high exhaust-gas-recirculation rate provides a remedy in this operating mode. The recirculated exhaust gases lower the combustion temperature and thereby reduce the temperature-dependent NO_X emissions.

Homogeneous lean
In the transitional range between stratified-charge and homogeneous mode, the engine can be run with a homogeneous lean air/fuel mixture ($\lambda > 1$). Since the pumping losses are lower due to "non-throttling", fuel consumption is lower in homogeneous lean mode than in standard homogeneous mode with $\lambda \leq 1$. However, this mode is accompanied by increased NO_X emissions since the three-way catalytic converter is unable to reduce these emissions in this range. Additional NO_X accumulator-type catalytic converters mean further efficiency losses as a result of the catalytic converter's regeneration phases.

Homogeneous stratified charge
In homogeneous stratified-charge mode, the complete combustion chamber is filled with a homogeneous lean basic mixture. This mixture is created by injecting a basic quantity of fuel during the induction stroke.

Fuel is injected a second time (dual injection) during the compression stroke. This leads to a richer zone forming in the area of the spark plug. This stratified charge is easily ignitable and can ignite the homogeneous lean mixture in the remainder of the combustion chamber with the flame along the same lines as torch ignition.

The homogeneous stratified-charge mode is activated for a number of cycles during the transition between stratified-charge and homogeneous modes. This enables the engine-management system to better adjust the torque during the transition. Due to the conversion to energy of the very lean basic mixture at $\lambda > 2$, the NO_X emissions are also reduced.

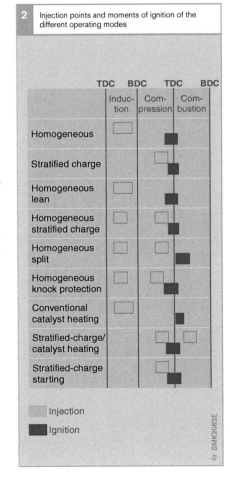

2 Injection points and moments of ignition of the different operating modes

	Induction	Compression	Combustion
	TDC BDC	TDC	BDC
Homogeneous	Injection		Ignition
Stratified charge		Injection	Ignition
Homogeneous lean	Injection		Ignition
Homogeneous stratified charge	Injection	Injection	Ignition
Homogeneous split	Injection	Injection	Ignition
Homogeneous knock protection	Injection	Injection	Ignition
Conventional catalyst heating	Injection		Ignition
Stratified-charge/ catalyst heating		Injection Injection	Ignition
Stratified-charge starting		Injection	Ignition

Injection

Ignition

SMK2082E

The distribution factor between the two injections is approximately 75 %, i.e., 75 % of the fuel is injected in the first injection, which is responsible for the homogeneous basic mixture.

Steady-state operation using dual injection at low engine speeds in the transitional range between stratified-charge and homogeneous modes reduces the soot emissions compared to stratified-charge mode, and lowers fuel consumption compared to homogeneous mode.

Homogeneous split

Homogeneous-split mode is a special application of homogeneous stratified-charge dual injection. It is used after the starting phase to bring the catalytic converter up to operating temperature as quickly as possible. Ignition can be significantly retarded (15...30° crankshaft after firing TDC) by the stabilizing effect of the second injection during the compression stroke. A large proportion of the combustion energy will then no longer influence an increase in torque, but rather increases exhaust-gas enthalpy. Due to this high exhaust-gas heat flow, the catalytic converter is already ready for operation just a few seconds after starting.

Homogeneous knock protection

In this operating mode, in view of the fact that charge stratification hinders knock, ignition-timing retardation as needed to avoid knocking can be dispensed with through the use of dual injection at full load. At the same time, the more favorable ignition point also leads to higher torque. In reality, the potential of this operating mode is very limited.

Stratified-charge/catalyst heating

Another form of dual injection makes it possible to heat up the exhaust system quickly, although the exhaust system must be optimized to accommodate this application. Here, in stratified-charge mode with a high level of excess air, injection is effected first during the compression stroke (as in stratified-charge mode) and then once again during the power stroke. This fuel is combusted very late and causes the engine's exhaust side and the exhaust manifold to heat up dramatically. When the engine is cold, however, this operating mode proves to be very limited in its potential for application to the extent that in this case homogeneous-split mode is significantly superior.

A further important application is for heating up the NO_X catalytic converter to temperatures in excess of 650 °C in order to initiate desulfurization of the catalytic converter. It is absolutely essential to use dual injection here since this high temperature cannot always be reached in all operating modes with conventional heating methods.

Stratified-charge starting

During stratified-charge starting, the quantity of fuel for starting is introduced during the compression stroke instead of conventional injection during the induction stroke. The advantage of this injection strategy is based on the fact that fuel is injected into air that is already compressed and therefore heated. In this way, more fuel evaporates in percentage terms than under cold ambient conditions, where a significantly larger proportion of the injected fuel remains as liquid wall-applied film in the combustion chamber and does not take part in combustion. The fuel quantity to be injected can therefore be reduced dramatically during stratified-charge starting. This results in greatly reduced HC emissions during starting. Because the catalytic converter cannot yet operate at the moment of starting, this is an important operating mode for developing low-emission concepts.

In order to facilitate mixture preparation in the short time available, stratified-charge starting is performed at fuel pressures of roughly 30...40 bar. These pressures can already be made available by the high-pressure pump by way of the starter revolutions.

Mixture formation

Function
It is the function of mixture formation to provide a combustible air/fuel mixture which is as homogeneous as possible.

Requirements
In homogeneous operating mode (homogeneous $\lambda \leq 1$ and also homogeneous lean), the mixture should be homogeneous throughout the entire combustion chamber. In stratified-charge mode, on the other hand, the mixture is only homogeneous within a limited area, while the remaining areas of the combustion chamber are filled with inert gas or fresh air.

All fuel must have evaporated before a gas mixture or gas/fuel-vapor mixture can be termed homogeneous. Evaporation is influenced by numerous factors, above all
- Combustion-chamber temperature
- Fuel-droplet size, and
- Time available for evaporation

Influencing factors
Depending on the engine's temperature, pressure and combustion-chamber geometry, a mixture containing gasoline is combustible within a range of $\lambda = 0.6...1.6$.

Temperature influence
The temperature has a decisive influence on fuel evaporation. At lower temperatures, fuel does not evaporate completely. More fuel must therefore be injected under these conditions in order to obtain a combustible mixture.

Pressure influence (fuel pressure)
The size of the droplets in the injected fuel is dependent on injection pressure and combustion-chamber pressure. Higher injection pressures result in smaller droplets, which then evaporate more quickly.

Geometry influence
With a constant combustion-chamber pressure and increasing injection pressure, the so-called penetration depth increases.

The penetration depth is defined as the distance traveled by an individual fuel droplet before it evaporates completely. The cylinder wall or the piston will be wetted with fuel if the distance needed for full evaporation exceeds the distance from the fuel injector to the combustion-chamber wall. If this wall-applied film fails to evaporate in good time by the point of ignition, it does not, or only incompletely, takes part in combustion.

The geometry of the engine (intake manifold and combustion chamber) is also responsible for the air flow and the turbulence in the combustion chamber. Both factors have a significant influence on mixture formation because they determine both mixture preparation and transportation of the ignitable mixture during charge stratification to the spark plug.

Mixture formation in homogeneous mode
The fuel should be injected as early as possible so that the maximum length of time is available for mixture formation. This is why the fuel is injected during the induction stroke in homogeneous mode. The intake air helps the fuel to evaporate quickly and ensures that the mixture is well homogenized. Mixture preparation is assisted above all by high flow velocities and their aerodynamic forces in the area of the opening and closing intake valve.

Wall interaction is not desired and the associated evaporation of wall-applied film plays a subordinate role here (Fig. 1).

Mixture formation in stratified-charge mode
The configuration of the combustible mixture cloud which is in the vicinity of the spark plug at the moment of ignition is crucial to stratified-charge mode. This is why the fuel is injected during the compression stroke so that a mixture cloud is created which can be transported to the vicinity of the spark plug by the air flows in the combustion chamber and by the upward stroke of the piston. The injection point is dependent on the engine speed and the requested torque.

In stratified-charge injection, mixture preparation profits from the higher temperature and the already increased pressure in the combustion chamber during the compression phase.

In the case of the wall-guided combustion process, condensation of fuel on the piston wall cannot be avoided, to such an extent that some mixture preparation takes place in the form of wall-applied-film evaporation (Fig. 2).

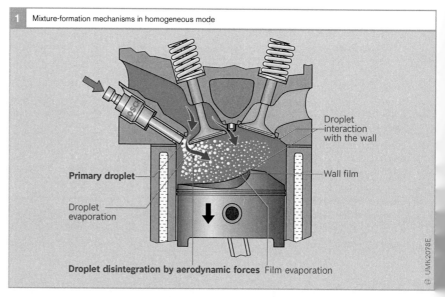

1   Mixture-formation mechanisms in homogeneous mode

Droplet interaction with the wall

Primary droplet

Wall film

Droplet evaporation

Droplet disintegration by aerodynamic forces Film evaporation

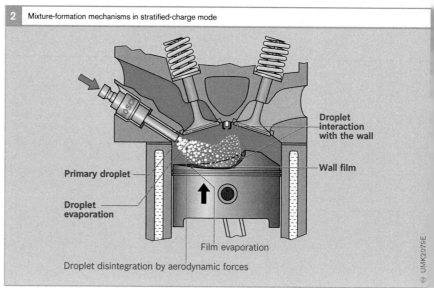

2   Mixture-formation mechanisms in stratified-charge mode

Droplet interaction with the wall

Primary droplet

Wall film

Droplet evaporation

Film evaporation

Droplet disintegration by aerodynamic forces

Ignition

Homogeneous mixtures

The ignition conditions for homogeneous mixtures with gasoline direct injection are to a large extent that same as those with manifold injection (see chapter entitled "Manifold injection").

Stratified-charge mixtures

When engine operation is dethrottled, from a general point of view, extremely lean air/fuel mixtures must be ignited in the lower part-load range. This is made possible by the creation of charge stratification in the area of the spark plug at the moment of ignition. Ideally, this stratified air/fuel charge is a virtually stoichiometric and thus easily ignitable mixture.

After ignition, a flame core is formed in the area of the stoichiometric stratified charge. The amount of energy released here is four times greater than that of an ignition spark. In this way, mixture in lean peripheral areas of the combustion chamber can also be ignited.

Wall/air-guided combustion process

In the case of the wall/air-guided combustion process, the injection time window must be chosen such that the piston directs the mixture cloud safely to the spark plug. This transportation movement is usually assisted by combustion-chamber flows. The mixture must be ignited when it reaches the spark plug, i.e. the moment of ignition is permanently linked in terms of time to the piston position – injection – mixture transportation sequence and is no longer available as a controlled variable of the combustion process.

When the piston surface is wetted with fuel, this surface acts as a mixture-formation component through its interaction with the liquid fuel. Here, the fuel that adheres to the surface evaporates at a relatively slow rate. When the flame front arrives at the piston surface, which is cold in comparison to the flame, the flame goes out as a result of high heat dissipation (quenching effect). The fuel that has still not evaporated at this time is not combusted and results in increased HC emissions.

The design of the piston surface, the combustion-chamber flow forming in the vicinity of the spark plug, and the injector quality thus have a direct influence on the ignition and burning performance.

Spray-guided combustion process

The close proximity of the spark plug and the injector ensures that an ignitable mixture concentration is available at the spark plug, even in the event of small injected fuel quantities. However, this also means that only a very small time window is available for evaporation and mixture preparation.

Fuel directly after it has left the injector is still unable to ignite since it has still not evaporated sufficiently and is mixed with surrounding air. Ignition significantly later than injection is barely possible since the mixture is increasingly removed from the spark plug and therefore becomes lean. Ideal ignition conditions are therefore only present during a relatively short period of time. Typically, a rapidly growing flame core in the now combustible air/fuel-vapor mixture initiated by the ignition spark is formed at the end of the injection process.

The following factors, among others, are crucial to safe ignition:
- An ignition spark lasting as long as possible
- The quality of mixture preparation
- The correct allocation of spark location and fuel spray
- Relatively precise adherence to the distance between spray and ignition location
- Constancy of the spray in relation to the combustion-chamber pressure
- Invariance of the spray over the engine's entire service life

High-pressure injector

Function

It is the function of the high-pressure fuel injector (HDEV) on the one hand to meter the fuel and on the other hand by means of its atomization to achieve controlled mixing of the fuel and air in a specific area of the combustion chamber. Depending on the desired operating mode, the fuel is either concentrated in the vicinity of the spark plug (stratified charge) or evenly distributed throughout the combustion chamber (homogeneous distribution).

Design and method of operation

The high-pressure injector (Fig. 1) comprises the following components:

- Inlet with filter (1)
- Electrical connection (2)
- Spring (3)
- Coil (4)
- Valve sleeve (5)
- Nozzle needle with solenoid armature (6), and
- Valve seat (7)

A magnetic field is generated when current passes through the coil. This lifts the valve needle off the valve seat against the force of the spring and opens the injector outlet passages (8). The system pressure now forces the fuel into the combustion chamber. The injected fuel quantity is essentially dependent on the opening duration of the injector and the fuel pressure.

When the energizing current is switched off, the valve needle is pressed by spring force back down against its seat and interrupts the flow of fuel.

Excellent fuel atomization is achieved thanks to the suitable nozzle geometry at the injector tip.

Requirements

Compared with manifold injection, gasoline direct injection differs mainly in its higher fuel pressure and the far shorter time which is available for directly injecting the fuel into the combustion chamber.

1   Design of HDEV5 high-pressure injector

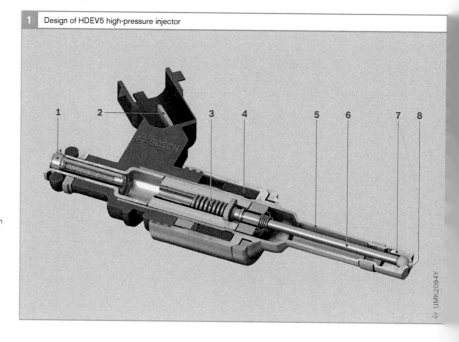

Fig. 1

1 Fuel inlet with filter
2 Electrical connection
3 Spring
4 Coil
5 Valve sleeve
6 Nozzle needle with solenoid armature
7 Valve seat
8 Injector outlet passages

Figure 2 underlines the technical demands made on the injector. In the case of manifold injection, two revolutions of the crankshaft are available for injecting the fuel into the manifold. This corresponds to an injection duration of 20 ms at an engine speed of 6000 rpm.

In the case of gasoline direct injection, however, considerably less time is available. In homogeneous operation, the fuel must be injected during the induction stroke. In other words, only a half crankshaft rotation is available for the injection process. At 6000 rpm, this corresponds to an injection duration of 5 ms.

With gasoline direct injection, the fuel requirement at idle in relation to that at full load is far lower than with manifold injection (factor 1:12). This results in an injection duration at idle of roughly 0.4 ms.

Actuation of HDEV high-pressure injector

The injector must be actuated with a highly complex current characteristic in order to comply with the requirements for defined, reproducible fuel-injection processes (Fig. 3). The microcontroller in the engine ECU delivers only a digital actuating signal (a). An output module (ASIC) uses this signal to generate the actuating signal (b) for the injector.

A DC/DC converter in the engine ECU generates the booster voltage of 65 V. This voltage is required in order to bring the current up as quickly as possible in the booster phase to a high current value. This is necessary in order to accelerate the injector needle as quickly as possible. In the pickup phase (t_{on}), the valve needle then achieves the maximum opening lift (c). When the injector is open, a small actuating current (holding current) is sufficient to keep the injector open.

With a constant injector-needle lift, the injected fuel quantity (d) is proportional to the injection duration.

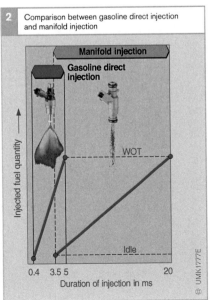

2 Comparison between gasoline direct injection and manifold injection

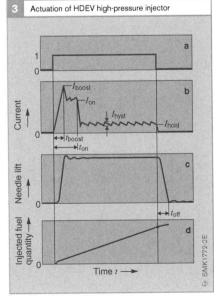

3 Actuation of HDEV high-pressure injector

Fig. 2
Injected fuel quantity
as a function of injection
duration

Fig. 3
a Actuating signal
b Current
   characteristic
   in injector
c Needle lift
d Injected fuel quantity

Operation of gasoline engines on natural gas

The Association of European Vehicle Manufacturers (ACEA) has undertaken a commitment to reduce average CO_2 emissions to 140 g/km by the year 2008. This represents a reduction of 25 % when set against the figures recorded back in 1995. Vehicle concepts based on CNG (Compressed Natural Gas) contribute to lowering CO_2 emissions. Because natural gas is not yet extensively available at filling stations, it should also be possible for the engine to be run on gasoline.

The EU Commission has plans by 2020 to replace 23 % of gasoline and diesel consumption with alternative fuels in Europe. The main contribution to this reduction – at 10 % – is to be made by natural gas (Fig. 1).

Germany currently (as at: July 2005) has roughly 30,000 natural-gas vehicles and 603 natural-gas filling stations. Because natural gas exhibits better environmental properties than gasoline and diesel, its use in motor vehicles is to be encouraged up to 2020 in Germany by means of a lowered tax on mineral oil. Thus, equivalent-energy natural gas will be offered at filling stations at prices which are roughly 50 % cheaper than gasoline.

Local efforts are also being made to use alternative fuels in South America and parts of Asia. Natural gas is leading the way in these local schemes.

Overview

Properties of natural gas

The primary constituent of natural gas is methane (CH_4). Natural gas therefore has the highest hydrogen content of all the fossil fuels. When it is burned, it creates approximately 25 % fewer CO_2 emissions than gasoline while providing the same amount of energy.

Natural gas is available throughout the world. Its composition, however, varies depending on where it originates. These variations influence density, calorific value and knock resistance. The use of natural gas in motor vehicles has hitherto not been subject to standardization.

Another advantage is that methane can also be regeneratively manufactured from biomass. In this way, the CO_2 cycle is completed and long-term availability can be increased.

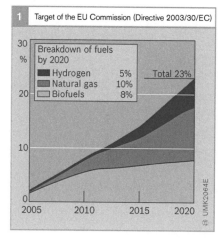

Fig. 1

Replacement of approx. 23 % of gasoline/diesel demand with alternative fuels

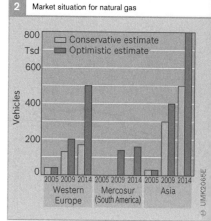

Properties of natural-gas drives

Because of the simpler molecular structure of methane and due to the fact that it is introduced into the engine in gaseous form, the untreated emissions (HC, NO_X) from a natural-gas engine are significantly lower than those from a gasoline engine. The emission of non-limited pollutants (aldehydes, aromatic hydrocarbons, etc.) and the emission of sulfur dioxide and particulates is virtually avoided completely with natural gas.

Natural gas has a very high knock resistance of up to 130 RON (by comparison, gasoline: 91...100). It is therefore possible, by comparison with a gasoline engine, to increase compression by approximately 20 % and thus raise efficiency. At the same time, the natural-gas engine is ideally suited to supercharging. In combination with a downsizing concept, in which the engine displacement is reduced and at the same time the engine is supercharged to its original power output, it is possible to obtain an additional improvement in efficiency and with it a further CO_2 reduction.

Because natural gas is low in density, it is more complicated to store it in a tank than it is to store gasoline and diesel. It is usually stored in the vehicle in gas form at an overpressure of 200 bar in steel or carbon-fiber tanks (hence the designation CNG = *Compressed Natural Gas*). It requires four times the conventional gasoline or diesel tank volume for the same energy content. It is nevertheless possible through optimized installation of the pressure accumulators (e.g., locating the tanks under the vehicle floorpan) to achieve ranges of currently roughly 400 km without additionally having to reduce the size of the luggage compartment.

Alternatively, natural gas can also be liquified at temperatures of $-162\,°C$ (LNG = *Liquefied Natural Gas*). However, the process of liquifying the gas expends a great deals of energy and the tanks are expensive. Today, almost exclusively CNG tanks are used in passenger-car applications.

Because of the advantage of lower CO_2 emissions and the possibility of adapting gasoline engines to natural gas at relatively little expense, natural gas fulfills a good many of the conditions to be able to experience a dramatic upturn in use in the short term.

Use of natural-gas vehicles

In Europe, natural-gas-powered vehicles have up to now predominantly been used in commercial fleets (e.g. vehicle pools of larger companies or city buses). The consistent buildup of the CNG filling-station network in Germany and agreements to provide natural gas at interstate-highway filling stations is encouraging the further spread of natural-gas vehicles, even in private transport.

Today, the largest fleet of natural-gas vehicles is in South America. This market has been determined up to now by conversion systems. However, a trend towards the series production of natural-gas vehicles can also be foreseen here. The greatest potential for growth is to be anticipated in Asia, because here, in addition to CO_2 emission aspects, underlying economic conditions promote the use of natural gas. Natural-gas-producing countries, such as, for instance, Iran, have a strong interest in using the gas in mobile applications as well.

In the NAFTA region (North America), there are currently no significant developments to speak of because the market-structure incentives are lacking. Here the trend is more towards gasoline hybrid vehicles.

Design and method of operation

Because of the limited number of natural-gas filling stations, today's natural-gas vehicles are primarily designed as bivalent vehicles (bifuel and monovalent-plus vehicles), i.e. as well as running on gas, they can also be run on conventional gasoline. The basis for the natural-gas system is the spark-ignition engine with manifold injection (Fig. 3). Additional components for supplying and injecting natural gas are required. The Bifuel-Motronic ECU controls both fuel operating modes.

Monovalent-plus vehicles are optimized for running on natural gas and only have a 15-l emergency gasoline tank. Today's natural-gas tanks are made from steel or fiber composites. For strength reasons, the shape of the tank cannot be freely selected. This often gives rise to problems of space in the vehicle. A combination of several tanks, sometimes even of different sizes, is therefore used in many vehicles.

Method of operation of fuel supply in a natural-gas system

The natural gas stored at approximately 200 bar in the tanks (Fig. 3, Pos. 13) flows through individual tank shutoff valves (12) to the pressure-regulator module (6). The electromagnetically actuated high-pressure shutoff valve on the tank ensures that when it is de-energized the tank is sealed off tight when the vehicle is stationary. In the event of a system failure, the pressure-limiting valve ensures that unacceptably high pressures in the system can be reduced.

The pressure regulator reduces the gas pressure from a tank pressure of roughly 200 bar to a constant system pressure of approximately 7 bar. A coolant port serves to heat the natural gas cooled by expansion. The high-pressure sensor enables the tank fill level (fuel gage) to be determined and can be called on for system diagnosis. In the interests of increasing the accuracy of the fill-level measurement, the pressure measurement can be combined with a temperature measurement.

Fig. 3

1   Carbon canister with canister-purge valve
2   Canister-purge valve
3   Fuel rail
4   Gasoline injector
5   Ignition coil with spark plug
6   Natural-gas pressure regulator
7   Natural-gas rail with natural-gas pressure and temperature sensor
8   Natural-gas injector
9   Fuel tank
10 Fuel-supply module with electric fuel pump
11 Filler neck for gasoline and natural gas
12 Natural-gas-tank shutoff valves
13 Natural-gas tank

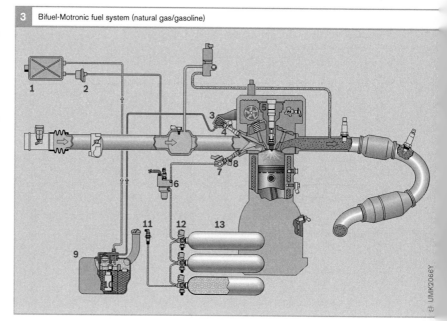

3 Bifuel-Motronic fuel system (natural gas/gasoline)

UMK2066Y

The gas is directed from the pressure-regulator module to the rail (7), which supplies one injector (8) per cylinder. Mixture formation is effected through the injection of fuel into the intake manifold. A combined low-pressure/temperature sensor (7) serves to correct the metering of the gas.

All the system components which may come into contact with natural gas must be certified for use in accordance with European Directive ECE-R110. In Germany, the necessary tests and inspections are carried out by TÜV (German Technical Inspection Agency).

Mixture formation

An unusual feature of the natural-gas engine is the injection of the fuel in gas form into the intake manifold (Fig. 4). This is effected along the same lines as gasoline by injection into the intake manifold ahead of the intake valves.

The natural-gas injectors are supplied via a common gas rail, which is connected to the pressure regulator. The system pressure regulated to 7 bar is monitored by a diagnostic function. For safety reasons, the gas-supply system features in addition to the injectors two gas shutoff valves on the tank and on

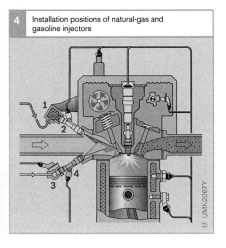

4 Installation positions of natural-gas and gasoline injectors

the pressure regulator which are electromagnetically actuated by the engine-management system. The shutoff valves are only opened when both the ignition is switched on and engine running is detected. This safety function is necessary so that the gas supply can be immediately and safely interrupted in the event of a malfunction or an accident.

As with gasoline injection, the system utilizes sequential multipoint injection, whereby the fuel is injected through an injector for each cylinder in sequence into the respective intake port. This process provides for efficient mixture preparation by means of precision-timed injection control. The injector can be either completely opened or completely closed. The injected gas quantity is adjusted solely by way of the injector's opening duration. The injector is opened once for every induction stroke of the engine.

In contrast to gasoline injection, gas injection involves a noticeable amount of fresh air being displaced by the natural gas. Due to the lower density of natural gas, at full load approximately 10 % by volume of the inducted air/fuel mixture consists of natural gas. At the same time, however, this means that natural-gas engines have a lower air delivery ratio. In naturally aspirated engines, this can cause a reduction in power compared with a gasoline engine. This can, however, be compensated for by higher compression and turbocharging. With these measures – combined with the high knock resistance of natural gas – it is possible even to achieve an increase in power compared with a gasoline engine.

During the injection of natural gas – comparable with gasoline injection – the injection duration is calculated while taking into account the injector constant. The injector constant here is dependent on the design of the injectors and defines the static through-flow $mfng_0$, which obtains under the standard condition and with flow above critical.

Fig. 4
1 Fuel rail
2 Gasoline injector
3 Natural-gas rail
4 Natural-gas injector

The gaseous mass flow through the injector is essentially calculated differently from a liquid fuel.

The density of natural gas is much lower than that of gasoline. In terms of gas-injector design, this results in larger opening cross-sections. Furthermore, the density ϱ of gases is dependent on temperature T and pressure p to a much greater extent than liquid fuels.
The density ϱ_{NG} of natural gas is:

$$\varrho_{NG} = \varrho_{NG0} \cdot \frac{p_{NG}}{p_0} \cdot \frac{T_0}{T_{NG}} \tag{1}$$

The index 0 denotes the status under the standard condition: $p_0 = 1013$ hPa, $T_0 = 273$ K

A defined flow velocity is obtained in the event of a gas flow above critical. This occurs when the pressure ratio at the injector is lower than 0.52. The gas then flows at the speed of sound. A natural-gas admission pressure in the gas rail of 7 bar (absolute pressure) ensures both the supply of the maximum natural-gas quantity required and a flow at the speed of sound at every engine operating point, even with supercharged engines. At the same time, it ensures that the gas is metered independently of the intake-manifold pressure.

The speed of sound is temperature-dependent. In natural gas, it is:

$$c_{NG} = c_{NG0} \sqrt{\frac{T_{NG}}{T_0}} \tag{2}$$

The obtained gas mass flow $mfng$ is then calculated with (1) and (2):

The natural-gas mass flow is dependent linearly on the pressure and with the index −1/2 on the temperature. The installation of a natural-gas pressure and temperature sensor in the natural-gas rail ensures that the variables which influence the mass flow are known. The injection duration is corrected accordingly and the injected gas quantity can thus be correctly introduced even under changing ambient conditions.

Two further corrections are needed for electrical actuation of the injectors. The opening delay of the injectors must be taken into account in the calculation of the opening duration; the opening delay is dependent on the battery voltage and also slightly on the admission pressure of the natural gas. Particularly in the case of injectors with metal/metal seals, the act of the injector closing can cause the valve needle to rebound as it contacts the valve seat, a motion which results in an undesired increase in the injected gas quantity. A correction based on the battery voltage and the natural-gas admission pressure compensates for these effects.

To ensure optimal combustion, it is necessary in addition to correct metering for the correct moment of injection to be determined as well. Generally, the fuel is injected into the intake manifold while the intake valves are still closed. The end of injection is determined by the pre-intake angle, the reference point of which is the closing of the intake valve. The pre-intake angle is specified as a function of the engine operating point. The start of injection can be calculated from the injection duration by means of the engine speed.

$$mfng = mfng_0 \cdot \frac{c_{NG}}{c_{NG0}} \cdot \frac{\varrho_{NG}}{\varrho_{NG0}} = mfng_0 \cdot \sqrt{\frac{T_{NG}}{T_0}} \cdot \frac{p_{NG}}{p_0} \cdot \frac{T_0}{T_{NG}} = mfng_0 \cdot \sqrt{\frac{T_0}{T_{NG}}} \cdot \frac{p_{NG}}{p_0}$$

Natural-gas injector NGI2

Development

Bosch has been making available an electro-magnetic natural-gas injector on the CNG market for many years now. This injector, bearing the designation EV1.3A, is based on the gasoline injector and has been adapted for gas metering with an increased needle lift and a stronger magnetic circuit.

Bosch in the meantime has developed a new generation of gas injectors, bearing the designation NGI2 (Natural-Gas Injector 2). The NGI2 shares only its external shape and electrical actuation with its original source, the current EV14 gasoline injector for manifold injection. All the functional components have been specifically conceived for use in modern natural-gas vehicles and are entirely new in terms of design. The know-how from the EV14 and above all the knowledge and experience gained with the previous EV1.3A gas injector are reflected in the NGI2's design. New findings and discoveries stemming from advance development in the related field of components for hydrogen drives have also played a role. Thus, the NGI2 brings together a series of technical innovations which sets new standards in the field of gas metering.

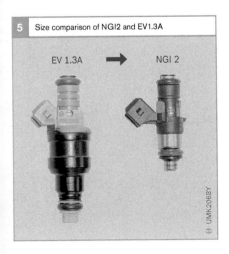

5 Size comparison of NGI2 and EV1.3A

EV 1.3A ➡ NGI 2

UMK2068Y

Design criteria

The maximum required natural-gas mass flow through the injectors, based on a defined engine spectrum, is specified for the development of the injectors. This maximum required natural-gas mass flow is derived from the inducted air-mass flow of the internal-combustion engine, the desired air/fuel ratio (lambda value) and the gas quality. The inducted air-mass flow is in turn dependent on the effective engine displacement, on the engine speed, on the mixture temperature and on the mixture pressure in the combustion chamber when the intake valves close.

A high percentage of CNG applications in the passenger-car field including turbocharged engines can be covered with a maximum natural-gas mass flow of 7.5 kg/h.

In order to supply the internal-combustion engine with this mass flow of gaseous fuel, it is necessary to meter through the gas injectors much more gas in terms of volume than gasoline in a conventional gasoline engine. This requirement places specific demands on the design of the gas injector, which must be adapted to the greater gas volume in its cross-sections. Even the high flow velocities that are encountered call for a special form of flow routing in order to reduce pressure losses ahead of the throttling point. Configuration for operation above critical (speed of sound in the narrowest cross-section) makes it possible even at higher intake-manifold pressures, such as, for example, on supercharged engines, to deliver a characteristic curve that is not dependent on the intake-manifold pressure.

In highly supercharged engines, the intake-manifold pressure can rise up to 2.5 bar (absolute). In order to suppress the influence of the intake-manifold pressure on the mass flow, it is necessary for the admission pressure at the narrowest cross-section accepted as the nozzle above critical to be at least twice as high as the maximum intake-manifold pressure. While taking into account possible pressure losses, this produces a minimum system pressure of 7 bar (absolute).

This design quantity forms the basis of the NGI2, since all applications are to be covered with a single design (principle of effective action).

Increasing the mass flow by raising the admission pressure is basically possible. At the same time, however, the opening force required in the injector increases with the result that the possible system pressure is limited in the upward direction by the restricted magnetic force. The NGI2 has been optimized for a system pressure of 7 bar (absolute) to its maximum mass flow. The injector can also be configured by simple modification to higher admission pressures with an identical mass flow. Even lowering the system pressure with smaller mass-flow demands is basically possible.

Design and method of operation

Individual parts and operation

The operating principles of the NGI2 are similar to those of the EV14 gasoline injector. The direction of fuel flow (top feed), the connections and the form of electrical actuation are identical. However, the individual parts have been adapted for use in the natural-gas system.

The solenoid armature (Fig. 6, Pos. 9) is guided in a sleeve (6). The armature has fuel flowing through it on the inside and has an elastomer seal at the discharge end. This seal closes on the flat seat (10) and thereby seals off the fuel supply from the intake manifold. When energized, the solenoid coil (7) effects the necessary force to lift the solenoid armature and open the metering cross-section (throttling point in the valve seat). When the coil is de-energized, the NGI2 is held closed by a resetting spring (8).

Size and weight

The outer shape of the NGI2 is the same as that of the EV14 gasoline injector. In comparison with established gas injectors, the NGI2 is extremely light and compact. These qualities make it easy to integrate in existing intake-manifold geometries.

Flow-optimized geometry

With regard to the routing of the flow in the NGI2, the pressure loss has been minimized ahead of the throttling point in order to ensure the greatest possible mass flow. Furthermore, the narrowest cross-section and thus the throttling point has been intentionally situated at the discharge end after the seal. The speed of sound is obtained here such that the injector conforms approximately to the physical description of an ideal nozzle. The injector is designed for operation above critical in order to minimize to the greatest possible extent the influence of the intake-manifold pressure on the mass flow. Thanks to a double routing of flow, the layout of the

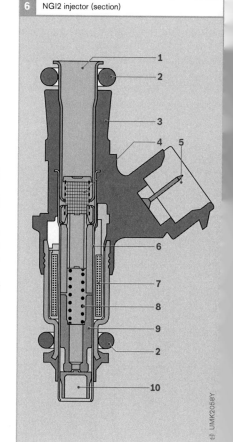

6   NGI2 injector (section)

Fig. 6
1 Pneumatic port
2 O-ring
3 Valve housing
4 Filter strainer
5 Electrical connection
6 Sleeve
7 Solenoid coil
8 Valve spring
9 Solenoid armature
   with elastomer seal
10 Valve seat

valve seat allows a large cross-section to open with relatively low opening force.

Variant range

The NGI2 is available in different lengths and with different plug connectors. Furthermore, variants have been developed for different system pressures and with different flow rates. Thus, as is the case with gasoline injectors, a range of variants is available for different engines.

Sealing geometry

The NGI2 is fitted with an elastomer seal and is similar in terms of its seal-seat geometry to shutoff valves for pneumatic applications. Thus, the NGI2 leaks much less than the EV1.3A. Damping in the elastomer also prevents "rebounding", i.e., a repeated, unwanted opening of the solenoid armature during the closing operation, and thus increases metering precision.

Noise

The optimized routing of flow greatly reduces the armature stroke. This in turn reduces the speed of the solenoid armature as it reaches its stops. When combined with the damping properties of the elastomer seal in the seat stop, the NGI2 has an overall sound-pressure level which is 2 dB lower than its predecessor.

Solenoid coil

At 12 ohms, the NGI2's solenoid coil has the same resistance as the EV14 gasoline injector. Whereas, previously, complicated actuation arrangements, such as parallel-switched output stages or peak-and-hold control, were used to operate low-resistance gas injectors, a standard switching output stage can be used with the NGI2.

Variable oil content

Natural gas is becoming increasingly more widespread as a fuel and the number of natural-gas filling stations is growing constantly in Europe. Especially newer filling stations are equipped with modern compressors which inject less oil into the compressed gas than older compressor types. In future, therefore, it can no longer be assumed that natural gas will have a lubricating and thus wear-reducing effect in the injector.
A solid-lubricant layer on the surface of the solenoid armature ensures that the NGI2 is able to cope with variable oil content in natural gas right down to the detection limit with minimal wear.

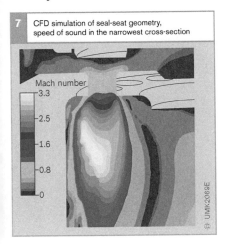

7   CFD simulation of seal-seat geometry, speed of sound in the narrowest cross-section

Mach number
- 3.3
- 2.5
- 1.6
- 0.8
- 0

UMK2069E

Natural-gas rail

The function of the rail is to supply the gas injectors with natural gas in uniform quantities and with minimal pulsations. It is made from stainless steel or aluminum. Design and construction (volume, dimensions, weight, etc.) are engine- and system-specific. The gas is usually supplied to the rail via a flexible low-pressure line.

The rail has a screwed connection, through which it is supplied with gas in the middle or from the side (low-pressure side of the fuel-supply system).

The injectors are held in place with retaining clips in the injector receptacles on the rail. The rail is provided optionally with an attachment point at which a gas pressure and temperature sensor can be attached to the rail.

Combined natural-gas pressure and temperature sensor

Function

Monolithic silicon pressure sensors are high-precision measuring elements for determining absolute pressure. They are particularly suitable for use under rough ambient conditions, such as, for example, for measuring the absolute natural-gas pressure in the rail of CNG-powered vehicles.

The combined low-pressure/temperature sensor (DS-K-TF) measures pressure and temperature in the natural-gas rail and controls exact gas metering by means of the ECU.

Design and method of operation

The sensor consists of the following main components:
- Plug housing (Fig. 8, Pos. 1) with electrical connection (6)
- Sensor cell (2) with silicon chip (9) and etched-in pressure diaphragm (8)
- NTC sensor element (5)
- Fitting (4)
- Outer O-ring (3)

A change in the gas pressure causes the silicon-chip pressure diaphragm to elongate; this elongation is recorded by way of changes in resistance by resistors situated on the silicon chip. The evaluator circuit is likewise integrated together with the electronic compensating elements on the silicon chip.

The silicon chip with glass base is soldered to a metal base with the pressure connecting tube (10). The gas pressure (CNG) is routed through this tube to the lower side of the pressure diaphragm. Underneath the cap (7) welded to the metal base is a reference

Fig. 8
a   Overall view
b   Sensor cell

1   Plug housing
2   Sensor cell
3   O-ring (outer)
4   Fitting
5   NTC (temperature sensor)
6   Electrical connection (plug)
7   Cap
8   Etched-in silicon-chip diaphragm
9   Silicon chip with glass base
10 Pressure connecting tube
11 Reference vacuum

p   Gas pressure

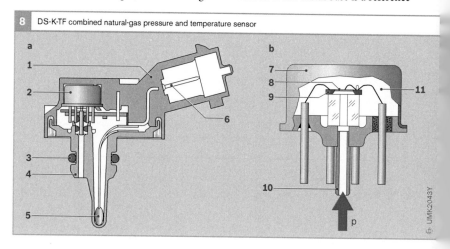

8   DS-K-TF combined natural-gas pressure and temperature sensor

vacuum (11), which enables the absolute pressure to be measured and simultaneously protects the upper side (circuit side) of the chip diaphragm against harmful environmental influences. The finish-compensated sensor cell is mounted in a plug housing with an electrical connection.

The sensor incorporates an NTC sensor element for recording the gas temperature. The fitting is glued tightly onto the plug housing.

The sensor is sealed, for example from the gas rail, by a natural-gas-resistant O-ring.

Signal evaluation

The combined low-pressure/temperature sensor delivers an analog pressure output signal which is ratiometric to the supply voltage. An RC low-pass filter in the input section of the subsequent electronic circuitry ensures that potentially disruptive harmonic waves are suppressed.

The integrated temperature sensor consists of an NTC thermistor and must be operated with a corresponding series resistor as a voltage divider.

DS-HD-KV4 high-pressure sensor

Function

The high-pressure sensor in natural-gas-powered spark-ignition engines is integrated in the pressure-regulator module. Its function is to measure the pressure of the natural gas in the tank.

Design and method of operation

The core of the sensor is a steel diaphragm, which is welded tight on a threaded fitting (Fig. 9). The sensor is mounted in the pressure-regulator module by means of the thread. Strain gages are integrated in a bridge circuit on the upper side of the steel diaphragm.

When pressure is applied, the steel diaphragm elongates and the bridge circuit is detuned. The resulting bridge voltage is proportional to the applied pressure. It is directed via bonded wires to the evaluator circuit, amplified and converted into an output voltage of 0.5 V...4.5 V. From this signal, the engine ECU calculates the current tank pressure using a characteristic curve.

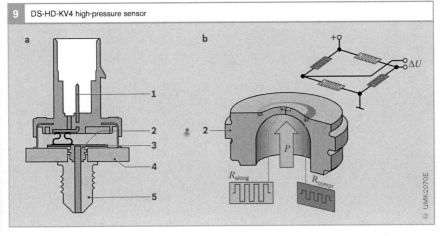

9   DS-HD-KV4 high-pressure sensor

Fig. 9

a   Sectional drawing
b   Measurement
     principle

1   Plug
2   Sensor element
3   Circuit holder
4   Housing base
5   Threaded fitting

TV-NG1 tank shutoff valve

Function

The TV-NG1 tank shutoff valve is screwed directly into the natural-gas tank and serves as the interface to the fuel system in the vehicle. The function of the TV-NG1 is to open up and shut off the gas flow. A solenoid shutoff valve (SOV-NG1) is integrated in the TV-NG1 for this purpose.

In addition, various service and safety devices are mounted on the TV-NG1:

- The gas flow can be interrupted for repairs with a mechanical shutoff valve.
- A flow limiter ensures that the contents of the tank are drained under throttled conditions if the natural-gas high-pressure line is severed in the event of an accident.
- A fusible link provides protection in the event of fire. At a temperature of approximately 110 °C, the fuse blows and ensures that the contents of the tank are discharged under controlled conditions to atmosphere.
- A pressure-limiting valve or a temperature sensor can also be optionally installed.
- It is possible with the aid of the optional temperature sensor to measure the tank contents more precisely when compared with pure pressure measurement.

Design

Two different types of tank shutoff valve may be used: internal and external. On an external tank shutoff valve, the individual attachment parts are mounted outside the gas bottle, as on a conventional gas fitting. On an internal tank shutoff valve, all the devices are integrated in the valve block and project into the gas tank. From the outside, only a plate containing the connections can be seen. This design provides for enhanced crash safety when compared with an external tank shutoff valve, and at the same time the reduced height makes it possible to use longer tanks and thereby optimize the tank volume.

Method of operation

Fig. 10 shows the external TV-NG1, consisting of the valve block and the SOV-NG1 modular shutoff valve (2). The SOV-NG1 is a two-stage solenoid valve for shutting off natural gas and is closed when de-energized. The closing process is initiated after the current is deactivated by a spring, which forces the sealing element onto the seal seat. The valve is also held closed with the assistance of the system pressure.

The SOV-NG1 operates according to a two-stage opening principle, i.e., pressure equalization is established in the first opening stage (small cross-section) and only then is the full throughflow cross-section opened. From a system point of view, a peak-hold actuation proves to be effective. In this case, after actuation, the valve switches back with a high opening current to a lower holding current in order to reduce electrical power loss. However, the valve can also be permanently operated with the opening current. A trapezoidal plug serves as the electrical interface.

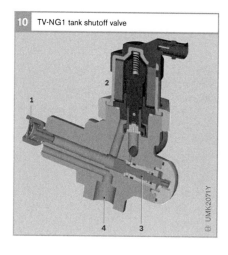

10   TV-NG1 tank shutoff valve

Fig. 10
1   Flow limiter
2   Electromagnetic valve
3   Manual shutoff valve
4   Port, safety valve

PR-NG1 pressure-regulator module

Function

The function of the PR-NG1 pressure-regulator module is to reduce the pressure of the natural gas from tank pressure to the nominal operating pressure. At the same time, the operating pressure must be kept constant within specific tolerances through all operating states. The operating pressure of present-day systems is usually about 7...9 bar (absolute). There are also systems which operate at pressures starting from 2 bar ranging up to 11 bar.

Design

Today, mainly diaphragm- or plunger-type pressure regulators are used. Pressure reduction is effected by means of throttle action and can occur either in one single stage or in several stages.

Figure 11 shows the sectional view of a single-stage diaphragm-type pressure regulator. A 40 µm sinter filter, a shutoff valve (SOV-NG1), and a high-pressure sensor are provided on the high-pressure side. The sinter filter is designed to retain solid particulates in the gas flow, while the SOV-NG1 serves to shut off the gas flow. A pressure sensor is incorporated to determine the fuel level in the tank.

A pressure-relief valve is mounted on the pressure regulator on the low-pressure side. In the event of a fault in the pressure regulator, this pressure-relief valve prevents damage to components in the low-pressure system. When the gas expands, the PR-NG1 cools down sharply in accordance with the Joule-Thompson effect. The PR-NG1 is therefore connected to the vehicle's heating circuit to prevent it from freezing up.

The operating pressure is preset by selecting the appropriate types of diaphragm and compression spring. An adjusting screw which is preset and sealed at the factory is used for fine adjustment of the spring preload.

Method of operation

The gas flows from the high-pressure side through a variable throttling orifice (5) into the low-pressure chamber (9), where the diaphragm (8) is situated. The diaphragm controls the opening cross-section of the throttling orifice (5) via a control rod (6). When the pressure in the low-pressure chamber is low, the diaphragm is forced by the spring (7) in the

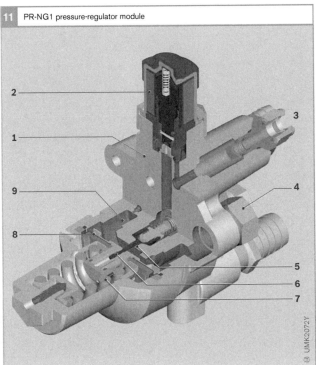

11   PR-NG1 pressure-regulator module

2
1
9
8
3
4
5
6
7

UMK2072Y

Fig. 11
1   Regulator housing
2   SOV-NG1 electromagnetic shutoff valve
3   CNG inlet
4   CNG outlet
5   Throttling orifice for pressure regulation
6   Control rod
7   Spring
8   Diaphragm
9   Low-pressure chamber

direction of the throttling orifice, which opens to allow the pressure to increase on the low-pressure side. In the event of excessive pressure in the low-pressure chamber, the spring is compressed more sharply and the throttling orifice closes. The decreasing cross-section of the throttling orifice reduces the pressure on the low-pressure side. In stationary operation, the system levels out at a specific throttle opening and keeps the pressure in the low-pressure chamber constant.

If now the gas demand in the system increases, e.g., when the accelerator pedal is pressed, at first more gas is discharged from the pressure regulator than can follow up through the throttling orifice. As a result, the pressure in the low-pressure area drops briefly until the throttling orifice opens to such an extent as to re-establish a constant pressure for the increased throughflow. Minimal system-pressure fluctuations in the event of load changes testify to the quality of the pressure regulator.

When the pressure in the low-pressure chamber exceeds a specific value, e.g., because no gas is discharged into the system, the throttling orifice closes completely. This pressure is known as the lock-off pressure. The throttling orifice reopens when the pressure drops. The process of the throttling orifice opening and closing is accompanied

by a buildup of noise and wear. In the interests of minimizing this, pressure regulators are normally designed in such a way that the lock-off pressure is so far above the system pressure that the throttling orifice always remains open in normal operation.

An ideal pressure regulator keeps the pressure constant, regardless of the throughflow. In reality, however, pressure regulators deviate from this ideal behavior on account of side effects. Figure 12 depicts a typical throughflow curve for a single-stage pressure regulator. The output pressure drops as throughflow increases, and hysteresis is also encountered between increasing and decreasing throughflow. This hysteresis arises on account of frictional and flow losses. Basically, the effects shown occur most clearly when the throughflow is high and the pressure regulator has a compact design.

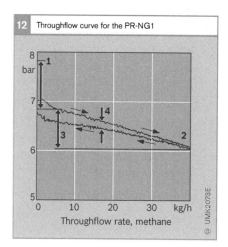

Fig. 12
1   Lock-off pressure
2   Setting pressure
3   Maximum pressure
    drop
4   Hysteresis

Manufacturers must furnish fuel-consumption data for their vehicles. The official figures are calculated based on the composition of the exhaust gases monitored during emissions testing. Emissions testing is conducted based on a standardised test procedure, or driving cycle. The standardized procedure provides emissions figures suitable for comparison among vehicles.

Motorists can make a major contribution to improved fuel economy by adopting a suitable driving style. Potential fuel savings vary according to a variety of factors.

By adopting the practices listed below, the "economy-minded" motorist can achieve fuel savings of 20...30% compared to the "average" driver. The latitude available for enhancing fuel economy depends upon a number of factors. Especially significant among these are operating environment (urban traffic or long-distance cruising, etc.) and general traffic conditions. This is why attempts to quantify the precise savings potential represented by each individual factor are not always logical.

Increasing fuel economy
- Tire pressures: Remember to increase inflation pressures when vehicle is loaded to capacity (savings: roughly 5%)
- Accelerate at wide throttle openings and low engine speeds, upshift at 2000 rpm
- Drive in the highest possible gear: even at engine speeds below 2000 rpm it is possible to apply full throttle
- Plan ahead to avoid continuous alternation between braking and acceleration
- Exploit the potential of the trailing-throttle fuel cutoff
- Switch off engine during extended stops, such as at traffic lights with extended red phases and at railroad crossings, etc. (3 minutes of idling consumes as much fuel as driving 1 kilometer)
- Use full-synthetic engine oils (savings of approximately 2% according to the manufacturer)

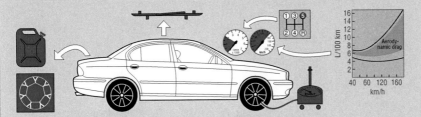

Negative influences on fuel economy
- Added vehicle weight caused by unnecessary ballast in the luggage compartment (adds roughly 0.3 liters/100 km)
- High driving speeds
- Increased aerodynamic drag from roof-mounted racks and luggage carriers
- Activation of supplementary electrical accessories such as rear-screen defroster, fog lamps (approximately 1 liter/1 kW load)
- Contaminated air filter and worn spark plugs (observe service intervals)

SMK1827E

Ignition systems over the years

The gasoline, or spark-ignition, engine is an internal-combustion machine that relies on an external source of ignition-energy to run. An ignition spark ignites the air/fuel mixture compressed in the combustion chamber to initiate the combustion process. This ignition spark is generated by a flashover between the electrodes of a spark plug extending into the combustion chamber. The ignition system must generate adequate levels of high-voltage energy to generate the flashover at the spark plug while also ensuring that the ignition spark is triggered at precisely the right instant.

Overview

Development history of Bosch ignition systems

Magneto

Ignition in gasoline engines posed a big problem in the early years of the automobile. It was only when Robert Bosch developed the low-voltage magneto that an ignition system became available which was deemed sufficiently reliable for the conditions obtaining at the time. The magneto generated by means of magnetic induction in a wound armature an ignition current which, when interrupted, triggered an ignition spark at the arcing mechanism. This spark was able to ignite the mixture in the combustion

chamber. However, the limits of this technology were soon to become apparent.

High-voltage magneto ignition was able to satisfy the demands of faster-running engines. This magneto also generated a voltage by means of magnetic induction. This voltage was transformed to such an extent that it was able to trigger a flashover at the electrodes of the spark plug which was now in common use.

Battery ignition

The demand for more cost-efficient ignition system led to the development of battery ignition; this gave rise to conventional coil ignition with a battery serving as the supplier of energy and an ignition coil serving as the energy storage medium (Fig. 2). The coil current was switched via the breaker point. A mechanical governor and a vacuum unit served to adjust the ignition angle.

Development did not stop there. Electronic components began to be used and gradually the amount of electronic components increased. First of all, with transistorized ignition, the coil current was switched via a transistor in order to prevent contact erosion at the breaker points and thereby to reduce wear. In further transistorized ignition variants, the breaker contact, which still served as the control element for activating the ignition coil, was replaced. This function was now taken over by Hall generators or induction-type pulse generators.

The next step was electronic ignition. The load- and speed-dependent ignition angle was now stored in a program map in the ECU. Now it was possible to take into account further parameters, such as, for example, the engine temperature, for determining the ignition angle. In the final step, with the arrival of distributorless semiconductor ignition, even the mechanical distributor has now been dispensed with.

Figure 1 shows this development process. Since 1998 only Motronic systems, which have integrated the functionality of distributorless semiconductor ignition in the engine-management systems, have been used.

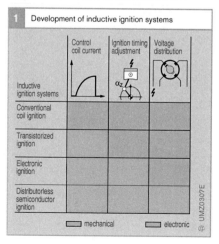

1 Development of inductive ignition systems

	Control coil current	Ignition timing adjustment	Voltage distribution
Inductive ignition systems		α_z	
Conventional coil ignition			
Transistorized ignition			
Electronic ignition			
Distributorless semiconductor ignition			

☐ mechanical   ☐ electronic

UMZ0307E

A training chart from 1969 showing Bosch battery ignition

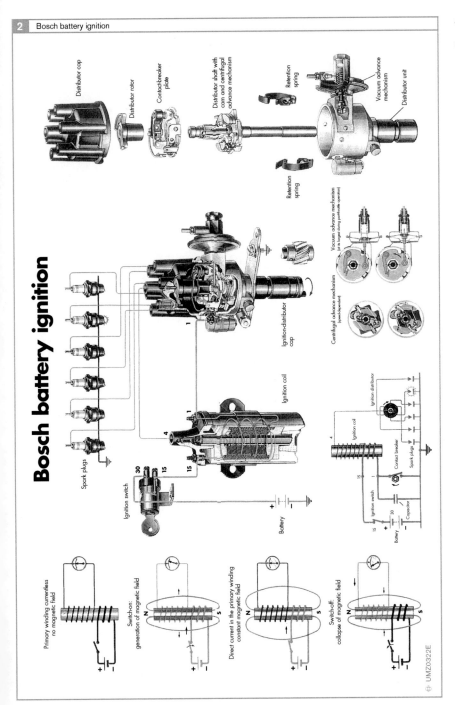

2 | Bosch battery ignition

Bosch battery ignition

Distributor cap
Distributor rotor
Contact-breaker plate
Distributor shaft with cam and centrifugal advance mechanism
Retention spring
Vacuum advance mechanism
Distributor unit
Retention spring

Vacuum advance mechanism (at its largest during part-throttle operation)

Centrifugal advance mechanism (speed-dependent)

Ignition-distributor cap

Spark plugs

Ignition coil

Ignition switch

Battery

Ignition distributor
Ignition coil
Contact breaker
Spark plugs
Ignition switch
Capacitor
Battery

Primary winding currentless no magnetic field

Switch-on: generation of magnetic field

Direct current in the primary winding constant magnetic field

Switch-off: collapse of magnetic field

UMZ0322E

Early ignition evolution

The Volta pistol combined two basic elements of engine technology: It used a mixture of air and gas, and relied on an electrical spark. It is here that the story of electric ignition begins.

Long before the first engines appeared at the end of the 19th century, inventors were engaged in efforts to evolve internal-combustion machines suitable for replacing the steam engines which were widely used at the time.

The first known attempt to create a thermal-energy machine to replace boiler, burner and steam with internal combustion was undertaken by Christiaan Huygens in the year 1673. The fuel used in this powder machine (Fig. 1) was gunpowder (1), which was ignited with a fuse (2). Following ignition, the combustion gases escape through non-return valves (4) from the tube (3), in which a vacuum is then created. Atmospheric pressure forces the piston (5) downwards, and a weight G (7) is lifted.

Because the machine had to be reloaded after each ignition, it could not serve as a true engine by providing continuous power.

Over 100 years later, in 1777, Alessandro Volta experimented with igniting a mixture of air and marsh gas using sparks. Spark generation was provided by the electrophorous tube which he had invented in 1775. This effect was utilized in the *Volta pistol*.

In 1807 Isaak de Rivaz developed an atmospheric piston engine, in which he utilized the principle of Volta's gas pistol and ignited a combustible air/gas mixture with an electrical spark. Rivaz built an experimental vehicle (Fig. 2) based on his patent drawings, but soon abandoned his efforts in response to less than satisfactory results. Working along similar lines to Huygens' powder machine, a piston was blasted upwards by the explosion before being pulled back again by atmospheric pressure. The vehicle was thus able to move forward a few meters, but then fresh combustion mixture had to be admitted into the cylinder and ignited.

Mobile applications in a motor vehicle called for engines with continuous outgoing power. Igniting the combustible mixture in the cylinder proved to be the main problem here. Many engine builders were working on finding solutions, and various systems came into being at the same time.

High-voltage vibrator ignition

A concept for a battery-based ignition system had been available since 1860, when the Frenchman Etienne Lenoir constructed a "high-voltage vibrator ignition" system (Fig. 3) for his stationary gas engine. To generate the ignition current, a Ruhmkorff spark inductor (2) was used, which was supplied, for example, by a galvanic element

Fig. 1
1 Capsule with gunpowder
2 Fuse
3 Tube
4 Non-return valve
5 Piston
6 Idler pulley
7 Weight G

Fig. 2
1 Button for transmitting ignition spark
2 Cylinder
3 Piston
4 Bladder, filled with hydrogen

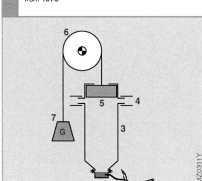

1 Concept of Christiaan Huygens' powder machine from 1673

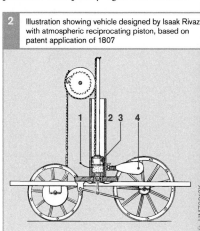

2 Illustration showing vehicle designed by Isaak Rivaz with atmospheric reciprocating piston, based on patent application of 1807

(voltaic pile) (battery ignition). Two insulated platinum wires (6) served as the electrodes to generate the flashover in the engine. Lenoir had thus invented the precursor of all spark plugs. Lenoir used a high-voltage distributor on contact rails (5) to control current flow to the two spark plugs on the dual-action engine.

In the Ruhmkorff spark inductor, a magnetic field builds up in the coil as soon as the circuit is completed. The current increases gradually. When it has reached a specific value, the armature (4) is attracted and the trembler contacts (3) open. The magnetic field collapses as a result of the broken circuit. The rapid magnetic-field change induces in the second coil a high induction voltage, which causes a flashover at the spark plug. The armature completes the circuit again and the process is repeated. Approximately 40 to 50 ignition processes were achieved with this high-tension vibrator ignition. The vibrator system emitted a characteristic buzzing sound during operation.

The following factors prevented this system from achieving widespread popularity in automotive applications.
- The system actually generated an entire series of sparks during the combustion stroke, which prevented efficient combustion at higher engine speeds.

- No option was available at the time for generating the required current while the vehicle was actually moving.

In 1886 Carl Benz further developed high-voltage vibrator ignition and was thereby able to achieve higher speeds than with his first vehicle engine (approximately 250 rpm). The electrical power source continued to pose problems, as the galvanic elements responsible for supplying current were ready for replacement after only 10 kilometers.

Hot-tube ignition

Increases in engine operating speeds were essential if the size of powerful gasoline engines for automotive applications was to be kept in check. Unfortunately, the control mechanisms employed for flame ignition, as were commonly used in stationary gas engines, were too slow to achieve higher speeds.

In 1883 the continuous-operation, hot-tube ignition system developed by Gottlieb Daimler was patented. This ignition system (Fig. 4) consisted of a passage which was connected to the combustion chamber in the cylinder. The passage was sealed gas-tight by a hot tube (2) which was permanently made to glow by a burner. During the compression stroke, the mixture was forced into the hot tube, where it ignited

Ignition was – as Carl Benz once observed – "the problem to end all problems".
"If there is no spark, then everything else has been in vain, and the most brilliant design is worthless".

It was not without reason that French drivers at the turn of the century bade each other not "Safe journey!" but "Safe ignition!" ("Bon Allumage!").

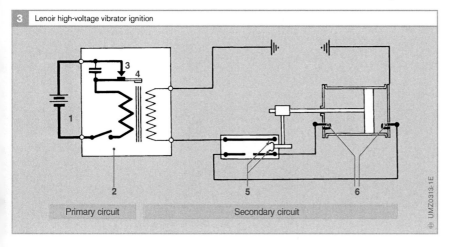

3   Lenoir high-voltage vibrator ignition

3	4
1	
2	

Primary circuit   Secondary circuit

UMZ0313-1E

Fig. 3
1 Battery (galvanic element)
2 Ruhmkorff spark inductor
3 Trembler contacts
4 Armature
5 Distributor with contact spring
6 Spark plug

and induced the remaining mixture in the combustion chamber to ignite. The hot tube had to be heated in such a way that ignition started only at the end of the compression stroke.

Hot-tube ignition enable engine speed to be increased dramatically. Depending on the system design, speeds as high as 700...900 rpm were possible.

For more than a decade, hot-tube ignition was the predominant type of ignition used by many engine manufacturers. The concept fostered widespread acceptance of both the Daimler engine and the motor vehicle in general. One disadvantage, however, lay in the fact that the hot tube always had to be adjusted to the correct heat. Furthermore, the flame was prone to go out in rainy or stormy conditions. If the burner was inexpertly handled, fire damage was a distinct possibility, which compelled the design engineer Wilhelm Maybach in 1897 to hypothesize in a memorandum that every automobile with hot-tube ignition would sooner or later be destroyed by fire. Even Daimler in the end turned to the principle of magneto ignition after this form of ignition had in the meantime proved to be workable.

Magneto-electric low-voltage snap-release ignition

In 1884 Nikolaus August Otto developed magneto-electric low-voltage snap-release ignition. A magneto-inductor with an oscillating double-T armature and rod-shaped permanent magnet generated a low-voltage ignition current (Fig. 5). Interrupting the current flow produced an opening ignition spark at the contact points in the cylinder. The armature drive's spring-loaded snap-release mechanism and the push rod controlling the ignition contact's trip lever were coordinated to open the circuit at precisely the instant when armature current peaked. This produced a powerful ignition spark at the moment of ignition.

The four-stroke engine developed by Otto in 1876 had up to that point been powered by municipal gas and had therefore only been suitable for stationary applications. Magneto-electric low-voltage snap-release ignition now allowed such an engine to be powered by gasoline. However, the engine speeds that could be achieved limited its use to slow-running, stationary engines only.

Magneto ignition

The ignition problem called out for a solution which would be more suitable for motor vehicles. In the end, this problem was addressed by a special company which did not build engines itself, but rather brought onto the market ignition devices for slow-running engines: This was Robert Bosch's Werkstätte für Feinmechanik und Elektrotechnik (Workshops for Light and Electrical Engineering), founded in 1886 in Stuttgart.

Bosch low-voltage magneto with snap-release mechanism
Bosch developed magneto-electric low-voltage devices for Otto's snap-release ignition (Fig. 5) in order to be able to offer them as accessory equipment to the manufacturers of stationary spark-ignition engines. The system's asset was its ability to operate without a battery. The high weight of the

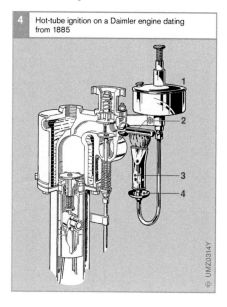

4 Hot-tube ignition on a Daimler engine dating from 1885

UMZ0314Y

Fig. 4
1 Gasoline reservoir
   for burner
2 Hot tube
3 Burner
4 Preheater bowl

armature and the slow ignition mechanism prevented its continued use in automotive engines.

Low-voltage magneto ignition

Bosch developed the slow snap-release ignition into faster and lighter make-and-break magneto ignition suitable for high-speed automotive engines.

Instead of allowing the heavy, wound armature to oscillate, the system now used a sleeve suspended between the pole shoes and the fixed armature (Fig. 6) to act as a conductor of the lines of flux. The sleeve was driven via bevel gears, which also served to adjust the moment of ignition. A cam rising slowly in the direction of rotation served to rotate the arcing mechanism. As soon as the mechanism sped through spring force away from the cam, the ignition lever was separated from the ignition pin in the cylinder, and the ignition spark was thereby generated.

The sleeve design of the magneto and the bevel-gear drive were immediately successful because this arrangement proved to be suitable for the speed range required at the time.

Daimler had one of these ignition systems installed in a vehicle in 1898, and then proceeded to road-test it by driving from Stuttgart to Tyrol, a trial which passed off successfully. Even the Daimler engine of the first Zeppelin airship operated with a Bosch make-and-break ignition system, since the flammability of the filling gas precluded the use of hot-tube ignition in the airship.

However, this ignition system was still a low-voltage magneto system, which required mechanically and later electromagnetically controlled arcing contacts in the combustion chamber to generate the opening ignition sparks via an arcing mechanism.

High-voltage magneto ignition

Higher engine speeds, compression ratios and combustion temperatures all combined to produce ignition demands that make-and-break ignition could not satisfy. Until problems with batteries could be resolved, magneto ignition using spark plugs instead of arcing contacts represented the only viable option. A source of high-voltage ignition current was essential for this purpose.

The double-T armature became the "Bosch armature", the symbol and logo of Robert Bosch GmbH.

| **5** | Design of the Bosch low-voltage magneto with snap-release mechanism and ignition flange dating from 1887 |

| **6** | Design of the Bosch low-voltage magneto with oscillating sleeve, 1897 version |

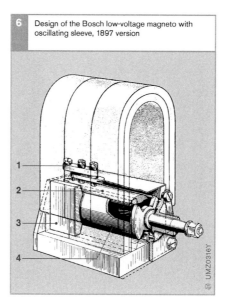

Fig. 5
a   Design
b   Block diagram
    (section)

1   Compression-spring
    arrangement
2   Ignition lever
3   Ignition pin
4   Ignition flange
5   Push rod
6   Double-T armature
7   Elbow lever
8   Control shaft
9   Terminal

Fig. 6
1   Terminal
2   Double-T armature
    (fixed)
3   Pole shoes
4   Sleeve (oscillating)

Robert Bosch assigned Gottlob Honold to design a magneto-based ignition system in which the arcing mechanism would be replaced by permanent ignition electrodes.

Honold's starting point was a low-voltage magneto with an oscillating sleeve, which he then proceeded to modify. The double-T armature received two windings; one consisted of a limited number of loops of thick wire, while the second comprised a larger number of loops of thin wire (Fig. 7). Rotating the sleeve generated initially generated a low voltage in the armature winding. The winding with the fewer number of loops was simultaneously shorted by a contact breaker (10). This produced a high current which was subsequently interrupted. This induced in the other winding with the larger number of loops a high, rapidly decaying voltage, which passed through the spark gap at the spark plug (16) to render it conductive. After this, a further voltage was induced in the same winding. Although substantially lower than the first voltage, it was sufficient to send a current through the now conductive spark gap and generate an arc familiar from make-and-break ignition.

The contact breaker was mechanically controlled by a cam (15) to enable it to complete or break the circuit of the low-voltage winding at a precisely defined time. A condenser was connected in parallel with the breaker points to inhibit arcing at the contact breaker.

The spark plugs also had to be redeveloped, since their electrodes eroded too quickly because of the hot, arc-like sparking by the new magneto. The development of Bosch spark plugs also dates back to this period. Contact breakers, which right from the start formed the heart of the high-voltage magneto, were developed further to make them more operationally reliable.

Yet another version of magneto ignition was developed by Ernst Eisemann. This system's high voltage was generated by a separate transformer fed by a low-voltage magneto. Initially, the winding of this magneto was shorted repeatedly during each current wave by a contactor which rotated synchronously with the armature. Later, Eisemann identified that just one short was sufficient. In Germany, Eisemann met with rejection. However, he enjoyed success in France,

Fig. 7

a Block diagram of high-voltage magneto
b Design of first series-manufactured high-voltage magneto

1 Pole shoe
2 Sleeve (rotating)
3 Double-T armature
4 Current collector with connecting bar to spark-plug terminals
5 Distributor disk with collector ring
6 Current conduction to distributor disk (secondary)
7 To ignition switch
8 Current conduction to contact breaker (primary)
9 Terminals to spark plugs
10 Contact-breaker lever
11 Breaker point
12 Condenser
13 Ignition-timing adjustment
14 Magnet
15 Cam
16 Spark plug

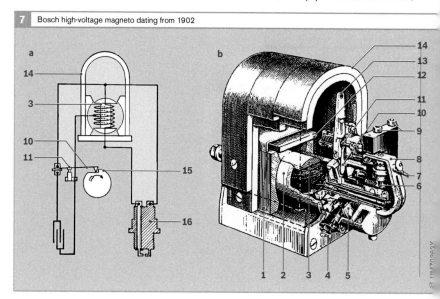

7   Bosch high-voltage magneto dating from 1902

where the engineer de la Valette secured the exclusive-marketing rights for Eisemann's magneto ignition. Later, Eisemann abandoned the separate coil in favor of the Bosch design featuring the familiar double-T armature with its two windings.

Battery ignition

When Robert Bosch AG introduced battery ignition in 1925, the automotive industry was dominated by magneto ignition, because it was the most reliable form of ignition. But vehicle manufacturers were demanding a less expensive system. After becoming established in series production in the US, battery ignition started to take hold on both motor cars and motorcycles within a few years in Europe too.

First series production in the US

By 1908 the American Charles F. Kettering had improved battery ignition to the point where it was ready for series production at Cadillac in 1910. Despite all its imperfections, it became increasingly popular during the First World War. The desire of the general population for affordable motor vehicles encouraged the success of the cheaper battery-ignition system. The vehicle's dependence on a battery came to be accepted because battery charging was now taken care of during vehicle operation by the installation of an alternator.

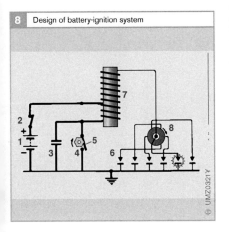

8 Design of battery-ignition system

European introduction of battery ignition by Bosch

In the initial years following World War I in Europe, motor cars were restricted to a small segment of the population, but the gradual rise in the demand for cars was accompanied by a desire for less expensive products, just as it had earlier in the US. In the 1920s conditions were ripe in Europe for the widespread breakthrough of battery-ignition systems. Bosch had long been in possession of the expertise required to design such a system for series production. Before 1914 Bosch was already supplying ignition coils – the core of a battery-ignition system – to the US market. Bosch was one of the first manufacturers to respond and in 1925 brought onto the European market a battery-ignition system, consisting of an ignition coil and an ignition distributor. Initially, they were only used in the Brennabor 4/25. But, by 1931, 46 of the 55 automotive models available in Germany were equipped with the system.

Design and method of operation

Battery ignition consisted of two separate devices: the engine-driven ignition distributor and the ignition coil (Fig. 8). The ignition coil (7) contained the primary and secondary windings, and the iron core. The distributor (8) comprised the stationary contact breaker (5), the rotating actuator cam (4), and a mechanism to distribute the secondary current. The ignition condenser (3) protected the points against premature wear by suppressing arcing.

The only moving parts in the system were the contact-breaker cam and the distributor shaft. The system also contrasted with magneto-based systems by requiring only negligible levels of motive force to sustain its operation.

Another difference relative to the magneto was that battery ignition obtained its primary current from the vehicle's electrical system. The high voltage was generated in a similar way to the magneto: the current, which built up a magnetic field in the primary winding, was interrupted by

Fig. 8
1 Battery
2 Ignition switch
3 Ignition condenser
4 Contact-breaker cam
5 Breaker point
6 Spark plugs
7 Ignition coil
8 Ignition distributor

a mechanically controlled contact breaker. The collapse of the magnetic field generated high voltage in the secondary winding.

Ignition-performance demands for "modern times"

The performance demands placed on ignition systems for internal-combustion engines increased dramatically and became more varied. Engines were operated with higher compression and leaner air/fuel mixtures. Even the maximum speed was increased. At the same time, demands, such as e.g., low noise, good idle performance, long service intervals, low weight, small dimensions, and low price, made rapid further development essential.

Higher compression ratios combined with more economical carburetor tuning meant that higher ignition voltages were needed to ensure safe and reliable flashover triggering. Meanwhile, wider spark-plug electrode gaps were required for smooth idling, and this also raised additional demands for ignition voltage. Voltage levels had to rise to more than twice their earlier level. This, in turn, had implications for the conductive elements in the high-voltage circuit, which had to be designed to resist arcing.

Also required was a way to adjust ignition timing to accommodate the expanded engine-speed range. Ignition timing had to adjusted through a larger range to compensate for the increased lag between firing point and flame-front propagation encountered at high engine speeds. In systems developed for multi-cylinder engines, the primary-current circuit breaker and the mechanism for distributing the high voltage supplied by the ignition coil were integrated in a single distributor housing, where they shared a common drive shaft. Ignition timing was regulated by shifting the position of the contact-breaker lever relative to the cam, an exercise initially performed from the driver's seat, and requiring both experience and some degree of mechanical sensitivity. Centrifugal timing adjusters operating in response to engine speed, found as early as 1910 in high-voltage magneto-ignition systems, were adopted in battery-ignition systems.

Fuel economy also became a progressively more important consideration, making it necessary to include the dependence on load of the combustion process in the timing adjustment. The answer was to install a diaphragm that responded to the intake-manifold pressure upstream from the throttle valve plate and generated actuating forces on the ignition distributor. This resulted in an ignition-angle correct acting in addition to the centrifugal timing adjuster. Bosch introduced this vacuum-controlled timing in its ignition distributors in 1936.

In developing the breaker points, Bosch was able to draw on experience already garnered while working with magnetos. All of battery-ignition components underwent improvement over the course of time. Eventually, technological advances – especially in the new field of semiconductor technology – paved the way for new ignition systems. While the basic concept mirrored that of the original battery-ignition system, the designs were radically different.

Bosch magneto ignition in motor racing

Bosch low-tension magneto ignition systems successfully absolved the acid test in the first car with the name Mercedes, which won three French races as well as achieving other victories in the course of 1901. One particularly significant event was the Irish Gorden Bennett race in 1903. With the Belgian driver Camille Jenatzy at the helm, the 60 HP Mercedes posted an impressive triumph – a success to which the reliability and superior performance of Bosch magneto ignition made a major contribution. By the time the 1904 Gorden Bennett rolled around, the five fastest cars were all equipped with Bosch ignition.

In June of 1902 a "light touring car" from Renault was the first to reach Vienna's Trabrennplatz at the culmination of the Paris to Vienna long-distance race. At the wheel was Marcel Renault, whose brother had already attracted considerable attention while at the same time laying the foundation for a major automotive marque with his "voiturette" in 1898. Renault's winning car was equipped with the new Bosch high-tension magneto ignition, an innovation still not available on standard vehicles at the time.

Camille Jenatzy as Bosch Mephisto on a Bosch advertising poster from 1911

In 1906, victory at the French Grand Prix also went to a vehicle equipped with the Bosch high-tension magneto system. This system soon found favored status as the system of choice among automotive manufacturers, resulting in a massive sales increase.

Magneto ignition in aircraft

It was in May, 1927, that postal aviator Charles Lindbergh embarked upon his historic flight across the Atlantic. His single-engine "Spirit of St. Louis" made the non-stop trip from New York to Paris in 33.5 hours. Trouble-free ignition during the journey was furnished by a magneto manufactured by Scintilla in Solothurn, Switzerland, now a member of the Bosch group.

In April, 1928, aviation pioneers Hermann Köhl, Günther Freiherr von Hünefeld and James Fitzmaurice achieved the first non-stop airborne traversal of the Atlantic from East to West in a

Junkers W33 featuring a fuselage of corrugated sheet metal. They took off from Ireland and landed 36 hours later in Greenly Island, Canada. They were unable to reach their original objective, New York, owing to violent weather. But: "the flight was successful with Bosch spark plugs and a Bosch magneto" (see illustration).

Battery ignition systems over the years

The period between the appearance of Bosch battery ignition in 1925 and the final versions of this system many years later was marked by constant change and continuous evolution.

There were no substantive changes in the basic concept behind battery ignition in this time. Most of the modifications focused on the mechanisms employed to adjust ignition timing. These were reflected in the changes to system components. Ultimately the only components remaining from the original battery ignition were the coil and the spark plug. Finally, at the end of the 1990s, control of ignition functions was incorporated in the Motronic engine-management system. Thus ignition systems with separate ignition control units – as described in the following section – are now history.

Conventional coil ignition (CI)

Conventional coil-ignition systems are controlled by contact-breaker points. The contact breakers in the distributor open and close the circuit to control current flow within the ignition coil. The contact is closed over a specific angle (dwell angle).

Design and operation

The components in the conventional coil-ignition system (Fig. 1) are the
- Ignition coil (3)
- Ignition distributor (4) with breaker point (6), ignition capacitor (5), centrifugal and vacuum advance mechanisms (7) and the
- Spark plugs (9)

During operation battery voltage flows through the ignition switch (2) on its way to the coil's Terminal 15. When the points close current flows through the ignition coil's primary winding (asphalt coil, refer to section ignition coils) and to ground. This flow produces a magnetic flux field in which ignition energy is stored. The rise in current flow is gradual owing to inductance and primary resistance in the primary winding. The time available for charging is determined by the dwell angle. The dwell angle, in turn, is defined by the contours of the distributor-cam lobes, which open and close the breaker points by pushing against the cam follower (Fig. 2b). At the end of the dwell period the cam lobe opens the contacts to interrupt current flow in the coil. The number of lobes on the cam corresponds the number of cylinders in the engine.

Points must be replaced at regular intervals owing to wear on the cam follower as well as burning and pitting on the contact surfaces.

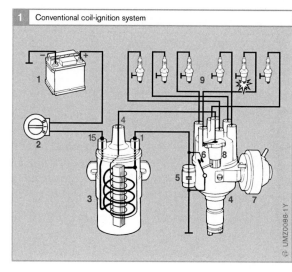

1   Conventional coil-ignition system

UMZ0088-1Y

Fig. 1
1. Battery
2. Ignition/starter switch
3. Ignition coil
4. Distributor
5. Capacitor
6. Contact-breaker points
7. Vacuum advance mechanism
8. Rotor
9. Spark plug

1, 4, 15 Terminals

Current, dwell time and the number of secondary windings in the coil are the primary determinants of the ignition voltage induced in the coil's secondary circuit.

A capacitor in parallel with the points prevents arcing between the contact surfaces, which would allow current to continue flowing after they open.

The high-tension voltage induced in the ignition coil's secondary winding is conducted to the distributor's centre contact.
As the rotor (Fig. 1, Pos. 8) turns it establishes an electrical path between this center contact and one of the peripheral electrodes. The current flows through each electrode in sequence, conducting high voltage to the cylinder that is currently approaching the end of its compression stroke to generate an arc at the spark plug. The distributor must remain synchronized with the crankshaft for its operation to remain in rhythm with the pistons in the individual cylinders. Synchronization is assured by a positive mechanical link between the distributor and either the camshaft or another shaft coupled to the crankshaft at a 2:1 step-down ratio.

Ignition advance adjustment

Because of the positive mechanical coupling between distributor shaft and crankshaft, it is possible to adjust the ignition timing to the specified angle by rotating the distributor housing.

Centrifugal advance adjustment
The centrifugal advance mechanism varies ignition timing in response to shifts in engine speed. Flyweights (Fig. 2a, Pos. 4) are mounted in a support plate (1) that rotates with the distributor shaft. These flyweights spin outward as engine and shaft speed increase. They shift the base plate (5) along the contact path (3) to turn it opposite the distributor shaft's (6) direction of rotation. This shifts the relative positions of the point assembly and distributor cam by the adjustment angle α. Ignition timing is advanced by this increment.

Vacuum advance adjustment
The vacuum-advance mechanism adjusts ignition timing in response to variations in the engine's load factor. The index of load factor is manifold vacuum, which is relayed via hose to the two aneroid capsules (Fig. 2b).

Falling load factors are accompanied by higher vacuum levels in the advance unit which pull the diaphragm (11) and its advance/retard arm (16) to the right. In doing so, the arm turns the breaker-point assembly's base plate (8) in the opposite direction to that of the distributor shaft's rotation and thus increases the ignition advance.

Vacuum in the retard unit, for which the manifold vacuum connection is behind the throttle plate instead of in front of it, moves the annular diaphragm (15) and its advance/retard arm to the left to retard the timing. This spark retardation system is used to improve engine emissions under certain operating conditions (idle, trailing throttle, etc.). The vacuum advance is the priority system.

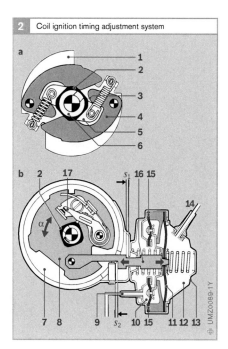

2   Coil ignition timing adjustment system

Fig. 2
a Centrifugal advance mechanism (illustrated in passive state)
b Vacuum advance and retard mechanism

1 Support plate
2 Distributor cam
3 Contact path
4 Flyweight
5 Base plate
6 Distributor shaft
7 Distributor
8 Breaker-point base plate
9 Manifold connection for retard unit
10 Retard unit
11 Diaphragm (ignition advance system)
12 Advance unit
13 Aneroid unit
14 Manifold connection for advance unit
15 Annular diaphragm (retard system)
16 Advance/retard arm
17 Contact-breaker points

s_1 Total timing advance
s_2 Total timing retardation
α Timing adjustment angle

Breaker-triggered transistorized ignition
Design and method of operation
The distributors used in transistorized
breaker-triggered ignition systems are identical to those employed with coil ignition. The
difference is in the control of the primary
ignition circuit. Instead of being opened and
closed by contact-breaker points, the circuit
is now controlled by a transistor – installed
along with supplementary electronics in the
ignition trigger box. In this system only the
control current for the transistorized ignition
system is switched by the breaker points.
Thus ultimate control of the system still re-
sides with the points. Figure 3 compares the
two designs.

When the breaker points (7) are closed,
control current flows to the base B, making
the path between the emitter E and the collector C on the transistor conductive. This
charges the coil. When the breaker points
open, no current flows to the base, and the
transistor blocks the flow of primary current.

The ballast resistors (3) limit the primary
current to the low-resistance, fast-charging
coil used in this ignition system. During starting, compensation for the reduced battery
voltage is furnished by bypassing one of these
resistors at the starter's Terminal 50.

Advantages over coil ignition
Two major assets distinguish breaker-triggered transistorized ignition from conventional coil systems. Because there is only
minimal current flow through the points,
their service life is increased dramatically. Yet
another advantage is the fact that the transistor can control higher primary currents than
mechanical contact breakers.
This higher primary current increases the
amount of energy stored in the coil, leading
to improvements in all high-voltage data,
including voltage levels, spark duration and
spark current.

Transistorized ignition with Hall-effect trigger
Design
In this transistorized ignition system the
contact breakers that were still present in the
breaker-triggered system are replaced by a
Hall-effect sensor integrated within the distributor assembly. As the distributor shaft
turns, the rotor's shutters (Fig. 4a, Pos. 1)
rotate through the gap (4) in the magnetic
triggering unit. There is no direct mechanical
contact. The two soft-magnetic conductive
elements with the permanent magnets (2)
generate a flux field. When the gap is vacated
the flux field penetrates the Hall IC (3). When
the shutters enter the gap, most of the magnetic flux is dissipated around them instead
of impacting on the IC. This process produces a digital voltage signal (Fig. 4b).

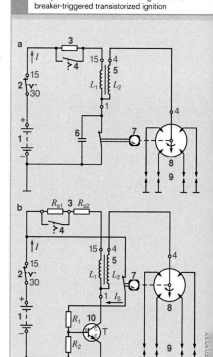

3 Comparison of conventional coil ignition and breaker-triggered transistorized ignition

Since the number of shutters corresponds to the number of cylinders, this voltage signal thus corresponds to the signal from the contact breaker in the breaker-triggered transistorized ignition system. One system relies on the distributor shaft's cam lobe to define the dwell angle, while the other uses the pulse factor of the voltage signal produced by the shutters. Depending on the particular ignition trigger box, the width b of the individual shutters can determine the maximum dwell angle. This angle thus remains constant throughout the Hall sensor's entire life, at least on systems without separate dwell-angle control. Dwell adjustments of the kind required with contact-breaker points thus become redundant.

Current and dwell-angle control
The application of rapid-charging, low-resistance coils made it necessary to limit primary current and power losses. The corresponding functions are integrated within the ignition system's trigger box.

Current control
The primary current is regulated to restrict flow within the coil and limit energy build-up to a defined level. Because the transistor enters its active range in its current-control phase, the voltage loss through the transistor is greater than in the switching mode. The result is high power loss in the circuit.

Dwell-angle control
An arrangement to regulate dwell to a suitable duration period is needed to minimize this power loss. Because it is possible to execute control operations by shifting the voltage threshhold using analog technology, the Hall-effect trigger's square-wave signal is converted to ramp voltage by charging and discharging a capacitor (Fig. 4c).

The ignition point defined by the distributor's adjustment angle lies at the end of the shutter width, correlating with 70 %. The dwell-angle control is set to provide a current control period $t_1{}^*$ that gives exactly the phase lead required for dynamic operation. The t_1 parameter is used to generate a voltage for comparison with the ramp's falling ramp. The primary current is activated to initiate the dwell period at the "ON" intersection. This voltage can be varied to shift the intersection on the ramp voltage curve to adjust the dwell period's start for any operating conditions.

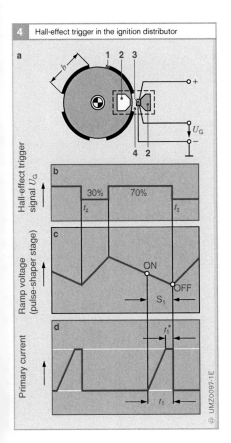

4 Hall-effect trigger in the ignition distributor

Fig. 4
a Schematic illustration of rotor design
b Hall sensor voltage output
c Ramp voltage for dwell control
d Primary current in coil

1 Shutter with width b
2 Soft-magnetic conductive element with permanent magnet
3 Hall IC
4 Gap

t_1 Dwell period
$t_1{}^*$ Current reduction period
t_z Ignition point

Transistorized ignition with induction-type pulse generator

Only minor differences distinguish transistorized ignition with a distributor containing an inductive trigger from the system with a Hall-effect sensor (Fig. 5a). The permanent magnet (1), inductive winding and core (2) on the inductive pulse generator form a fixed unit, the stator. A reluctor or "rotor" located opposite this stationary arrangement rotates to trigger the pulses. The rotor and core are manufactured in soft-magnetic material and feature spiked ends (stator and rotor spikes).

The operating concept exploits the continuous change in the gap between the rotor and stator spikes that accompanies rotation. This variation is reflected in the magnetic-flux field. The change in the flux field induces AC voltage in the inductive winding (Fig. 5b). Peak voltage varies according to engine speed: approximately 0.5 V at low rpm, and roughly 100 V at high revs. The frequency f is the number of sparks per minute.

Control of current and dwell angle with inductively triggered ignition are basically the same as with Hall-effect transistorized ignition. In this case no generation of a ramp voltage is required, as the AC induction voltage can be used directly for dwell-angle control.

Electronic ignition

As demands for precise engine management grew, the very basic ignition timing curves offered by the centrifugal and vacuum mechanisms in conventional distributors proved unable to satisfy the requirements. In the early 1980s the introduction of automotive microelectronics opened up new options for ignition-system design.

Design and operation

Electronic ignition requires neither centrifugal nor vacuum-based timing adjustment. Instead, sensors monitor engine speed and load factor and then convert these into electrical signal data for processing in the ignition control unit. The microcontroller is essential for achieving the functionality associated with electronic ignition.

Engine speed is registered by an inductive pulse sensor than scans the teeth of a reluctor mounted on the crankshaft. An alternative is to monitor rpm using a Hall-effect sensor in the ignition distributor.

A hose connects the atmosphere within the intake manifold to a pressure sensor in the control unit. If the engine is equipped with electronic injection then the load signal employed to govern the mixture-formation process can also be tapped for ignition purposes.

The control unit uses these data to generate the control signal for the ignition's coil driver. The corresponding circuitry can be integrated within the control unit or mounted externally on the ignition coil, etc.

The most pronounced asset of electronic ignition is its ability to use a program map for ignition timing. The program map contains the ideal ignition timing for range of engine operating coordinates as defined by engine rpm and load factor; the timing is defined to provide the best compromise for each performance criterion during the engine's design process (Fig. 6a). Ignition timing for any given operating coordinates is selected based on

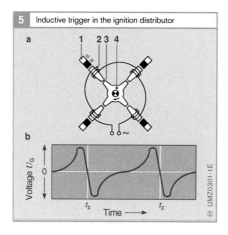

Fig. 5
a Design concept
b Inductive voltage
   curve

1 Permanent magnet
2 Inductive winding
   with core
3 Variable gap
4 Rotor

t_z Ignition point

5 Inductive trigger in the ignition distributor

a   1 2 3 4

b

Voltage U_G

0

t_z t_z

Time

UMZ0301-1E

- Torque
- Fuel economy
- Exhaust-gas composition
- Margin to knock limit
- Driveability, etc.

Designs assign priority to specific individual parameters based on the optimization criteria. This is why 3D representations of program maps for systems with electronic control show a craggy and variegated landscape, as opposed to the smooth slopes of mechanical timing-adjustment systems (Fig. 6b).

A map based on engine speed and battery voltage is available for dwell angle. This ensures that the energy stored in the ignition coil can be regulated just as precisely as with separate dwell control.

A number of other parameters can also have an effect upon the ignition angle, and if these are to be taken into account this entails the use of additional sensors to monitor
- Engine temperature
- Intake-air temperature (optional)
- Throttle-plate aperture (at idle and at WOT)

It is also possible to monitor battery voltage – important as a correction factor for dwell angle – without a sensor. An analog-digital converter transforms the analog signals into digital information suitable for processing in the microcontroller.

Advantages of electronic ignition-timing adjustment
The step from mechanically-adjusted ignition timing to systems featuring electronic control brought decisive assets:
- Improved adaptation of ignition timing
- Improved starting, more stable idle and reduced fuel consumption
- Extended monitoring of operational data (such as engine temperature)
- Allows integration of knock control

Distributorless (fully-electronic) ignition
Fully-electronic ignition includes the functionality of basic electronic systems. As a major difference, the distributor used for the earlier rotating high-voltage distribution has now been deleted in favour of stationary voltage distribution governed by the control unit. The fully-electronic ignition system generates a separate, dedicated control signal for the individual cylinders, each of which must be equipped with its own ignition coil. Dual-spark ignition uses one coil for two cylinders.

Advantages
The advantages of distributorless ignition are
- Substantially reduced electromagnetic interference, as there are no exposed sparks
- No rotating parts
- Less noise
- Lower number of high-tension connections and
- Design benefits for the engine manufacturer

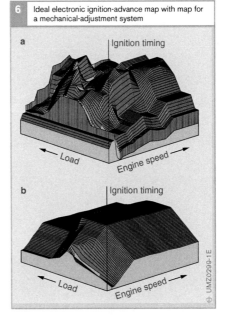

6 Ideal electronic ignition-advance map with map for a mechanical-adjustment system

a — Ignition timing — Load — Engine speed

b — Ignition timing — Load — Engine speed

UMZ0299-1E

Fig. 6
a Ignition-advance map for electronic ignition
b Ignition-advance response with conventional coil ignition

Inductive ignition system

Ignition of the air/fuel mixture in the gasoline engine is electric; it is produced by generating a flashover between the electrodes on a spark plug. The ignition-coil energy converted in the spark ignites the compressed mixture immediately adjacent to the spark plug, creating a flame front which then spreads to ignite the mixture in the entire combustion chamber. The inductive ignition system generates in each power stroke the high voltage required for flashover and the spark duration required for ignition. The electrical energy drawn from the vehicle electrical system battery is temporarily stored in the ignition coil for this purpose.

The most significant application for the inductive ignition system is in passenger cars with gasoline engines. The most commonly used are four-stroke engines with four cylinders.

Design

Figure 1 shows the basic design of the ignition circuit of an inductive ignition system using the example of a system with distributorless (stationary) voltage distribution – as is used in all current applications – and single-spark ignition coils. The ignition circuit comprises the following components:
- Ignition driver stage (5), which is integrated in the Motronic ECU or in the ignition coil
- Ignition coils (3), designed as pencil coils or as a compact coil to generate one spark (as illustrated) or two sparks
- Spark plugs (4), and
- Connecting devices and interference suppressors

Older ignition systems with rotating high-voltage distribution require an additional high-voltage distributor. This ensures that the ignition energy generated in the ignition coil is directed to the correct spark plug.

Fig. 1
Illustration of a cylinder of an inductive ignition system with distributorless voltage distribution and single-spark ignition coils
1   Battery
2   AAS diode (Activation Arc Suppression), integrated in ignition coil
3   Ignition coil
4   Spark plug
5   Ignition driver stage (integrated in engine ECU or in ignition coil)
6   Engine ECU Motronic

Term. 15, Term. 1, Term. 4, Term. 4a
Terminal designation
⊓ Actuation signal for ignition driver stage

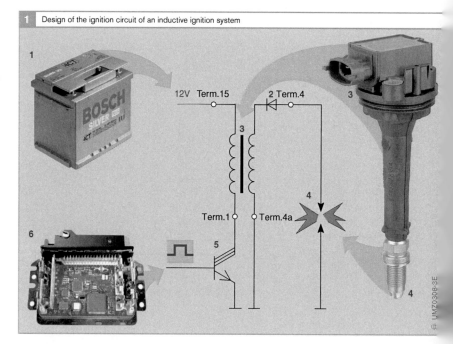

1   Design of the ignition circuit of an inductive ignition system

12V   Term.15   2 Term.4   3

Term.1   Term.4a

5

6

4

UMZ0308-3E

Function and method of operation

It is the function of the ignition to ignite the compressed air/fuel mixture and thus initiate its combustion. Safe combustion of the mixture must be guaranteed in the process. To this end, sufficient energy must be stored in the ignition coil prior to the moment of ignition and the ignition spark must be generated at the correct moment of ignition.

All the components of the ignition system are adapted in terms of their designs and performance data to the demands of the overall system.

Generating the ignition spark

A magnetic field is built up in the ignition coil when a current flows in the primary circuit. The ignition energy required for ignition is stored in this magnetic field.

Interrupting the coil current at the moment of ignition causes the magnetic field to collapse. This rapid magnetic-field change induces a high voltage (Fig. 2) on the secondary side of the ignition coil as a result of the large number of turns (turns ratio approx. 1:100). When the ignition voltage is reached, flashover occurs at the spark plug and the compressed air/fuel mixture is ignited.

The current in the primary winding only gradually attains its setpoint value because of the induced countervoltage. Because the energy stored in the ignition coil is dependent on the current ($E = 1/2 LI^2$), a certain amount of time (dwell period time) is required in order to store the energy necessary for ignition. This dwell period is dependent on, among others, the vehicle system voltage. The ECU program calculates from the dwell period and the moment of ignition the cut-in point, and cuts the ignition coil in via the ignition driver stage and out again at the moment of ignition.

Flame-front propagation

After the flashover, the voltage at the spark plug drops to the spark voltage (Fig. 2). The spark voltage is dependent on the length of the spark plasma (electrode gap and deflection due to flow) and ranges between a few hundred volts and well over 1 kV. The ignition-coil energy is converted in the ignition spark during the ignition-spark period; this spark duration lasts between 100 μs to over 2 ms. Following the breakaway of the spark, the attenuated voltage decays.

The electrical spark between the spark-plug electrodes generates a high-temperature plasma. When the mixture at the spark plug is ignitable and sufficient energy is supplied by the ignition system, the flame core that is created develops into an automatically propagating flame front.

Moment of ignition

The instant at which the spark ignites the air/fuel mixture within the combustion chamber must be selected with extreme precision. It is usually specified as an ignition angle in °cks (crankshaft) referred to Top Dead Center (TDC). This variable has a crucial influence on engine operation and determines

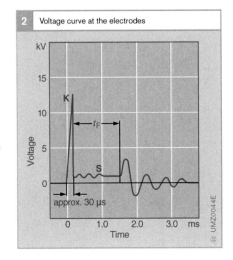

2   Voltage curve at the electrodes

Fig. 2
K   Spark head
S   Spark tail
t_F Spark duration

- The delivered torque
- The exhaust-gas emissions, and
- The fuel consumption

The moment of ignition is specified in such a way that all requirements are met as effectively as possible. However, continuous engine knocking must not develop during operation.

The influencing variables that determine the moment of ignition are engines speed and engine load, or torque. Additional variables, such as, for example, engine temperature, are also used to determine the optimal moment of ignition. These variables are recorded by sensors and then relayed to the engine ECU (Motronic). The moment of ignition is calculated from program maps and characteristic curves, and the actuation signal for the ignition driver stage is generated.

Knock control

Knock is a phenomenon which occurs when ignition takes place too early (Fig. 3). Here, once regular combustion has started, the rapid pressure increase in the combustion chamber leads to auto-ignition of the unburnt residual mixture which has not been reached by the flame front. The resulting abrupt combustion of the residual mixture leads to a considerable local pressure increase. The pressure wave which is generated propagates, strikes the cylinder walls, and can be heard as combustion knock.

If knock continues over a longer period of time, the engine can incur mechanical damage caused by the pressure waves and the excessive thermal loading. To prevent knock on today's high-compression engines, no matter whether of the manifold-injection or direct-injection type, knock control is now a standard feature of the engine-management system. With knock control, knock sensors (structure-borne-noise sensors) detect the start of knock and the ignition timing is retarded at the cylinder concerned (Fig. 4). The pressure increase after the mixture has ignited therefore occurs later, which reduces the tendency to knock. When the knocking stops, the ignition-timing adjustment is reversed in stages. To obtain the best-possible engine efficiency, therefore, the basic adaptation of the ignition angle (ignition map) can be located directly at the knock limit.

Fig. 3

Pressure curves at different moments of ignition

1 Ignition Z_a at correct moment
2 Ignition Z_b too advanced (combustion knock)
3 Ignition Z_c too retarded

Fig. 4

K1...3 Occurrence of knock at cylinders 1...3, no knock at cylinder 4
a Dwell time before timing retardation
b Drop depth
c Dwell time before reverse adjustment
d Timing advance

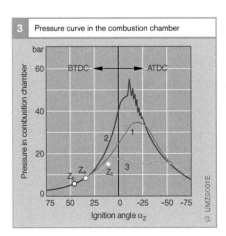

3 Pressure curve in the combustion chamber

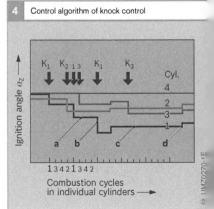

4 Control algorithm of knock control

Ignition parameters

Moment of ignition

Engine-speed and load dependence

Once ignition has been initiated by the ignition spark, it takes a few milliseconds for the air/fuel mixture to burn completely. This period of time remains roughly constant as long as the mixture composition remains unchanged. The moment of ignition point must be selected so that main combustion, and the accompanying pressure peak in the cylinder, takes place shortly after TDC. As engine speed increases, the ignition angle must therefore be advanced.

The cylinder charge also has an effect on the combustion curve. The flame front propagates at a slower rate when the cylinder charge is low. For this reason, with a low cylinder charge, the ignition angle must also be advanced.

In the case of gasoline direct injection, the range for variation of the moment of ignition in stratified-charge mode is limited by the end of injection and the time needed for mixture preparation during the compression stroke.

Basic adaptation of ignition angle

In electronically controlled ignition systems, the ignition map (Fig. 5) takes into account the influence of engine speed and cylinder charge on the ignition angle. This map is stored in the engine-management system's data memory, and forms the basic adaptation of the ignition angle.

The map's x and y axes represent the engine speed and the relative air charge. A specific number of values, typically 16, forms the data points of the map. One ignition angle is stored for each pair of values. The map therefore contains 256 adjustable ignition-angle values. By applying linear interpolation between two data points, the number of ignition-angle values is increased to 4096.

Using the ignition-map principle for electronic control of the ignition angle means that, for every engine operating point, it is possible to select the best-possible ignition angle. These maps are ascertained on the engine test stand, or dynamic power analyzer, where demands pertaining to, for example, noise, comfort and component protection are also taken into account.

Additive ignition-angle corrections

Different impacting factors on the moment of ignition are taken into account through additive corrections of the basic ignition angle, such as, for instance, knock control or warming-up after the starting phase. The engine temperature has a further influence on the selection of the ignition angle (e.g., shifting of the knock limit when the engine is hot).

Temperature-dependent ignition-angle corrections are therefore also necessary. Such corrections are stored in the data memory in the form of fixed values or characteristic curves (e.g., temperature-dependent correction). They shift the basic ignition angle by the specified amount. The ignition-angle correction can be either an advance or a retardation.

5 Ignition map

Ignition angle

Relative air charge

Engine speed

UMZ0030-1E

Ignition angles for specific operating
conditions
Specific operating states, e.g., starting or
stratified-charge mode with gasoline direct
injection, require ignition angles that deviate
from the ignition map. In such cases, access
is obtained to special ignition angles stored
in the data memory.

Dwell period

The energy stored in the ignition coil is dependent on the level of the primary current
at the moment of ignition (cut-out current)
and the inductance of the primary winding.
The level of the cut-out current is essentially
dependent on the cut-in period (dwell period) and the vehicle system voltage. The
dwell periods for obtaining the desired cut-
out current are stored in voltage-dependent
curves or program maps. Changing the
dwell period by way of the temperature
can also be compensated for.

In order not to thermally overload the
ignition coil, it is essential to adhere rigidly
to the time required to generate the required
ignition energy in the coil.

Ignition voltage

The ignition voltage at the point where
flashover between the spark-plug electrodes
occurs is the ignition-voltage demand.
It is dependent, among other things, on
- The density of the air/fuel mixture in the
  combustion chamber, and thus on the
  moment of ignition
- The composition of the air/fuel mixture
  (excess-air factor, lambda value)
- The flow velocity and turbulence
- The electrode geometry
- The electrode material, and
- The electrode gap

It is vital that the ignition voltage supplied
by the ignition system always exceed the
ignition-voltage demand under all conditions.

Ignition energy

The cut-out current and the ignition-coil
parameters determine the amount of energy
that the coil stores for application as ignition
energy in the spark. The level of ignition energy has a decisive influence on flame-front
propagation. Good flame-front propagation
is essential to delivering high-performance
engine operation coupled with low levels
of toxic emissions. This places considerable
demands on the ignition system.

Energy balance of an ignition
The energy stored in the ignition coil is
released as soon as the ignition spark is
initiated. This energy is divided into two
separate components.

Spark head
In order that an ignition spark can be generated at the spark plug, first the secondary-
side capacitance C of the ignition circuit
must be charged, and this is released again
on flashover. The energy required for this
increases quadratically with the ignition
voltage U ($E = \frac{1}{2} CU^2$). Figure 6 shows the
component of this energy contained in the
spark head.

Spark tail
The energy still remaining in the ignition
coil after flashover (inductive component)
is then released in the course of the spark
duration. This energy represents the difference between the total energy stored in the
ignition coil and the energy released during
capacitive discharge. In other words: The
higher the ignition-voltage demand, the
greater the component of total energy contained in the spark head, and the less energy
is converted in the spark duration, i.e., the
shorter the spark duration. When the ignition-voltage demand is high, due for instance to badly worn spark plugs, the energy
stored in the spark tail may no longer be
enough to completely burn an already ignited mixture or to re-ignite a spark that
has broken away.

Further increases in the ignition-voltage demand lead to the ignition-miss limit being reached. Here, the available energy is no longer enough to generate a flashover, and instead it decays in a damped oscillation (ignition miss).

Energy losses

Figure 6 shows a simplified representation of the existing conditions. Ohmic resistance in the ignition coil and the ignition cables combined with the suppression resistors cause losses, which are then unavailable as ignition energy.

Additional losses are produced by shunt resistors. While these losses can result from contamination on the high-voltage connections, the primary cause is soot and deposits on the spark plugs within the combustion chamber.

The level of shunt losses is also dependent on the ignition-voltage demand. The higher the voltage applied at the spark plug, the greater the currents discharging through the shunt resistors.

Mixture ignition

Under ideal (e.g., laboratory) conditions, the energy required to ignite an air/fuel mixture with an electrical spark for each individual injection is approximately 0.2 mJ, provided the mixture in question is static, homogeneous and stoichiometric. Under such conditions, rich and lean mixtures require in excess of 3 mJ.

The energy that is actually required to ignite the mixture is only a fraction of the total energy in the ignition spark, the ignition energy. With conventional ignition systems, energy levels in excess of 15 mJ are needed to generate a high-voltage flashover at the moment of ignition at high breakdown voltages. This additional energy is required to charge the capacitance on the secondary side. Further energy is required to maintain a specific spark duration and to compensate for losses, due for instance to contamination shunts at the spark plugs. These requirements amount to ignition energies of at least 30...50 mJ, a figure which corresponds to an energy level of 60...120 mJ stored in the ignition coil.

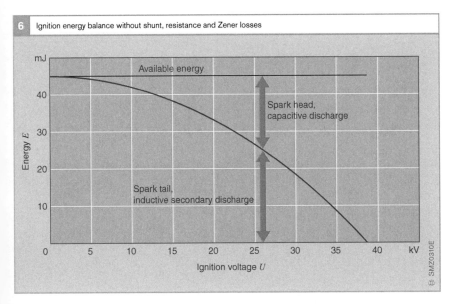

6   Ignition energy balance without shunt, resistance and Zener losses

Fig. 6
The energy figures are for a sample ignition system with a coil capacitance of 35 pF, an external load of 25 pF (total capacitance $C = 60$ pF) and secondary inductance of 15 H.

Turbulence within the mixture of the kind encountered when engines with gasoline direct injection are operated in stratified-charge mode can deflect the ignition spark to such an extent that it breaks away (Fig. 7). A number of follow-up sparks is then needed to ignite the mixture, and this energy must also be provided by the ignition coil.

The ignition tendency decreases in the case of lean mixtures. A particularly high level of energy is therefore required to be able to cover the increased ignition-voltage demand and at the same time to ensure an effectively long spark duration.

If inadequate ignition energy is available, the mixture will fail to ignite. No flame front is established, and combustion miss occurs. This is why the system must furnish adequate reserves of ignition energy: To ensure reliable detonation of the air/fuel mixture, even under unfavorable external conditions. It may be enough to ignite just a small portion of the mixture directly with the spark plug. The mixture igniting at the spark plug then ignites the remaining mixture in the cylinder and thereby initiates the combustion process.

Factors affecting ignition performance

Efficient preparation of the mixture with unobstructed access to the spark plug improves ignition performance, as do extended spark durations and large spark lengths or large electrode gaps. Mixture turbulence can also be an advantage, provided enough energy is available for follow-up ignition sparks should these be needed. Turbulence supports rapid flame-front distribution in the combustion chamber, and with it the complete combustion of the mixture in the entire combustion chamber.

Spark-plug contamination is also a significant factor. If the spark plugs are very dirty, energy is discharged from the ignition coil through the spark-plug shunt (deposits) during the period in which the high voltage is being built up. This reduces the high voltage whilst simultaneously shortening spark duration. This affects exhaust emissions, and can even lead to ignition misses under extreme conditions, as when the spark plugs are severely contaminated or wet.

Ignition misses lead to combustion misses, which increase both fuel consumption and pollutant emissions, and can also damage the catalytic converter.

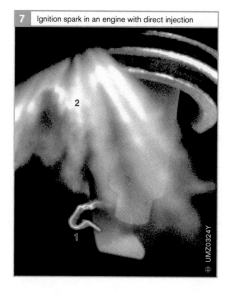

7   Ignition spark in an engine with direct injection

2

1

UMZ0324Y

Fig. 7

Photograph of an ignition spark: taken in a transparent engine using a high-speed camera

1   Ignition spark
2   Fuel spray

▶ Danger of accident

All electrical ignition systems are high-voltage systems. To avoid potential dangers, always switch off the ignition or disconnect the power source before working on any ignition system. These precautions apply to, e.g.,

- Replacing components, such as ignition coils, spark plugs, ignition cables, etc.
- Connecting engine testers, such as timing stroboscope, dwell-angle/speed tester, ignition oscilloscope, etc.

When checking the ignition system, remember that dangerously high levels of voltage are present within the system whenever the ignition is on. All tests and inspections should therefore only be carried out by qualified professional personnel.

Voltage distribution

Rotating high-voltage distribution
The high voltage generated in the ignition coil (Fig. 8a, Pos. 2) must be applied at the correct spark plug at the moment of ignition. In the case of rotating high-voltage distribution, the high voltage generated by this single ignition coil is mechanically distributed to the individual spark plugs (5) by an ignition distributor (3).
The rotation speed and the position of the distributor rotor, which establishes the electrical connection between the ignition coil and the spark plug, are coupled to the camshaft.

This form of distribution is no longer of any significance to new, modern-day engine-management systems.

Distributorless (stationary) voltage distribution
The mechanical components have been dispensed with in the distributorless, or stationary, voltage-distribution system (Fig. 8b). Voltage is distributed on the primary side of the ignition coils, which are connected directly to the spark plugs. This permits wear-free and loss-free voltage distribution. There are two versions of this type of voltage distribution.

System with single-spark ignition coils
Each cylinder is allocated an ignition driver stage and an ignition coil. The engine ECU actuates the ignition driver stages in specified firing order.

Since there are no distributor losses, these ignition coils can be very small in design. They are preferably mounted directly over the spark plug.

Distributorless voltage distribution with single-spark ignition coils can be used with any number of cylinders. There are no limitations on the ignition-timing adjustment range. In this case, the spark plug of the cylinder which is at firing TDC is the one that fires. However, the system does also

have to be synchronized by means of a camshaft sensor with the camshaft.

System with dual-spark ignition coils
One ignition driver stage and one ignition coil are allocated to every two cylinders. The ends of the secondary winding are each connected to a spark plug in different cylinders. The cylinders have been chosen so that when one cylinder is in the compression stroke, the other is in the exhaust stroke (applies only to engines with an even number of cylinders). Flashover occurs at both spark plugs at the moment of ignition. Because it is important to prevent residual exhaust gas or fresh induction gas from being ignited by the spark during the exhaust stroke additional spark, the latitude for varying ignition timing is limited with this system. However, it does not need to be synchronized with the camshaft. Because of these limitations, dual-spark ignition coils cannot be recommended.

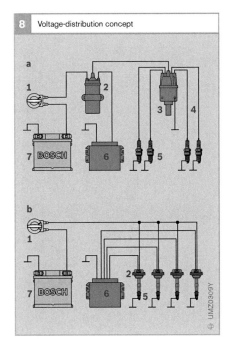

8 Voltage-distribution concept

Fig. 8
a   Rotating distribution
b   Distributorless
    (stationary)
    distribution with
    single-spark
    ignition coils

1   Ignition lock
2   Ignition coil
3   Ignition distributor
4   Ignition cable
5   Spark plug
6   ECU
7   Battery

Ignition driver stage

Function and method of operation

The function of the ignition driver stage is to control the flow of primary current in the ignition coil. It is usually designed as a three-stage power transistor with BIP technology (Bosch Integrated Power, bipolar technology). The functions of primary-voltage limitation and primary-current limitation are integrated as monolithic components on the ignition driver stage, and protect the ignition components against overload.

During operation, the ignition driver stage and the ignition coil both heat up. In order not to exceed the permissible operating temperatures, it is necessary that appropriate measures be taken to ensure that the heat losses are reliably dissipated to the surroundings even when outside temperatures are high. In order to avoid high power loss in the ignition driver stage, the function of primary-current limitation is only to limit the current in the event of a fault (e.g., short circuit).

In the future, the three-stage circuit-breakers will be superseded by the new IGBTs (Insulated Gate Bipolar Transistors, hybrid form on field-effect and bipolar transistors), also for ignition applications. The IGBT has some advantages over BIP:

- Virtually power-free actuation (voltage instead of current)
- Low saturation voltage
- Higher load current
- Lower switching times
- Higher clamp voltage
- Higher holding temperature
- Protected against polarity reversal in the 12 V vehicle electrical system

Design variations

Ignition driver stages are categorized into internal and external driver stages. The former are integrated on the engine ECU's printed-circuit board, and the latter are located in their own housing outside the engine ECU. Due to the costs involved, external driver stages are no longer used on new developments.

Furthermore, it is becoming increasingly common for driver stages to be incorporated in the ignition coil. This solution avoids cables in the wiring harness which carry high currents and are subjected to high voltages. In addition, the power loss incurred in the Motronic ECU is accordingly lower. Stricter demands with regard to actuation, diagnostic capability and temperature load are made of the driver stages integrated in the ignition coil. These demands are derived from the installation circumstances directly on the engine with higher ambient temperatures, ground offsets between ECU and ignition coil, and the additional expenditure involved in transmitting diagnostic information from the ignition coil to the ECU either via an additional cable or through the intelligent use of the control line to include the return transmission of diagnostic information.

Fig. 9

a BIP ignition driver stage (monolithic integrated)

a IGBT ignition driver stage (monolithic integrated)

1 Base resistor
2 Triple Darlington transistor
3 Basic emitter resistors
4 Emitter current regulator
5 Collector-voltage limitation
6 Current-recording resistor
7 Inverse diode
8 Polysilicon protective-diode chain
9 Gate resistor
10 Polysilicon clamp-diode chain for collector-voltage limitation
11 Gate emitter resistor
12 IGBT transistor
13 Resistor (omitted from standard IGBT)

B Base
E Emitter
C Collector
G Gate

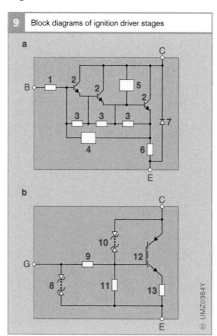

9   Block diagrams of ignition driver stages

Connecting devices and interference suppressors

Ignition cables

The high voltage generated in the ignition coil must be delivered to the spark plug. For this purpose, plastic-insulated, high-voltage-proof cables with special connectors at their ends for contacting the high-voltage components are used with ignition coils which are not mounted directly on the spark plug (e.g., dual-spark ignition coils).

Since, for the ignition system, each high-voltage cable represents a capacitive load which reduces the available secondary voltage, the ignition cables must be kept as short as possible.

Interference-suppression resistors, screening

Each flashover is a source of interference due to its pulse-shaped discharge. Interference-suppression resistors in the high-voltage circuit limit the peak current during discharge. In order to minimize the interference radiation from the high-voltage circuit, the suppression resistors should be installed as close as possible to the source of interference.

Normally, the suppression resistors are integrated in the spark-plug connectors and cable connectors. Spark plugs are also available which feature an integral suppression resistor. However, increasing resistance on the secondary side leads to additional energy losses in the ignition circuit, with lower ignition energy at the spark plug as the ultimate result.

Interference radiation can be even further reduced by partially or completely screening the ignition system. This screening includes the ignition cables. This effort is justified only in special cases (official government and military vehicles, radio equipment with high transmitting power).

---

10 Ignition cables

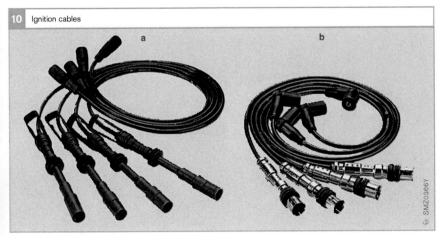

Fig. 10

a   Cable set with straight connectors and unscreened spark-plug connectors

b   Cable set with elbow connectors and partially screened spark-plug connectors

SMZ0366Y

Ignition coils

Within the inductive ignition system, the ignition coil is the component responsible for converting the low battery voltage into the high voltage required to generate flashover at the spark plug. The ignition coil operates on the basis of electromagnetic induction: The energy stored in the magnetic field of the primary winding is transmitted by magnetic induction to the secondary side of the coil.

Function

The high voltage and ignition energy required to ignite the air/fuel mixture must be generated and stored prior to flashover. The coil acts as a dual-function device by serving as both transformer and energy accumulator. It stores the magnetic energy built up in the magnetic field generated by the primary current and then releases this energy when the primary current is deactivated at the moment of ignition.

The coil must be precisely matched to the other components in the ignition system (ignition driver stage, spark plugs). Essential parameters are:
- The spark energy W_{sp} available to the spark plug
- The spark current I_{sp} applied to the spark plug at the flashover point
- The duration of the spark at the spark plug t_{sp}, and
- An ignition voltage U_{ig} adequate for all operating conditions

Important considerations in designing the ignition system include the interactions of individual system parameters with the ignition driver stage, the ignition coil and the spark plug, as well as the specific demands associated with the engine's design concept.

Fig. 1
1. Module with three single-spark coils
2. Module with four single-spark coils
3. Single-spark coil (compact coil)
4. Single-spark coil (pencil coil)
5. Dual-spark coil (one magnetic circuit)
6. Dual-spark coil with two magnetic circuits (four high-voltage domes)
7. Module with two single-spark coils

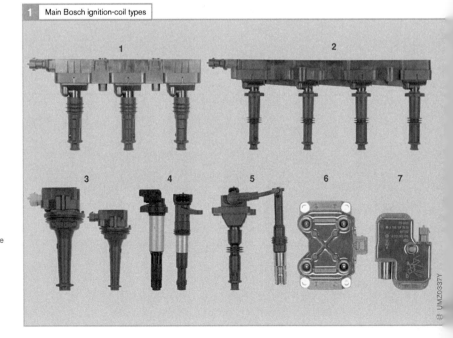

1   Main Bosch ignition-coil types

Examples:
- To ensure secure and reliable ignition of the mixture under all conditions, turbocharged engines need more spark energy than manifold-injection engines; engines with gasoline direct injection have the highest energy requirement of all.
- Spark current has a relatively limited effect on the service life of modern-day spark plugs.
- Turbo- and supercharged engines need consistently higher ignition voltages than non-charged engines.
- The ignition driver stage and the ignition coil must be mutually matched for correct configuration of the operating point (primary current).
- The connection between the ignition coil and the spark plug must be designed for safe and reliable performance under all conditions (voltage, temperature, vibration, resistance to aggressive substances).

Areas of application
Ignition coils made their debut in Bosch ignition systems when battery-based ignition replaced magneto ignition in the 1930s. Since then, they have been subject to ongoing improvements while being adapted to various new areas of applications. Coils are used in all vehicles and machines equipped with inductive ignition systems.

Requirements

Emission-control legislation imposes limits on pollutant emissions from internal-combustion engines. Ignition misses and incomplete mixture combustion, which lead to rises in HC emissions, must be avoided. It is thus vital to have coils that consistently provide adequate levels of ignition energy throughout their service lives.

In addition to these considerations, coils must also suit the geometry and design configuration of the engine. Earlier ignition systems with rotating high-voltage distribution (distributor, [asphalt] ignition coil, ignition cables) featured standardized coils in for mounting on the engine or the vehicle body.

The ignition coil is subject to severe performance demands – electrical, chemical and mechanical – yet still expected to provide fault- and maintenance-free operation for the entire life of the vehicle. Depending on where they are installed in the vehicle – often directly in the cylinder head – today's ignition coils must be able to operate under the following conditions:
- Operating-temperature range of $-40...+150\,°C$
- Secondary voltage up to 30,000 V
- Primary current between 7 and 15 A
- Dynamic vibration loading up to 50 g
- Durable resistance to various substances (gasoline, oil, brake fluid, etc.)

Design and method of operation

Design

Primary and secondary windings

The ignition coil (Fig. 1, Pos. 3) operates in accordance with the principle of a transformer. Two windings surround a shared iron core.

The primary winding consists of thick wire with a relatively low number of turns. One end of the winding is connected to the battery's positive terminal (1) via the ignition switch (terminal 15). The other end (terminal 1) is connected to the ignition driver stage (4) to control the flow of primary current.

Although contact-breaker points were still being used to control primary current as late as the end of the 1970s, this arrangement is now obsolete.

The secondary winding consists of thin wire with a larger number of turns. The turns ratio usually ranges between 1:50 and 1:150.

In the basic economy circuit (Fig. 2a), one terminal from the primary winding is connected to one terminal on the secondary winding, and these are both linked to terminal 15 (ignition switch). The other end of the primary winding is connected to the ignition driver stage (terminal 1). The secondary winding's second terminal (terminal 4) is connected to the ignition distributor or to the spark plug. The autotransformer principle makes the coil less expensive thanks to the common terminal at terminal 15. But because there is no mutual electrical isolation between the two electric circuits, electrical interference from the coil can be propagated into the vehicle's electrical system.

The primary and secondary windings are not interconnected in Figs. 2b and 2c. On the single-spark coil, one side of the secondary winding is connected to ground (terminal 4a), while the other side (terminal 4) leads directly to the spark plug. Both of the secondary-winding connections on the dual-spark ignition coil (terminals 4a and 4b) lead to a spark plug.

Fig. 1
1 Battery
2 AAS diode (integrated in ignition coil)
3 Coil with iron core and primary and secondary windings
4 Ignition driver stage (alternatively integrated in Motronic ECU or in ignition coil)
5 Spark plug

Term. 1, Term. 4, Term. 4a, Term. 15   Terminal designations

Fig. 2
a Single-spark coil in economy circuit (AAS diode not required on ignition systems with rotating high-voltage distribution)
b Single-spark ignition coil
c Dual-spark ignition coil

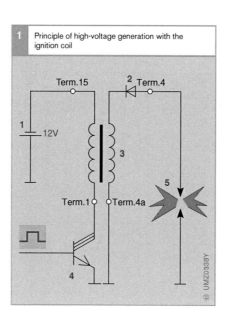

1   Principle of high-voltage generation with the ignition coil

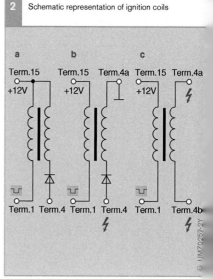

2   Schematic representation of ignition coils

Operating principle

High-voltage generation

The engine ECU activates the ignition driver stage for the calculated dwell period. During this period, the coil's primary current climbs to the setpoint level to generate a magnetic field.

The level of the primary current and the coil's primary inductance determine the amount of energy stored in the magnetic field.

At the moment of ignition (ignition point), the ignition driver stage interrupts the current flow. The resulting shift in the magnetic field induces secondary voltage in the coil's secondary winding. The maximum possible secondary voltage (secondary-voltage supply) is dependent on the energy stored in the ignition coil, the winding capacitance, the coil's turns ratio, the secondary load (spark plug), and the primary-voltage limitation (clamp voltage) of the ignition driver stage.

The secondary voltage must in any case exceed the voltage level required for flashover at the spark plug (ignition-voltage demand). The spark energy must be sufficiently high to ignite the mixture even when follow-up sparks are generated. Follow-up, or secondary, sparks occur when the ignition spark is deflected by turbulence in the mixture and breaks away.

When the primary current is activated, an undesired voltage of roughly 1...2 kV is induced in the secondary winding (switch-on voltage); its polarity is opposed to that of the high voltage. A flashover at the spark plug (switch-on spark) must be avoided.

In systems with rotating high-voltage distribution, the switch-on spark is effectively suppressed by the upstream distributor spark gap. In systems with distributorless (stationary) voltage distribution with single-spark ignition coils, a diode (AAS diode, see Figs. 2 a and 2 b) suppresses the switch-on spark in the high-voltage circuit. This AAS diode can be installed on the "hot" side (facing toward the spark plug) or the "cold" side (facing away from the spark plug).

In systems with dual-spark coils, the switch-on spark is suppressed by the high flashover voltage in the series circuit feeding the two spark plugs without additional measures.

Deactivating the primary current produces a self-induction voltage of several hundred volts in the primary winding. To protect the driver stage, this is limited to 200...400 V.

Generating the magnetic field

A magnetic field is generated in the primary winding as soon as the driver stage completes the circuit. Self-induction creates an inductive voltage in this winding, which according to Lenz's law opposes the cause – i.e., the generation of the magnetic field. This rule explains why the rate at which the magnetic field is generated is always comparatively low (Fig. 3) in relation to the iron cross-section and the winding (inductance).

The primary current will continue to rise while the circuit remains closed; beyond a certain current flow, magnetic saturation occurs in the magnetic circuit. The actual level is determined by the ferromagnetic material used. Inductance falls and current flow rises more sharply. Losses within the ignition coil

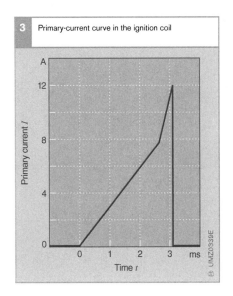

3 Primary-current curve in the ignition coil

also rise steeply. It is therefore sensible to have the operating point as far as possible below the magnetic-saturation level. This is determined by means of the dwell period.

Magnetization curve and hysteresis
The ignition coil's core consists of a soft-magnetic material (in contrast, permanent magnets are hard-magnetic material). This material displays a characteristic magnetization curve that defines the relationship between the magnetic field strength H and the flux density B within it (Fig. 4). Once maximum flux density is reached, the effect of additional increases in field strength on flux density will be minimal: saturation has occurred.

Yet another property of this material is hysteresis in the magnetization curve. This material property denotes a situation where the flux density (i.e., the magnetization) is dependent not only on the currently effective field strength but also on the earlier magnetic state. The magnetization curve assumes a different shape in the case of magnetization (increasing field strength) than it does in the case of demagnetization

(decreasing field strength). The intrinsic losses in the material used are proportional to the level of hysteresis. The area included by the hysteresis curve is a measure of the intrinsic losses.

Magnetic circuit
The material most commonly used in ignition coils is electrical sheet steel, processed in various layer depths and to various specifications. Depending on what is required of it, the material is either grain-oriented (high maximum flux density, expensive) or non-grain-oriented (low maximum flux density).

Sheet metal with layer depths of 0.3...0.5 mm is most commonly used. Mutually insulated plates are used to reduce eddy-current losses. The plates are stamped, combined in plate packs and joined together; this process provides the required thickness and geometrical shape.

The best possible geometry for the magnetic circuit must be defined to obtain the desired electrical performance data for an ignition coil from any given coil geometry.

Fig. 4
1   New curve (magnetization curve of demagnetized iron core)
2   Hysteresis curve

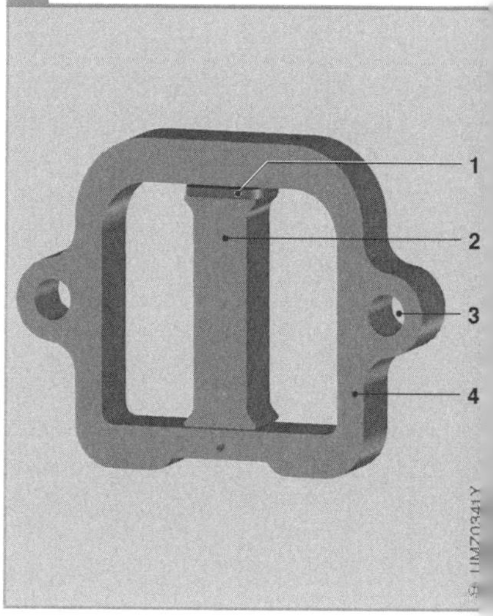

Fig. 5
1   Air gap or permanent magnet
2   I core
3   Fastening hole
4   O core

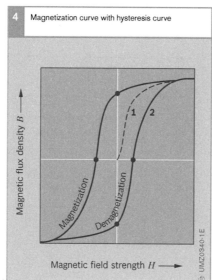

| 4 | Magnetization curve with hysteresis curve |

| 5 | Magnetic circuit in compact coil with O and I cores |

To meet the electrical requirements (spark duration, spark energy, secondary-voltage rise, secondary-voltage level), an air gap is needed which effects a shear in the magnetic circuit (Fig. 5, Pos. 1). A larger air gap (greater shear) permits a higher magnetic field strength in the magnetic circuit and thus leads to a higher magnetic energy that can be stored. This substantially raises the current levels at which magnetic saturation occurs in the magnetic circuit. Without this air gap, saturation would occur at low currents, and subsequent rises in current flow would produce only insignificant increases in levels of stored energy (Fig. 6).

What is important here is that the overwhelming proportion of the magnetic energy is stored in the gap.

In the coil-development process, FEM simulation is employed to define the dimensions for the magnetic circuit and the air gap that will provide the required electrical data. The object is to obtain ideal geometry for maximum storable magnetic energy for a given current flow without saturating the magnetic circuit.

It is also possible to respond to the requirements associated with limited installation space, especially important with pencil coils, by installing permanent magnets (Fig. 5, Pos. 1) to increase the magnetic energy available for storage. The permanent magnet's poles are arranged to allow it to generate a magnetic field opposed to the field in the winding. The advantage of this premagnetization lies in the fact that more energy can be stored in this magnetic circuit.

Switch-on sparks
Activating the primary current changes the current gradients to produce a sudden shift in magnetic flux in the iron core. This induces voltage in the secondary winding. Because the gradient for the current change is positive, the polarity of this voltage polarity is opposed to that of the induced high voltage when the circuit is switched off. Because this gradient is very small relative to the gradients that occur when the primary current is deactivated, the induced voltage is relatively low, despite the large turns ratio arising from the disparity in turn numbers between the two windings. It lies within a range of 1...2 kV, and could be enough to promote spark generation and mixture ignition under some conditions. To prevent possible engine damage, preventing a flashover (switch-on spark) at the spark plug is vital.

In systems with rotating high-voltage distribution, this switch-on spark is suppressed by the upstream distributor spark gap. The rotor-arm contact is not directly across from the cap contact when activation occurs.

In systems with distributorless (stationary) voltage distribution and single-spark ignition coils, the AAS diode (Activation Arc Suppression) suppresses the switch-on spark (see Fig. 1, Pos. 2). With dual-spark coils, the switch-on spark is suppressed by the high flashover voltage of the series circuit with its two spark plugs, and no supplementary measures are required.

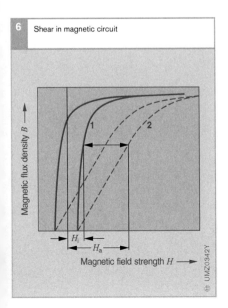

6   Shear in magnetic circuit

Magnetic flux density B →

Magnetic field strength H →

H_i

H_a

UMZ0342Y

Fig. 6

1   Hysteresis for iron core without air gap
2   Hysteresis for iron core with air gap

H_i Modulation for iron core without air gap
H_a Modulation for iron core with air gap

Heat generation in the coil

The efficiency, defined as the available secondary energy relative to the stored primary energy, is on the order of 50...60 %. Under certain boundary conditions, high-performance ignition coils for special applications can achieve efficiency levels as high as 80 %.

The difference in energy is primarily converted into heat through the resistance losses in the windings as well as remagnetization and eddy-current losses.

A driver stage integrated directly in the coil can represent yet another source of thermal loss. The primary current causes a voltage drop in the semiconductive material, leading to lost efficiency. A further and thoroughly significant energy loss is attributable to the switching response when the primary current is deactivated, especially when the driver stage is "slow" in its dynamic response.

High secondary voltages are usually limited by the restriction on primary voltage in the driver stage, where part of the energy stored in the coil is dissipated as thermal loss.

Capacitive load

Capacitance in the ignition coil, the ignition cable, the spark-plug well, the spark plug, and adjacent engine components is low in absolute terms, but remains a factor of not inconsiderable significance in view of the high voltages and voltage gradients. The increased capacitance reduces the rise in secondary voltage. Resistive losses in the windings are higher, high voltage is reduced. In the end, all of the potential secondary energy is not available to ignite the mixture.

Spark energy

The electrical energy available for the spark plug within the ignition coil is called spark energy. It is an essential criterion in ignition-coil design; depending on the winding configuration, it determines such factors as the spark current and the spark duration at the spark plug.

Spark energies of 30...50 mJ are the norm for igniting mixtures in naturally aspirated and turbocharged engines. A higher spark energy (up to 100 mJ) is needed for safe and reliable ignition at all engine operating points in engines with gasoline direct injection.

Ignition-coil types

Single-spark ignition coil

Each spark plug has its own ignition coil in systems with single-spark ignition coils.

The single-spark coil generates one ignition spark per power stroke via the spark plug. It is thus necessary to synchronize operation with the camshaft in these systems.

Dual-spark ignition coil

Single-spark ignition (one spark plug per cylinder)

The dual-spark coil generates ignition voltage for two spark plugs simultaneously. The voltage is distributed to the cylinders in such a way that

- The air/fuel mixture in the one cylinder is ignited at the end of the compression stroke
- The ignition spark in the other cylinder is generated during the valve overlap at the end of the exhaust stroke

The dual-spark coil generates a spark for every crankshaft rotation, corresponding to twice for each power stroke. This means that no synchronization with the camshaft is required with this ignition system. However, this ignition coil can only be used in engines with an even number of cylinders.

There is no compression within the cylinder at the point of valve overlap, and the flash-over voltage at the spark plug is therefore very low. This "additional or maintenance spark" therefore requires only very small amounts of energy for flashover.

Dual-plug ignition

In ignition systems with two spark plugs per cylinder, the ignition voltages generated by one ignition coil are distributed to two different cylinders. The resulting advantages are

- Emissions reductions
- A slight increase in power
- Two sparks at different points in the combustion chamber
- The option of using ignition offset to achieve "softer" combustion
- Good emergency-running characteristics when one ignition coil fails due to a fault

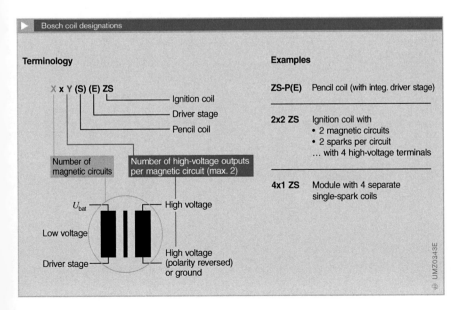

Bosch coil designations

Terminology

X x Y (S) (E) ZS
— Ignition coil
— Driver stage
— Pencil coil
— Number of high-voltage outputs per magnetic circuit (max. 2)
Number of magnetic circuits

U_{bat} — High voltage
Low voltage
Driver stage — High voltage (polarity reversed) or ground

Examples

ZS-P(E)   Pencil coil (with integ. driver stage)

2x2 ZS   Ignition coil with
- 2 magnetic circuits
- 2 sparks per circuit
... with 4 high-voltage terminals

4x1 ZS   Module with 4 separate single-spark coils

UMZ0343E

Bosch has introduced these designations to rationalize its type definitions

Types

Virtually all of the coils in the ignition systems being designed today are either

- Compact coils, or
- Pencil coils

It is also possible to integrate the ignition driver stage within the housing on some of the coil models described in the following.

Compact ignition coil

Design

The compact coil's magnetic circuit consists of the O core and the I core (Fig. 1), onto which the primary and secondary windings are plugged. This arrangement is installed in the coil housing. The primary winding (I core wound in wire) is electrically and mechanically connected to the primary plug connection. Also connected is the start of the secondary winding (coil body wound in wire). The connection on the spark-plug side of the secondary winding is also located in the housing, and electrical contacting is established when the windings are fitted.

Integrated within the housing is the high-voltage contact dome. This contains the contact section for spark-plug contacting, and also a silicone jacket for insulating the high voltage from external components and the spark-plug well.

Once the components have been assembled, impregnating resin is vacuum-injected into the inside of the housing, where it is allowed to harden. This process provides

- High resistance to mechanical loads
- Effective protection against environmental factors, and
- Excellent insulation against high voltage

The silicone jacket is then pushed onto the high-voltage contact dome for permanent attachment.

The ignition coil is ready for use after it has been tested to ensure compliance with all the relevant electrical specifications.

1   Compact-ignition-coil design

Remote and COP versions
The ignition coil's compact dimensions make it possible to implement the design shown above in Figure 1. This version is called COP (**C**oil **o**n **P**lug). The ignition coil is mounted directly on the spark plug, thereby rendering additional high-voltage connecting cables superfluous (Fig. 2a). This reduces the capacitive load on the coil's secondary winding. The reduction in the number of components also increases operational reliability (no rodent bites in ignition cables, etc.).

In the less common remote version, the compact coils are mounted within the engine compartment using screws. Attachment lugs or an additional bracket are provided for this purpose. The high-voltage connection is effected by means of a high-voltage ignition cable from the coil to the spark plug.

The COP and remote versions are virtually identical in design. However, the remote version (mounted on the vehicle body) is subject to fewer demands with regard to temperature and vibration conditions due to the fact that it is exposed to fewer loads and strains.

Other coil types
ZS 2x2
Rotating high-voltage distribution is being gradually superseded by distributorless (stationary) voltage distribution.

An uncomplicated means for converting an engine model to distributorless distribution is offered by the ZS 2x2 (Fig. 3) and the ZS 3x2 (German: <u>Z</u>ünd<u>s</u>pule = ignition coil, hence ZS). These ignition coils contain two (or three) magnetic circuits, and generate two sparks per circuit. They can thus be used to replace the distributors in four- and six-cylinder engines. Because the units can be mounted almost anywhere in the engine compartment, the vehicle manufacturer's adaptation effort is minimal, although the engine ECU has to be modified. Another factor is that high-voltage ignition cables are required in most cases for layouts with remote ignition coils.

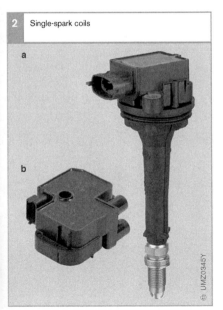

2	Single-spark coils

a

b

UMZ0345Y

3	2x2 ignition coil for converting from rotating to distributorless voltage distribution

UMZ0346Y

Fig. 2
a COP version of a single-spark compact coil
b Remote version: two single-spark coils in module, spark plugs connected via two ignition cables

Ignition-coil modules

Ignition-coil modules combine several coils in a shared housing to form a single assembly (Fig. 4). These coils continue to operate individually.

The advantages furnished by coil modules are

- Simplified installation (just a single operation for three or four ignition coils)
- Fewer threaded connections
- Connection to the engine wiring harness with just one plug
- Cost savings thanks to faster installation and simplified wiring harness

Disadvantages:

- It is necessary to adapt the module's geometry to fit the engine, and
- Modules must be designed to fit individual cylinder heads; no universal designs

Pencil coil

The pencil coil makes optimal use of the space available within the engine compartment. Its cylindrical shape makes it possible to use the spark-plug well as a supplementary installation area for ideal space utilization on the cylinder head (Fig. 5).

Because pencil coils are always mounted directly on the spark plug, no additional high-voltage connecting cables are required.

Design and magnetic circuit

Pencil coils operate like compact coils in accordance with the inductive principle. However, the rotational symmetry results in a design structure that differs considerably from that of compact coils.

Although the magnetic circuit consists of the same materials, the central rod core (Fig. 6, Pos. 5) consists of laminations in various widths stacked in packs that are virtually spherical. The yoke plate (9) that

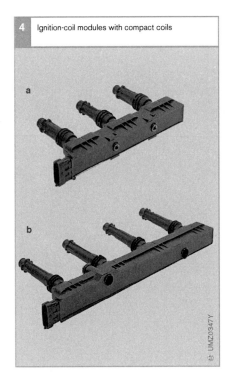

4   Ignition-coil modules with compact coils

a

b

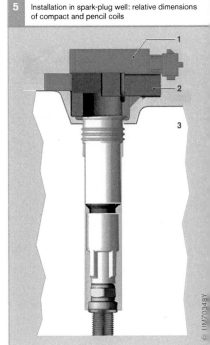

5   Installation in spark-plug well: relative dimensions of compact and pencil coils

1

2

3

Fig. 4
a   ZS 3x1M
b   ZS 4x1M

Fig. 5
1   Compact coil
2   Pencil coil
3   Cylinder head

provides the magnetic circuit is a rolled and slotted shell – also in electrical sheet steel, sometimes in multiple layers.

Another difference relative to compact coils is the primary winding (7), which has a larger diameter and is above the secondary winding (6), while the body of the winding also supports the rod core. This arrangement brings benefits in the areas of design and operation.

Owing to restrictions imposed by their geometrical configuration and compact dimensions, pencil coils allow only limited scope for varying the magnetic circuit (rod core, yoke plate) and windings.

In most pencil-coil applications, the limited space available dictates that permanent magnets be used to increase the spark energy.

The arrangements for electrical contact with the spark plug and for connection to the engine wiring harness are comparable with those used for compact coils.

Variants

An extended range of variants (e.g., different diameters and lengths) is available to provide pencil coils for assorted applications. The ignition driver stage can also be integrated within the housing as an option.

A typical diameter, as measured at the cylindrical center section (yoke plate, housing), is roughly 22 mm. This dimension is derived from the hole diameter of the spark-plug well within the cylinder head as used with standard spark plugs featuring a 16 mm socket fitting. The length of the pencil coil is determined by the installation space in the cylinder head and the required or potential electrical performance specifications. Extending the active section (transformer) is subject to limits, however, due to the parasitic capacitance and the deterioration of the magnetic circuit involved.

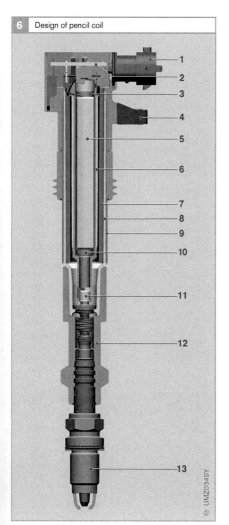

6   Design of pencil coil

Fig. 6
1 Plug connection
2 Printed-circuit board with ignition driver stage
3 Permanent magnet
4 Attachment arm
5 Laminated electrical-sheet-steel core (rod core)
6 Secondary winding
7 Primary winding
8 Housing
9 Yoke plate
10 Permanent magnet
11 High-voltage dome
12 Silicone jacket
13 Attached spark plug

Cavities filled with sealing compound

Ignition-coil electronics

In earlier designs, the ignition driver stage was usually incorporated within a separate module, and attached to the coil or the distributor within the engine compartment of a vehicle with rotating voltage distribution. The conversion to distributorless ignition combined with increasing miniaturization of electronic componentry to foster the development of ignition driver stages embedded in integrated circuits and thus suitable for incorporation in the ignition or engine ECUs.

The constantly increasing functional scope of engine ECUs (Motronic) and new engine concepts (e.g., gasoline direct injection) have increased thermal stresses (overall heat loss from driver stages) and reduced installation space. These factors have produced a trend toward remote driver stages located outside the ECU. One option is integration within the ignition coil, which also makes it possible to use a shorter primary wire to reduce line loss.

Design

The driver stage can be integrated in the housing of both compact and pencil coils. Figure 1 shows installation in a pencil coil. The driver-stage module – which can also incorporate additional functions – is mounted on a small printed-circuit board. SMD (Surface Mounted Device) components are used because of the restricted dimensions.

The driver-stage transistors (7) are integrated in standardized TO housings and connected to the circuit board or conductor rails. Additional functions for monitoring, diagnosis or other functions (e.g., closed-circuit current deactivation, customer-specific input circuit) can be optionally integrated in further electronic components (3). The primary connector (1) is connected directly to the circuit board. Below the board are the contacts for the coil's primary winding (4).

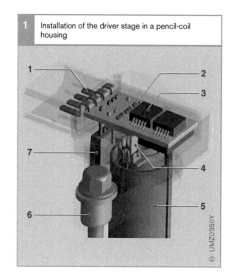

1 Installation of the driver stage in a pencil-coil housing

Fig. 1

1   Primary plug
2   SMD components
3   Electronics for
    ignition functions
4   Primary winding
    contacts
5   Pencil-coil
    transformer
6   Attachment lug
7   Driver stage

Electrical parameters

Inductance
Inductance is a physical variable which denotes the electromagnetic efficiency or self-induction capability of a coil or, in general, of an electrical conductor.

Inductance is determined by the material and cross-section of the permeated magnetic circuit, the number of windings, and the geometry of the copper winding.

An ignition coil includes primary and secondary inductance elements, with the secondary inductance being many times greater.

Capacitance
Three different types are encountered in an ignition coil: inherent capacitance, parasitic capacitance and load capacitance. Inherent capacitance of an ignition coil is essentially created by the winding itself. It is created from the fact that neighboring wires within the secondary winding form a capacitor.

Parasitic capacitance is a "harmful" capacitance within an electrical system. Part of the available or generated energy is needed to charge or recharge this parasitic capacitor and is therefore not available at the connections. In an ignition coil, parasitic capacitance is created, for example, by the small gap between the secondary and primary windings or by cable capacitance between ignition cable and neighboring components.

Load capacitance is essentially created by the spark plug. It is determined by the installation environment (e.g., metallic spark-plug well), the spark plug itself, and any high-voltage connecting cables that may be present. These factors are not usually subject to modification and must be taken into consideration in the design of the ignition coil.

Stored energy
The amount of magnetic energy that can be stored depends upon numerous factors such as the coil's design (geometry, material in the magnetic circuit, additional magnets) and the ignition driver stage used. Once a certain point is reached, any additional increases in primary current will deliver only minimal rises in stored energy, while losses grow disproportionately, ultimately leading to destruction of the coil within a short period of time.

While taking into account all the tolerances, the ideal coil thus operates just below the magnetic circuit's magnetic saturation point.

Resistance
The resistance of the windings is determined by the temperature-sensitive specific resistance of copper.

The primary resistance (resistance of the primary winding) is normally within the range of $0.4...0.7\,\Omega$. It should not be too high, because in the event of low vehicle system voltage (voltage dip during cold starting) the ignition coil would not reach its rated current, and would thus not be able to generate a lower spark energy.

The secondary resistance (resistance of the secondary winding) is in the range of several $k\Omega$; it differs from the primary resistance in the larger number of turns on the secondary winding (by a factor of $70...100$) and the small wire diameter (by a factor of approximately 10).

1	Parameters for ignition coils (series application)	
I_1	Primary current	6.5...9.0 A
T_1	Charging time	1.5...4.0 ms
U_2	Secondary voltage	29...35 kV
T_{sp}	Spark duration	1.3...2.0 ms
W_{sp}	Spark energy	30...50 mJ, up to 100 mJ for gasoline direct injection
I_{sp}	Spark current	80...115 mA
R_1	Resistance of primary winding	0.3...0.6	
R_2	Resistance of secondary winding	5...15 k|
N_1	Number of turns in primary winding	150...200
N_2	Number of turns in secondary winding	8000...22,000

Power loss
The losses in an ignition coil are determined by resistance in the windings, capacitive losses and remagnetization losses (hysteresis), as well as by construction-necessitated deviations from the ideal configuration for a magnetic circuit. At an efficiency level of 50...60 %, relatively high power losses are generated in the form of heat at high engine speeds. The losses are kept as low as possible by loss-minimized configurations, suitable design solutions, and materials subject to high thermal loads.

Turns ratio
The turns ratio is the ratio of the number of turns in the primary copper winding to the number in the secondary copper winding. On standard ignition coils, it is on the order of 1:50...1:150. Determination of the turns ratio is used in combination with driver-stage specifications to affect such factors as the level of spark current and – to some degree – the maximum secondary voltage.

High-voltage and spark-generation properties
The ideal coil remains relatively impervious to load factors while producing as much high voltage as possible within an extremely brief rise period. These properties ensure flashover at the spark plug for reliable mixture ignition under all conditions encountered in operation.

At the same time, the real-world properties of the windings, the magnetic circuit and the driver stage used all unite to impose limits on performance.

The polarity of the high voltage ensures that the spark plug's center electrode maintains negative potential relative to chassis ground. This negative polarity counteracts the tendency of the spark plug's electrodes to erode.

Dynamic internal resistance
Yet another important parameter is the coil's dynamic internal resistance (impedance). Because impedance combines with internal and external capacitance to help determine voltage rise times, it serves as an index of the amount of energy that can flow from the coil and through shunt-resistance elements at the moment of flashover. Low internal resistance is an asset when spark plugs are contaminated or wet. Internal resistance depends on secondary inductance.

Simulation-based development of ignition coils

Ever-increasing technical demands mean that the efficiency limits of conventional design and development methods are reached early. Product-development processes using CAE provide a solution. CAE (Computer Aided Engineering) is the generic term for computer-based engineering services. It includes all aspects of CAD (Computer Aided Design) as well as calculation routines.

The advantages offered by CAE are:
- Informed decisions at early stages in the development process (also without prototypes)
- Identification of specimens suitable for testing, and
- Enhanced understanding of physical interrelationships

Calculation programs are employed in various areas of simulation in ignition-coil development:
- Structural mechanics (analysis of mechanical and thermal stress factors)
- Fluid mechanics (analysis of fluidic charging processes)
- Electromagnetics (analysis of the system's electromagnetic performance)

Electromagnetic simulation tools are especially important in the development ignition coils. Two different types of simulation may be used here: geometry-orientated simulation and performance simulation.

The Finite Element Method (FEM) is used with geometry-orientated simulation. Here, the coil's geometry is modeled on the basis of a CAD model. This is provided with appropriate boundary conditions (current density, electrical potential, etc.) and then converted into an FEM model. Transfer to a calculation model and derivation of specifications from the corresponding equation series follow. The result is a clear calculated solution to the problem.

This method permits 100 % virtual, simulation-based ignition-coil design. Depending on the analysis objective, it is possible here to optimize the geometry of the ignition coil (magnetic optimization of the magnetic circuit, electrostatic optimization of electrically conductive contours).

Following geometry-orientated analysis, performance simulation can be employed to examine the coil's electrical characteristics within the overall system, consisting of driver stage, coil and spark plug, under conditions reflecting the actual, real-world environment. This calculation method provides the initial specification data for the coil. It also supports subsequent calculation of electrical parameters such as spark energy and spark current.

Electromagnetic simulation tools make "virtual" development of ignition coils possible. The simulation results define geometrical data and winding design to furnish the basis for specimen construction. The electrical performance of these specimen coils will approximate the simulation results. This substantially reduces the number of time-consuming recursion processes that occur when coils are produced using conventional product-development methods.

Spark plugs

The air/fuel mixture in the gasoline, or spark-ignition, engine is ignited electrically. Electrical energy drawn from the battery is temporarily stored in the ignition coil for this purpose. The high voltage generated within the coil produces a flashover between the spark-plug electrodes in the engine's combustion chamber. The energy contained in the spark then ignites the compressed air/fuel mixture.

Function

The function of the spark plug is to introduce the ignition energy into the gasoline engine's combustion chamber and to produce a spark between the electrodes to initiate combustion of the air/fuel mixture.

Spark plugs must be designed to ensure positive insulation between spark and cylinder head while also sealing the combustion chamber.

In combination with engine components, such as the ignition and mixture-formation systems, the spark plug plays a crucial role in determining operation of the gasoline engine. It must
- facilitate reliable cold starts,
- ensure consistent operation with no ignition miss throughout its service life, and
- not overheat under extended operation at or near top speed.

To ensure this kind of performance throughout the spark plug's service life, the correct plug concept must be established early in the engine-design process. Research investigating the ignition process is employed to determine the spark-plug concept that will provide the best emissions and most consistent engine operation.

An important spark-plug parameter is the heat range. The right heat range prevents the spark plug from overheating and inducing the thermal auto-ignition that could lead to engine damage.

1   Spark plug in a gasoline engine

UMZ0336Y

Usage

Areas of application

Bosch first used a spark plug in a passenger car in 1902, when it was installed in a system featuring magneto ignition. The spark plug then went on to become an unparalleled success in automotive technology.

Spark plugs are used in all vehicles and machinery powered by gasoline engines, both 2-stroke and 4-stroke. They can be found in

- Passenger cars
- Commercial vehicles
- Single-track vehicles (motorcycles, scooters, motor-assisted bicycles)
- Ships and small craft
- Agricultural and construction machinery
- Motor saws
- Garden appliances (e.g., lawnmowers), etc.

To accommodate the wide array of potential applications, more than 1200 different spark plug designs are available.

Because multi-cylinder passenger-car engines require at least one plug per cylinder, it is in this sector that most spark plugs are used.

Motorized machinery, because of the lower engine power, usually relies on a single-cylinder engine needing only a single spark plug.

Within Europe, most commercial vehicles – at least in heavy-duty applications – are powered by diesel engines, which limits the demand for spark plugs in this segment. In the US, however, gasoline engines are also the most prevalent powerplants in heavy vehicles.

Variety of types

The engine of 1902 delivered only about 6 HP for each 1000 cc of displacement. Today's comparable figure is 100 HP, with up to 300 HP available from racing engines. The technical resources invested in engineering and producing the spark plugs that allow this performance is enormous.

The first spark plug was expected to ignite 15 to 25 times per second. Today's spark plugs must ignite five times as often. The upper temperature limit as risen from 600 °C to approximately 900 °C, and the ignition voltage from 10,000 V to up to 30,000 V. Whereas today's spark plugs must function for at least 30,000 km, original spark plugs had to be replaced every 1000 km.

Although the spark plug's basic concept has changed little in the course of 100 years, in this period Bosch has designed more than 20,000 different types to meet the needs of various engine configurations.

The current spark-plug range continues to embrace a wide array of models. The spark plug is subject to immense demands in the areas of

- electrical and
- mechanical performance, as well as
- resistance to chemical and
- thermal loads.

In addition to satisfying these performance criteria, the spark plug must also be matched to the geometrical conditions defined by the individual engine (e.g, spark-plug length in the cylinder head). Combined with the extensive range of engines being manufactured, these requirements make it necessary to offer a wide variety of spark plugs. Bosch currently supplies more than 1250 different spark-plug types, all of which must be available to service workshops/garages and commercial distributors.

Requirements

Electrical-performance requirements

During operation in electronic ignition systems, spark plugs must handle voltages as high as 30,000 V with no disruptive discharge at insulator. Residue from the combustion process, such as soot, carbon and ash from fuel and oil additives, can be electrically conductive under certain thermal conditions. Yet under these same conditions it remains imperative that flashover through the insulator be avoided.

The insulator must continue to display adequate electrical resistance at up to 1000 °C with only very minor diminution of this figure throughout its service life.

Mechanical-performance requirements

The spark plug must be capable of withstanding periodic pressure peaks (up to about 100 bar) in the combustion chamber, while still providing an effective gas seal. High resistance to mechanical stresses is also required from the ceramic insulator, which is exposed to loads during installation as well as from the spark-plug connector and the ignition cable itself during operation. The shell must absorb the torque applied during installation with no permanent deformation.

Chemical-performance requirements

Because the spark plug extends into a combustion chamber hot enough to make its nose glow red, it is exposed to the chemical reactions that occur at extreme temperatures. Substances within the fuel can form aggressive residue deposits on the spark plug, affecting its performance characteristics.

Thermal-performance requirements

In operation the spark plug must alternately absorb heat from hot combustion gases and then withstand the cold incoming air/fuel mixtures in rapid succession. This is why insulators must display immense resistance to thermal shock.

The spark plug must also dissipate the heat absorbed in the combustion chamber to the engine's cylinder head with maximum efficiency; the terminal end of the spark plug should remain as cool as possible.

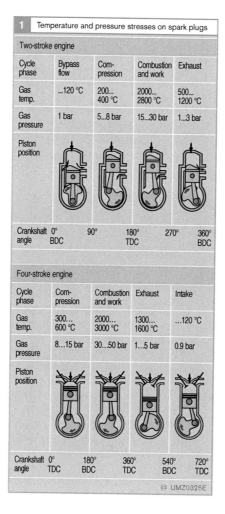

1   Temperature and pressure stresses on spark plugs

Two-stroke engine

Cycle phase	Bypass flow	Com-pression	Combustion and work	Exhaust
Gas temp.	...120 °C	200...400 °C	2000...2800 °C	500...1200 °C
Gas pressure	1 bar	5...8 bar	15...30 bar	1...3 bar
Piston position				

Crankshaft angle	0° BDC	90°	180° TDC	270°	360° BDC

Four-stroke engine

Cycle phase	Com-pression	Combustion and work	Exhaust	Intake
Gas temp.	300...600 °C	2000...3000 °C	1300...1600 °C	...120 °C
Gas pressure	8...15 bar	30...50 bar	1...5 bar	0.9 bar
Piston position				

Crankshaft angle	0° TDC	180° BDC	360° TDC	540° BDC	720° TDC

UMZ0325E

Design

The essential components of the spark plug
are (Fig. 1)
● Terminal stud (1)
● Insulator (2)
● Shell (3)
● Seal seat (6), and
● Electrodes (8, 9)

Terminal stud

The steel terminal stud is mounted gas-tight
in the insulator with an electrically conductive glass seal, which also establishes the connection to the center electrode. The terminal
end protruding from the insulator features
a thread for connecting the spark-plug connector of the ignition cable. In the case of
connectors designed to ISO/DIN standards,
a terminal nut (with the required outer
contour) is screwed onto the terminal-stud
thread, or the stud is equipped with a solid
ISO/DIN connection manufacture.

Insulator

The insulator is cast in a special ceramic material. Its function is to insulate the center
electrode and terminal stud from the shell.
The demand for a combination of good
thermal conductivity and effective electrical
insulation is in stark contrast to the properties displayed by most insulating substances.
Bosch uses aluminum oxide (Al_2O_3) along
with minute quantities of other substances.
Following firing, this special ceramic meets
all requirements for mechanical and chemical durability, while its dense microstructure
provides high resistance to disruptive discharge.

On air-gap spark plugs, the outer contour
of the insulator nose can also be modified to
improve heating for better response during
repeated cold starts.

The surface of the insulator's terminal end
is coated with a lead-free glaze. The glazing
helps prevent moisture and contamination
from adhering to the surface, which helps to
prevent tracking currents to a large extent.

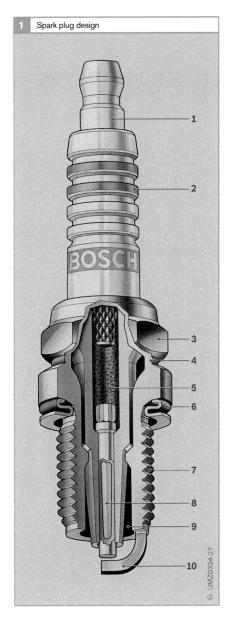

1 Spark plug design

Fig. 1
1 Terminal stud
   with nut
2 Al_2O_3 ceramic
   insulator
3 Shell
4 Heat-shrinkage zone
5 Conductive glass
6 Sealing ring
   (seal seat)
7 Thread
8 Composite center
   electrode (Ni/Cu)
9 Breathing space
   (air space)
10 Ground electrode
   (here Ni/Cu
   composite)

Shell

The shell is manufactured from steel in a cold-forming process. The shell castings emerge from the pressing tool with their final contours, limiting subsequent machining operations to just a few areas.

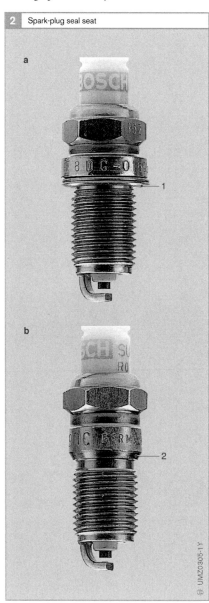

2	Spark-plug seal seat

a

b

1

2

Fig. 2
a	Flat seal seat with
	sealing ring
b	Conical seal seat
	without sealing ring

1	Sealing ring
2	Conical sealing
	surface

The bottom end of the shell includes threads (Fig. 1, Pos. 7), making it possible to install the plug in the cylinder head and then remove it after a specified replacement interval. Depending on the specific design, as many as four ground electrodes can be welded to the end of the shell.

An electroplated nickel coating is applied to the surface to protect the shell against corrosion and prevent it from seizing in the sockets of the aluminum cylinder heads.

To accommodate the spark plug-wrench, the upper section of the conventional shell has a 6-point socket fitting; newer shell designs may use a 12-point fitting. The 12-point fitting makes it possible to reduce the socket attachment's size to 14 mm without modifying insulator-head geometry. This reduces the spark plug's space demands in the cylinder head and allows the engine designer greater freedom in locating the cooling passages.

The top end of the spark-plug shell is flanged after the plug core (comprising insulator with reliably mounted center electrode and terminal stud) has been inserted, and secures the plug core in position. The subsequent shrink-fitting process – inductive heating under high pressure – produces a gas-tight connection between insulator and shell to ensure effective thermal conductivity.

Seal seat

Depending on engine design, either a flat or conical seal seat (Fig. 2) effects the seal between the spark plug and the cylinder head.

In the case of a flat seal seat, a sealing ring (1) is used as the sealing element. This captive sealing ring is permanently attached to the spark-plug shell. Its special contours adapt to form a durable yet flexible seal when the spark plug is installed. In the case of a conical seal seat, a conical, or tapered, surface (2) on the spark-plug shell mates directly with the cylinder head to provide a seal without the use of a sealing ring.

Electrodes

During flashover and high-temperature operation, the electrode material is subjected to such strong thermal load that the electrodes become worn – the electrode gap widens accordingly. To satisfy demands for extended replacement intervals, electrode materials must effectively resist erosion (burning by the spark) and corrosion (wear due to aggressive thermochemical processes). These properties are achieved primarily through the use of temperature-resistant nickel alloys.

Center electrode

The center electrode (Fig. 1, Pos. 8), which includes a copper core for improved heat dissipation, is anchored at one end in the conductive glass seal.

In "long-life" spark plugs, the center electrode serves as the base material for a noble-metal pin, which is permanently connected to the base electrode by means of laser welding. Other spark plug designs rely on electrodes formed from a single thin platinum wire, which is then sintered to the ceramic base for good thermal conductivity.

Ground electrodes

The ground electrodes (10) are attached to the shell and usually have quadrilateral cross-sections. Available arrangements include the front electrode and the side electrode (Fig. 3b). The ground electrode's fatigue strength is determined by its thermal conductivity. As with center electrodes, composite materials can be used to improve heat dissipation, but it is the length and the end surface that will ultimately determine the ground electrode's temperature, and thus its resistance to wear.

Spark-plug life can be extended through the use of greater end-surface areas and multiple ground electrodes.

3   Electrode shapes

Fig. 3

a   Front electrode
b   Side electrodes
c   Surface-gap spark plug without ground electrode (special application for racing engines)

Electrode materials

As a basic rule, pure metals conduct heat better than alloys. Yet pure metals – such as nickel – are also more sensitive than alloys to chemical attack from combustion gases and solid combustion residues. Manganese and silicon can be added to nickel to produce alloys with enhanced resistance to aggressive chemical, especially sulfur dioxide (SO_2, sulfur is a constituent of lube oil and fuel). Aluminum and yttrium additives enhance resistance to scaling and oxidation.

Compound electrodes

Corrosion-resistant nickel alloys are now the most widely used option in spark-plug manufacture. A copper core can be used for further increases in heat dissipation, producing compound electrodes that satisfy exacting demands for high thermal conductivity and high corrosion resistance (Fig. 1).

The ground electrodes, which must be flexible enough to bend when the gap is set, may also be manufactured from a nickel-based alloy or from a composite material.

Silver-center electrodes

Silver has the best electrical and thermal conductivity of any material. It also displays extreme resistance to chemical attack, provided that it is not exposed to either leaded fuels or to high temperatures in reducing atmospheres (rich air/fuel mixture).

Composite particulate materials with silver as their basic substance can substantially enhance heat resistance.

Platinum electrodes

Platinum (Pt) and platinum alloys display high levels of resistance to corrosion, oxidation and thermal erosion. This is why platinum is the substance of choice for use in "long-life" spark plugs.

In some spark-plug types, the Pt pin is cast in the ceramic body early in the manufacturing process. In the subsequent sintering process, the ceramic material shrinks onto the Pt pin to permanently locate it in the plug core.

In other spark-plug types, thin Pt pins are welded onto the center electrode (Fig. 2). Bosch relies on continuous-operation lasers to produce a durable bond.

Fig. 1
a   With front electrode
b   With side electrode

1   Conductive glass
2   Air gap
3   Insulator nose
4   Composite center electrode
5   Composite ground electrode
6   Ground electrodes

Fig. 2
1   Compound electrode (Ni/Cu)
2   Laser-welded seam
3   Platinum pin

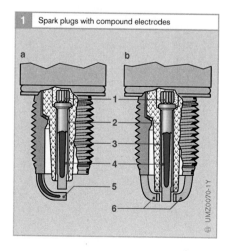

1   Spark plugs with compound electrodes

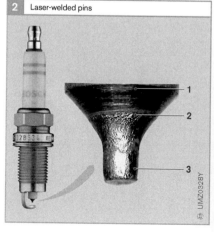

2   Laser-welded pins

Spark-plug concepts

The mutual arrangement of the electrodes and the locations of the ground electrodes relative to the insulator determine the type of spark-plug concept (Fig. 1).

Air-gap concept
Center and ground electrodes are configured to produce a linear spark to ignite the air-fuel mixture located within the space between them.

Surface-gap concept
As a result of the defined position of the ground electrodes relative to the ceramic, the spark travels initially from the center electrode across the surface of the insulator nose before jumping across a gas-filled gap to the ground electrode. Because the ignition voltage required to produce discharge across the surface is less than that needed to produce discharge across an air gap of equal dimensions, a surface-gap spark can bridge wider electrode gaps than an air-gap spark with an identical ignition-voltage demand. This produces a larger flame core for more effective creation of a stable flame front.

The surface-gap spark also promotes self-cleaning during repeated cold starts, preventing soot deposits from forming on the insulator nose. This improves performance on engines exposed to frequent cold starts at low temperatures.

Surface-air-gap concepts
On these spark plugs, the ground electrodes are arranged at a specific distance from the center electrode and the end of the ceramic insulator. This produces two alternate spark paths, which facilitate both forms of discharge – air gap and surface-air gap – and different ignition-voltage-demand values. Depending on operating conditions and spark-plug condition (wear), the spark travels as an air-gap or surface-air-gap spark.

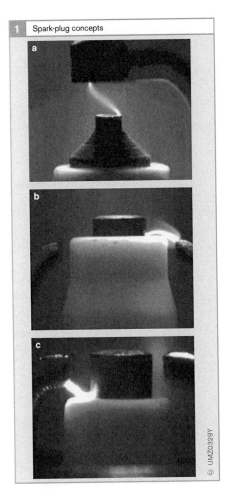

1   Spark-plug concepts

a

b

c

Fig. 1
a   Air-gap spark
b   Surface-gap spark
c   Surface-air-gap
     spark

Electrode gap

As the shortest distance between the center and ground electrodes, the electrode gap determines, among others, the length of the spark (Fig. 1). The smaller the electrode gap, the lower the voltage that is required to generate an ignition spark.

An excessively small gap produces only a small flame core in the electrode area. Because this flame core loses energy through the electrode contact surfaces (quenching), the rate at which the flame core propagates is only very slow. Under extreme conditions, the energy loss can be high enough to produce ignition miss.

As electrode gaps increase (e.g., due to electrode wear), lower quenching losses lead to improved conditions for ignition, but larger gaps also increase the ignition-voltage demand (Fig. 2). The reserves afforded by any given level of ignition voltage in the ignition coil are reduced and the danger of ignition miss increases.

Engine manufacturers use various test procedures to determine the ideal electrode gap for each engine. The first step is to conduct ignition tests at characteristic engine operating points to determine the minimum electrode gap. Salient considerations include exhaust emissions, smooth operation and fuel consumption.

In subsequent extended test runs, the wear performance of these spark plugs is determined and then evaluated with regard to ignition-voltage demand. The specified electrode gap is then defined at a point providing an adequate safety margin to the miss limit. Gap specifications are quoted in vehicle owner's manuals as well as in Bosch spark-plug sales documentation.

Bosch spark plugs are set to the correct electrode gap at the factory.

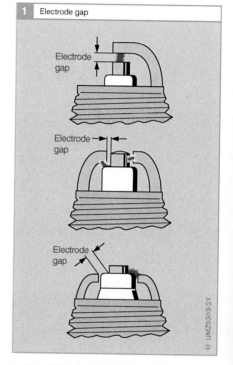

1 Electrode gap

Electrode gap

Electrode gap

Electrode gap

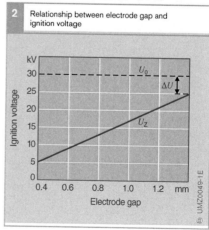

2 Relationship between electrode gap and ignition voltage

Fig. 1
a   Spark plug with front electrode (air-gap spark)
b   Spark plug with side electrode (air-gap or surface-air-gap spark)
c   Surface-gap spark plug

Fig. 2
U_0 Available ignition voltage
U_Z Ignition voltage
ΔU Ignition-voltage reserve

Spark position

The spark position (Fig. 1a) is the location of the spark gap relative to the walls of the combustion chamber. Spark position has a substantial effect on combustion in modern engines (especially direct-injection engines). The criterion for defining the quality of the combustion process is the engine's operating consistency, or smoothness, which is in turn based on a statistical evaluation of the indicated mean effective pressure. The extent of the standard deviation or of the variation coefficient ($cov = s/p_{ime} \cdot 100\,[\%]$) is an index of the uniformity of the combustion. These values also provide information on any major effects that delayed or missed combustion will have on engine operation. A value of 5% is defined for cov as the measure of the operation limit.

Figure 1 illustrates the effects of leaning out the air/fuel mixture and varying ignition timing on operational consistency at two different spark-plug positions. The lines describe constant consistency levels, while the 5% limit is shown in bold blue. Values above this curve (< 5% range) correlate with smooth engine operation – the combustion process of the individual working cycles is uniform and free from major fluctuations. Values below this curve (> 5% range) correspond to poor engine operation – combus-

tion is not always uniform, with isolated misses or delayed combustion occurring in extreme situations.

Comparison of the two diagrams indicates that on this engine projecting the spark position further into the combustion chamber would substantially improve ignition, as the ignition-timing range increases above the 5% curve and the operating limit is pushed toward higher excess-air factors.

However, extending the length of the ground electrodes leads to higher temperatures, which in turn produce rises in electrode wear. The self-resonant frequency also falls, which can lead to ruptures and fissures from vibration. When the spark position is shifted forward, a number of other measures are needed to ensure adequate service life:

- Extending the spark-plug shell inward beyond the combustion-chamber wall. The shoulder reduces the danger of electrode rupture.
- Inserting copper cores into the ground electrodes. Placing copper in direct contact with the spark-plug shell can reduce temperatures by approximately. 70 °C.
- Using highly heat-resistant electrode materials.

Fig. 1
a   Definition of spark position f
b   Diagram for $f = 3\,mm$
c   Diagram for $f = 7\,mm$

Curves indicate operating points with constant cov values

$cov = s/p_{ime} \cdot 100\,[\%]$
s Standard deviation
p_{ime} Indicated mean effective pressure

5% curve: operation limit
< 5% range: good running consistency
> 5% range: poor running consistency

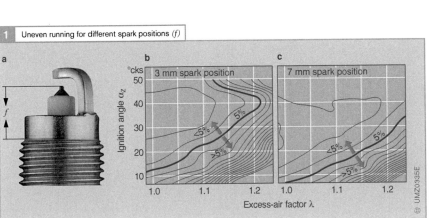

1   Uneven running for different spark positions (f)

a

b   °cks   3 mm spark position
c   7 mm spark position

Ignition angle α_z (vertical axis): 50, 40, 30, 20, 10

Excess-air factor λ (horizontal axis): 1.0, 1.1, 1.2, 1.0, 1.1, 1.2

<5%, >5%, 5%

UMZ0335E

Spark-plug heat range

Spark-plug operating temperatures

Operating range

Engines run on a rich air/fuel mixture when cold. This can lead to incomplete combustion and formation of soot deposits on spark plugs and combustion-chamber surfaces. These deposits contaminate the insulator nose to form a partially conductive link between the center electrode and the spark-plug shell (Fig. 1). This "shunting" effect allows a portion of the ignition energy to escape as "shunt current", reducing the overall energy available for ignition. As contamination increases, so does the probability that no spark will be produced.

The tendency for combustion residues to form deposits on the insulator nose is heavily dependent on its temperature and takes place predominantly at temperatures below approximately. 500 °C. At higher temperatures, the carbon-based residues are burned from the insulator nose, i.e., the spark plug cleans itself.

The objective is therefore to heat the insulator nose to an operating temperature which is above the "self-cleaning limit" of roughly 500 °C (for unleaded fuel) and is obtained shortly after starting.

An upper temperature limit of approximately 900 °C should not be exceeded. Above this limit, the electrodes are subject to heavy wear due to oxidation and hot-gas corrosion.

If temperatures rise even further, the risk of auto-ignition can no longer be ruled out (Fig. 2). In this process, the air/fuel mixture ignites on the hot spark-plug components to produce uncontrolled ignition events; these can damage or even destroy the engine.

Thermal loading capacity

When the engine is running, the spark plug is heated by the temperatures generated in the combustion process. Some of the heat absorbed by the spark plug is dissipated to the fresh gas. Most is conducted via the center electrode and the insulator to the spark-plug shell, from which point it is dissipated to the cylinder head (Fig. 3). The ultimate

Fig. 1
- - - Shunt current

Fig. 2
1   Spark plug with correct heat-range code number
2   Spark plug with heat-range code number too low (cold plug)
3   Spark plug with heat-range code number too high (hot plug)

The temperature in the operating range should be 500...900 °C at the insulator, varying according to engine power

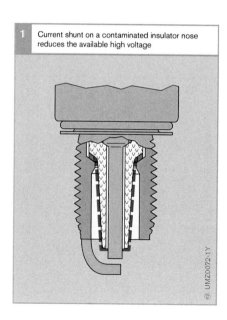

1 Current shunt on a contaminated insulator nose reduces the available high voltage

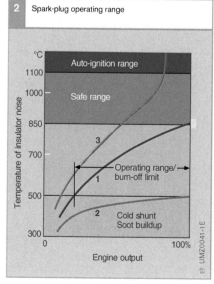

2 Spark-plug operating range

Temperature of insulator nose

°C
1100 — Auto-ignition range
1000 — Safe range
850
700
3
500 — 1 Operating range/burn-off limit
300 — 2 Cold shunt / Soot buildup
0   100%
Engine output

operating temperature is the point at which absorption of heat from the engine and its dissipation to the cylinder head reach a state of equilibrium.

The amount of heat supplied is dependent on the engine. Engines with high specific power output generally operate with higher combustion-chamber temperatures than those with low specific power output.

The design of the insulator nose is the primary determinant of heat dissipation. The size of the insulator surface determines heat absorption, while the cross-sectional area and the center electrode affect heat dissipation.

The spark plug's heat-absorption capacity must therefore be matched to the individual engine type. The index indicating a spark plug's thermal loading capacity is its heat range.

Heat range and heat-range code number

The heat range of a spark plug is determined relative to calibration spark plugs and described with the aid of a heat-range code number. A low code number (e.g., 2…5) indicates a "cold" spark plug with low heat absorption through a short insulator nose. A high code number (e.g., 7…10) indicates a "hot" spark plugs with high heat absorption through a long insulator nose. These code numbers form an integral part of the spark-plug designation so that spark plugs with different heat ranges can be easily distinguished and allocated to different engines.

The correct heat range is determined in full-load measurements because it is at these very operating points that thermal loading of spark plugs is at its greatest. During operation, the spark plugs should never become so hot as to represent a source of thermal auto-ignition. The heat-range recommendation is always defined with a safety margin relative to this auto-ignition limit to accommodate production variations in both plugs and engines. This margin is also important in view of the fact that an engine's thermal

properties can vary over the course of time. One example is the potential increase in compression ratio caused by ash deposits in the combustion chamber, which in turn results in higher temperatures for the spark plug. If no malfunctions occur with sooted spark plugs in the subsequent cold-starting tests with this heat-range recommendation, then the correct heat range for the engine is determined.

Because vehicle engines display a wide range of different properties with regard to operating loads, method of operation, compression, engine speed, cooling and fuel, it is impossible to use just one spark plug for all engines. A plug that overheats in one engine would run at relatively cold temperatures in another.

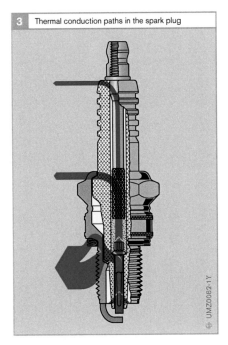

3   Thermal conduction paths in the spark plug

UMZ0082-1Y

Fig. 3
A large proportion of the heat absorbed from the combustion chamber is dissipated by thermal conduction (small contribution to cooling of approximately 20 % from flow of fresh induction mixture is not included)

Adaptation of spark plugs

Bosch works together with engine manufacturers in jointly defining the ideal spark plugs for each engine.

Temperature measurement

Thermocouple spark plugs specially designed and produced for temperature monitoring (Fig. 1) provide initial information on the right choice of plug. A thermocouple (2) embedded in the center electrode (3) makes it possible to record the temperatures in the individual cylinders as a function of engine speed and load. This process represents a simple means of identifying the hottest cylinder and operating conditions for subsequent measurements as well as assisting in reliable designation of the correct plug for any specific application.

Ionic-current measurement

The Bosch ionic-current measurement procedure employs the combustion process as a factor for determining the heat-range requirement. The ionizing effect of flames enables the progress in terms of time of combustion to be assessed by measuring the conductivity in the spark gap (Fig. 2). Because the electrical ignition spark produces a large number of charged particles in the spark gap, the ionic current rises sharply at the moment of ignition. Although the current flow falls once the ignition coil is discharged, the number of charged particles maintained by the combustion process is large enough to allow continued monitoring. Simultaneous monitoring of combustion-chamber pressure provides a record of normal combustion with a uniform pressure increase, peaking after ignition TDC. If the spark plug's heat range is varied during these measurements, the combustion process displays characteristic shifts with the thermal loading of a spark plug as a function of the heat range (Fig. 4).

Fig. 1

1 Insulator
2 Thermocouple sleeve
3 Center electrode
4 Measuring point

Fig. 2

1 High voltage from ignition coil
2 Ionic-current adapter
2a Break-over diode
3 Spark plug
4 Ionic-current device
5 Oscilloscope

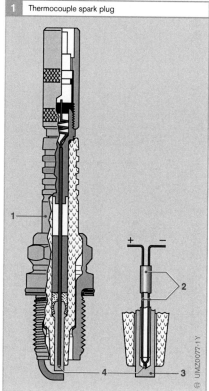

1 Thermocouple spark plug

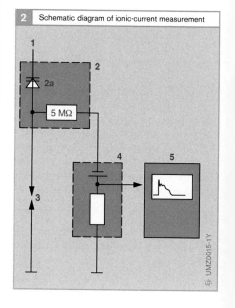

2 Schematic diagram of ionic-current measurement

The advantage of this method over measurements focusing exclusively on temperatures in the combustion chamber is that it indicates ignition probability, which is dependent not only on temperature, but also on the design parameters of the engine and the spark plug.

Definition of terminology

Terminology and definitions for uncontrolled ignition of air/fuel mixtures for heat-range adaptation of spark plugs have been defined in an international agreement (ISO 2542 – 1972, Fig. 3).

Thermal auto-ignition

Auto-ignition is defined as a process that results in ignition of the air/fuel mixture without an ignition spark, usually starting on a hot surface (e.g., on the excessively hot insulator-nose surface of a spark plug with too high a heat range). These events can be classified in one of two categories, according to the point at which they occur relative to the moment of ignition.

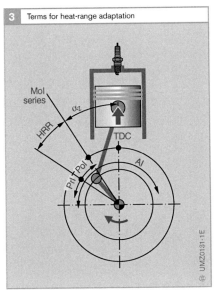

3 Terms for heat-range adaptation

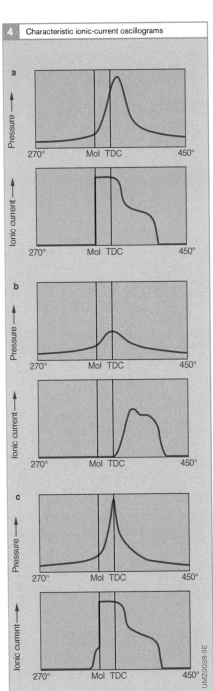

4 Characteristic ionic-current oscillograms

Fig. 3
AI Auto-Ignition
TDC Top Dead Center
Prl Pre-Ignition
Pol Post-Ignition
HRR Heat-Range Reserve in °cks
Mol Moment of Ignition in °cks before TDC
α_z Ignition angle

Fig. 4
a Normal combustion
b Scanned ignition with post-ignition
c Pre-ignition

Post-ignition

Post-ignition occurs after the moment of electrical ignition, but is not a critical factor in practical engine operation in that electrical ignition always takes place earlier. Conducting measurements to determine whether the spark plug is producing thermal auto-ignition entails suppressing the electrical spark. When post-ignition occurs, the sharp rise in ionic current does not occur until after the moment of ignition. But, because it initiates a combustion process, a pressure rise and therefore a torque output are also registered (Fig. 4b).

Pre-ignition

Pre-ignition occurs before the moment of electrical ignition (Fig. 4c) and can cause serious engine damage due to its uncontrolled progression. Premature initiation of the combustion process shifts the pressure peak relative to TDC while also increasing maximum pressure levels in the combustion chamber, promoting additional thermal load on the components in the combustion chamber. It is thus essential when adapting spark plugs to ensure that no pre-ignition will take place.

Assessment of measurement results

The Bosch ionic-current measurement procedure can be used to detect both types. However, the ignition spark must be suppressed at specific intervals for the purpose of detecting post-ignition. The point at which post-ignition occurs relative to the moment of ignition combines with the percentage of post-ignition events relative to the scan rate to provide information on the stresses to which the spark plug is being subjected within the engine. Because spark plugs with extended insulator noses (hot spark plugs) absorb more heat from the combustion chamber and dissipate that heat less effectively, they are more likely to induce post-ignition or even pre-ignition than spark plugs with shorter insulator noses. The application measurements employed to select the correct heat range for the

respective engine thus rely on mutual comparisons of spark plugs with various heat ranges and analysis of their tendency to produce pre-ignition or post-ignition.

The preferred environments for conducting spark-plug adaptation measurements are thus the engine test stand and the chassis dynamometer. For reasons of safety, measurement test runs to determine the hottest operating point at full load over an extended period of time on public highways are not permitted.

Spark-plug selection

The object of an adaptation is to select a spark plug which can be operated without pre-ignition and which has an adequate heat-range reserve, i.e., pre-ignition should not occur with spark plugs that are not hotter by at least two heat-range numbers.

As this section has indicated, selection and use of spark plugs is a finely-tuned process. The procedure for choosing the ideal spark plug generally includes close cooperation between the spark-plug manufacturer and the engine manufacturer.

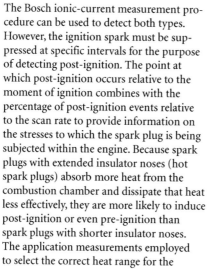

Fig. 5

Spark-plug selection:
The engine manufacturer's specifications and the recommendations contained in the Bosch sales documents are binding on drivers. Bosch supplies the ideal spark plug for every engine –
You can find the right plug in this catalog

5   Bosch Spark-Plug Catalog

Spark plugs   2006

BOSCH

SMZ0354E

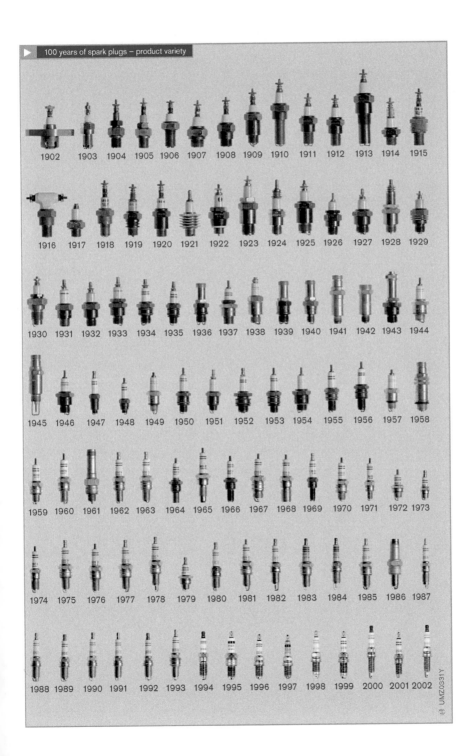

100 years of spark plugs – product variety

Spark-plug performance

Changes in the course of service life

Because spark plugs operate within an aggressive atmosphere, sometimes at extremely high temperatures, electrodes are subject to wear, which increases the ignition-voltage demand. When the situation finally reaches the point at which the ignition-voltage demand can no longer be covered by the ignition coil, then ignition misses occur.

Spark-plug operation can also be detrimentally affected by changes in an aging engine and by contamination. As engines age, blowby and leakage increase, raising the amount of oil in the combustion chamber. This, in turn, leads to more deposits of soot, ash and carbon on the spark plug, which can give rise to shunting, ignition misses, and in extreme cases auto-ignition. Yet another factor is the use of antiknock additives in fuels, which can form deposits, become conductive at high temperatures, and produce hot shunts. The ultimate result is ignition miss, characterized by a substantial increase in pollutant emissions along with potential damage to the catalytic converter. This is why spark plugs should be replaced at regular intervals.

Electrode wear

Electrode wear is synonymous with electrode erosion, a material loss which causes the gap to grow substantially over the course of time. This phenomenon essentially arises from two sources:
- Spark erosion, and
- Corrosion in the combustion chamber

Spark erosion and corrosion

Flashover of electrical sparks causes the electrodes to heat up to their melting point. The minute, microscopic particles deposited on surfaces react with the oxygen or the other constituents of the combustion gases. This results in material erosion, widening the electrode gap and raising the ignition-voltage demand (Fig. 1).

Electrode wear is minimized by using materials with high temperature stability (e.g., platinum and platinum alloys). It is also possible to reduce erosion without limiting service life using suitable electrode geometry (e.g., smaller diameter, thin pins) and alternate spark-plug designs (surface-gap plugs).

The electrical resistance effected in the conductive glass seal also reduces erosion and wear.

Abnormal operating states

Abnormal operating states can destroy both the spark plugs and the engine. Such states include:
- Auto-ignition
- Combustion knock, and
- High oil consumption (ash and carbon deposits)

1 Wear on center and ground electrodes

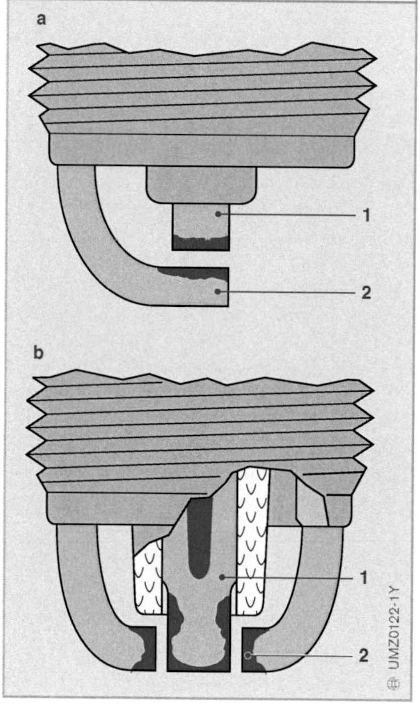

Fig. 1
a Spark plug with front electrode
b Spark plug with side electrodes

1 Center electrode
2 Ground electrode

Engine and spark plugs can also be damaged by incorrect ignition-system settings, spark plugs with the wrong heat range for the engine, and unsuitable fuels.

Auto-ignition

Auto-ignition is an uncontrolled ignition process accompanied by increases in combustion-chamber temperatures severe enough to cause serious damage to both spark plugs and engine.

Full-load operation can produce localized hot spots and induce auto-ignition in the following areas:

- At the spark-plug's insulator nose
- On exhaust valves
- On protruding sections of cylinder-head gaskets, and
- On flaking deposits

Combustion knock

Knocking is characteristic of an uncontrolled combustion process with very sharp rises in pressure. Knock is caused by spontaneous ignition of the mixture in areas which the advancing flame front, initiated by the usual electrical spark, has not yet reached. Combustion proceeds at a considerably faster rate than normal. High-frequency pressure pulsations with extreme pressure peaks are then superimposed on the normal pressurization curve (Fig. 3). The severe pressure gradients expose components (cylinder head, valves, pistons and spark plugs) to extreme thermal loads capable of damaging one or numerous components.

The damage is similar to that associated with cavitation damage from ultrasonic flow currents. On the spark plug, pitting on the ground electrode's surface is the first sign of combustion knock.

2 Ground electrode damaged from excessive knock

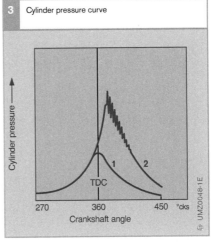

3 Cylinder pressure curve

270   360   450   °cks

TDC

Crankshaft angle

Cylinder pressure →

Fig. 3
1   Normal combustion
2   Combustion knock

Types

SUPER spark plug

SUPER spark plugs (Fig. 1) make up the majority of the Bosch spark-plug range, and serve as the basis for various derivative spark-plug types and concepts. A suitable version with precisely the right heat range is available for virtually every engine and application.

A cutaway view of the SUPER spark plug is shown in the section entitled "Design" (Fig. 1). The most significant characteristics of the SUPER spark plug are:

- A composite center electrode consisting of nickel-chromium alloy and featuring a copper core
- Optionally, a composite ground electrode for reducing ground-electrode wear by reducing the maximum temperature at the electrode, and
- An electrode gap that is preset for the relevant engine the factory

Various spark-plug profiles are employed to satisfy specific individual demands. The spark plug illustrated in Figure 2b is a current version that varies in a number of details from the classic SUPER (Fig. 2a). The spark position projects further into the combustion chamber, while optimized insulator-nose geometry and a thinner center electrode offer improved performance in repeated cold starts.

The version in Figure 2c features a laser-welded noble-metal pin. This not only extends service life, but also improves ignition and flame propagation thanks to its small diameter.

SUPER 4 spark plug

Design

The special features that distinguish the Bosch SUPER 4 spark plug from conventional SUPER plugs include

- Four symmetrically arranged ground electrodes (Fig. 3)
- A silver-plated center electrode, and
- A preset electrode gap requiring no adjustment during the plug's service life

Fig. 1
1   Composite center electrode with copper core

Fig. 2
a   Front electrode
b   Front electrode and forward spark
c   Front electrode and platinum center electrode

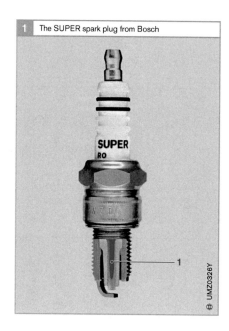

1   The SUPER spark plug from Bosch

SUPER
RO

1

UMZ0326Y

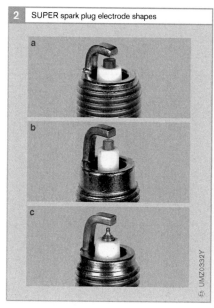

2   SUPER spark plug electrode shapes

a

b

c

UMZ0332Y

Method of operation

The four ground electrodes are manufactured from a thin profile section to ensure good ignition and flame-front propagation. The defined gap separating them from the center electrode and the insulator nose allows the spark – depending on the operating conditions – to jump either as an air-gap spark or as a surface-air-gap spark. The result is a total of eight potential spark gaps. Which of these spark gaps is selected is dependent on the operating conditions and the density of the air/fuel mixture at the moment of ignition.

Uniform electrode wear

Because the probability of spark propagation is the same for all electrodes, the sparks are evenly distributed across the insulator nose. In this way, even the wear is evenly distributed across all four electrodes.

Operating range

The silver-plated center electrode provides effective heat dissipation. This reduces the risk of auto-ignition due to overheating and extends the safe operating range. These assets mean that each SUPER 4 has a heat range corresponding to at least two ranges in a conventional spark plug. In this way, a wide range of vehicles can be refitted during servicing with relatively few spark-plugs types.

Spark-plug efficiency

The SUPER 4's thin ground electrodes absorb less energy from the ignition spark than the electrodes on conventional spark plugs. The SUPER 4 thus offers higher operating efficiency by providing up to 40 % more energy to ignite the air/fuel mixture (Fig. 5).

Ignition probability

Higher excess air (lean mixture, $\lambda > 1$) reduces the probability that the energy transferred to the gas will be sufficient to ignite the mixture reliably. In laboratory tests, the SUPER 4 has demonstrated the ability to ignite reliably mixtures as lean as $\lambda = 1.55$, whereas more than half of all ignition attempts failed under these conditions when a standard spark plug was used (Fig. 5).

Performance in repeated cold starts

Surface-gap sparking ensures effective self-cleaning, even at low temperatures. This means that up to three times as many cold starts (starting without warming up the engine) are possible as with conventional plugs.

3 Electrodes on Bosch SUPER 4 spark plug

UMZ0282-1Y

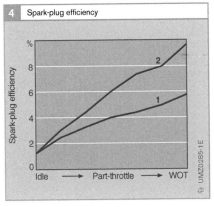

4 Spark-plug efficiency

Spark-plug efficiency

Idle ⟶ Part-throttle ⟶ WOT

UMZ0285-1E

Fig. 4
1 Conventional spark plug
2 Bosch SUPER 4 spark plug

Environmental and catalytic-converter protection

Improved cold-start performance and more reliable ignition, under all conditions including the warm-up phase, reduce the amount of unburnt fuel and thereby the HC emissions.

Advantages

The improved properties that set the SUPER 4 apart from conventional spark plugs include:
- Greater ignition reliability thanks to eight potential spark gaps
- Self-cleaning thanks to surface-gap technology, and
- Extended heat range

Platinum+4 spark plug

Design

The Platinum+4 spark plug (Fig. 6) is a surface-gap spark plug designed for extended replacement intervals. It is distinguished from conventional spark plugs by
- Four symmetrically arranged ground electrodes with double curvatures (9)
- A thin sintered center electrode made from platinum (8)
- A geometrically improved contact pin (7) made from a special alloy
- A ceramic insulator (2) with high breakdown resistance, and
- An insulator nose redesigned for improved performance

Method of operation

Ignition reliability

The extended electrode gap of 1.6 mm lends the Platinum+4 the capacity to deliver outstandingly reliable ignition, while the four earth electrodes assume an ideal position in the combustion chamber to ensure that the ignition spark has unobstructed access to the mixture. This allows the flame core to spread into the combustion chamber with virtually no interference, ensuring complete ignition of the entire air/fuel mixture.

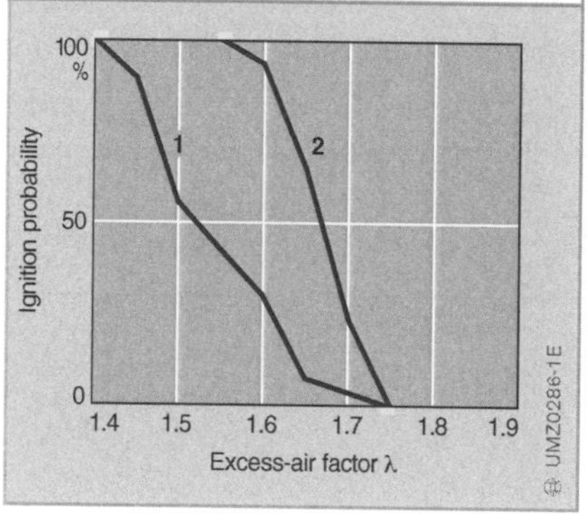

Fig. 5
1 Conventional spark plug
2 Bosch SUPER 4 spark plug

Fig. 6
1 Terminal stud
2 Insulator
3 Shell
4 Heat-shrinkage zone
5 Sealing ring
6 Conductive glass seal
7 Contact pin
8 Platinum pin (center electrode)
9 Ground electrodes (only two of four electrodes shown)

5 Effect of mixture composition on ignition probability

Ignition probability vs *Excess-air factor λ*

100 %

50

0

1.4   1.5   1.6   1.7   1.8   1.9

UMZ0286-1E

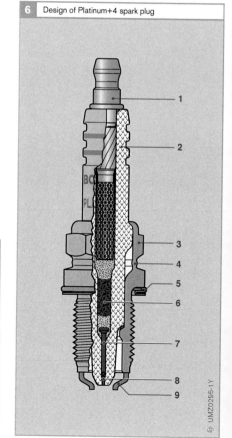

6 Design of Platinum+4 spark plug

UMZ0295-1Y

Response to repeated cold starts
The surface-gap concept provides substantial improvements over air-gap plugs in repeated cold starting.

Electrode wear
There are also advantages in respect of electrode wear, thanks to the erosion-resistant platinum pin in the center electrode and improved materials in the four ground electrodes. The resistance in the conductive glass seal reduces capacitive discharge, making a further contribution to reduced spark erosion.

The comparison in Figure 7 shows the rise in demand for ignition energy over a period of engine operation of 800 hours on an engine test stand (corresponding to 100,000 km of highway use). The Platinum+4 spark plug's lower electrode wear delivers substantial reductions in the rate at which voltage demand increases relative to conventional spark plugs. Figures 8 and 9 show the profiles of a Platinum+4 spark plug when new and after a period of engine operation of 800 hours; the minimal electrode wear at the end of the endurance test is clear to see.

Advantages of the Platinum+4 spark plug
The Platinum+4 spark plug is characterized by a host of properties which make it ideal for extended-duty applications:
● Durable electrodes and ceramic components extend the plug-replacement intervals to up to 100,000 km
● Higher numbers of repeated cold starts possible
● Extremely good ignition and flame-front propagation for major improvements in smooth engine running

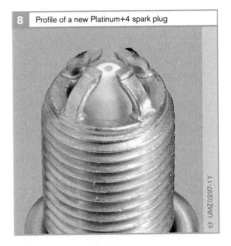

8   Profile of a new Platinum+4 spark plug

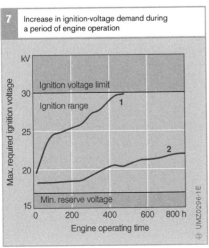

7   Increase in ignition-voltage demand during a period of engine operation

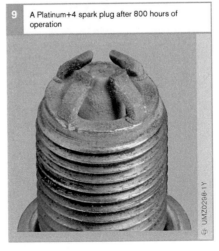

9   A Platinum+4 spark plug after 800 hours of operation

Fig. 7
1   Spark plug with air-gap spark (gap = 0.7 mm)
2   Platinum+4 spark plug with surface-gap spark (gap = 1.6 mm)

Spark plugs for direct-injection gasoline engines

In direct-injection engines, the fuel is introduced in stratified-charge mode via the high-pressure injector directly into the combustion chamber during the compression stroke. The design of the intake manifold and the piston crown generates a swirl- or tumble-like charge movement with which the fuel is transported to the spark plug. Because both the mass and direction of the flow vary at the engine's different operating points, a spark position projecting far into the combustion chamber is very advantageous to mixture ignition. This forward-spark concept has a negative effect on the temperature of the ground electrode to the extent that measures need to be take to reduce the temperature. By extending the shell into the combustion chamber, it is possible to reduce further the length of the ground electrode and thereby its temperature so that workable spark-plug concepts are possible.

Because of the numerous possible spark gaps, surface-gap concept offer a greater degree of reliability with regard to ignition misses. The improved self-cleaning performance by the surface-gap spark marks this spark-plug concept out for the wall- and air-guided combustion processes.

If the flow velocity at the spark location is not too great, even air-gap plugs can deliver good ignition results. This is because
- The spark is not so sharply deflected
- Breakaway and re-ignition are avoided, and
- The ignition energy can be transferred to generate a stable flame core

In the wall- and air-guided combustion processes, stratified mixture formation is closely linked to piston stroke to the extent that adjustment of combustion to the optimum efficiency cannot always be guaranteed. In addition, soot is caused by the intensive contact of the spray with the cylinder wall and the piston. For this reason, combustion processes which do not manifest these disadvantages have taken hold in recent years. By injecting the fuel during the induction stroke, the air/fuel mixture is set to $\lambda = 1$ and the engine is operated under homogeneous conditions. The homogeneous combustion processes place similar

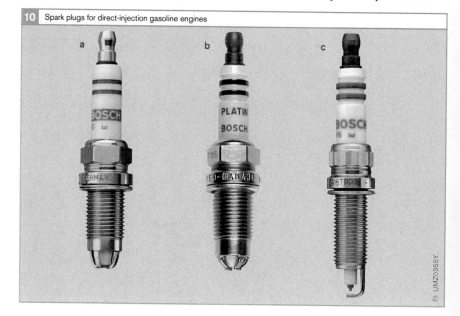

10   Spark plugs for direct-injection gasoline engines

a   b   c

Fig. 10
a   Surface-gap spark
    plug without noble
    metal
b   Surface-gap spark
    plug with platinum
    center electrode
c   Air-gap spark plug
    with platinum on
    center electrode

demands on the ignition performance of the spark plugs, as is the case with manifold-injection engines. However, these engines are often operated with exhaust-gas turbochargers in order to achieve higher power figures, i.e., at the moment of ignition the air/fuel mixture has a higher density and therefore also a higher ignition-voltage demand. Here, air-gap plugs with noble-metal pins are generally used on the center electrode in order to be able to reliably satisfy the service-life requirements after 60,000 km and more.

Spray-guided combustion processes
In contrast, the demands placed on spark plugs are significantly greater in more recent developments pertaining to spray-guided combustion processes. Due to the fact that the spark plug is located close to the fuel injector, long, narrow plugs are preferred because this shape allows addition cooling passages to be accommodated between the injector and the spark plug. The alignment of the spark plug to the injector must be determined in extensive tests. In this way, the spark is drawn into the peripheral area of the spray by the flow of the injection jet (entrainment flows), and thereby ignition of the mixture is ensured.

In these combustion processes, it is extremely important for the spark always to jump at the same location. By configuring the geometry of the spark plugs on the combustion-chamber side, it possible to prevent the spark from disappearing in the breathing space (air space between the spark-plug shell and the insulator on the combustion-chamber side) so that it remains available for ignition. But reversing the ignition polarity (center electrode as the anode, ground electrode as the cathode) is another way of avoiding surface-gap sparking into the spark-plug shell (Fig. 11). It is also necessary to check whether restricted axial/radial position tolerances are needed in order to reduce the reciprocal action between the injector and the spark plug.

If the spark plug is situated too closely to the injector, the peripheral zone of the spray will not yet be sufficiently prepared such that ignition problems may arise due to over-rich mixture zones. If the spark plug is situated too far away from the injector, this may already give rise in the peripheral zones of the spray to leaning-out effects, which in turn are not conducive to a stable ignition phase.

In the case of a close spray-cone tolerance, it is also necessary to keep the spark location constant. If the spark position is too deep, the spark plug projects into the spray and is saturated with fuel; this may cause damage to the spark plug and sooting on the insulator. If the spark position is pulled back too far towards the combustion-chamber wall, the spray might no longer be drawn into the mixture by the spray-induced flow, resulting in ignition misses.

From this, it is possible to deduce that close coordination and cooperation is required between the design engineers responsible for spark-plug development and combustion-process in order to ensure reliable functioning in the spray-guided combustion processes.

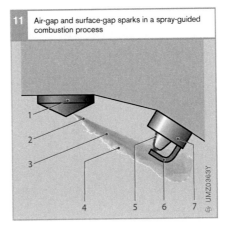

11   Air-gap and surface-gap sparks in a spray-guided combustion process

Fig. 11
The air-gap spark can ignite the air/fuel mixture, the surface-gap spark is generated outside the mixture cloud
1   High-pressure fuel injector
2   Fuel spray
3   Rich area
4   Lean area
5   Surface-gap spark
6   Air-gap spark
7   Spark plug

Special-purpose spark plugs

Applications

Special-purpose spark plugs are available for use in certain applications. These plugs feature unique designs dictated by the operating conditions and installation environments in individual engines.

Spark plugs for motor-sport applications

Constant full-load operation subjects the engines in competition vehicles to extreme thermal loads. The spark plugs produced for this operating environment usually have noble-metal electrodes (silver, platinum) and a short insulator nose. The heat absorption of these spark plugs is very low through the insulator nose, while heat dissipation through the center electrode is high (Fig. 12).

Spark plugs with resistors

A resistor can be installed in the supply line to the spark plug's spark gap to suppress transmission of interference pulses to the ignition cable and thereby educe interference radiation. The reduced current in the ignition spark's arcing phase also leads to lower electrode erosion. The resistor is formed by the special conductive glass seal between the center electrode and the terminal stud. Appropriate additives lend the conductive glass seal the desired level of resistance.

Fully-shielded spark plugs

Shielded spark plugs may be required in applications characterized by extreme demands in the area of interference suppression (radio equipment, car phones).

In fully shielded spark plugs, the insulator is surrounded by a metal shielding sleeve. The connection is inside the insulator. A union nut attaches the shielded ignition cable to the sleeve. Fully shielded spark plugs are also watertight (Fig. 13).

Fig. 12
1 Silver center electrode
2 Short insulator

Fig. 13
1 Special conductive glass seal (interference-suppression resistor)
2 Ignition-cable connection
3 Shielding sleeve

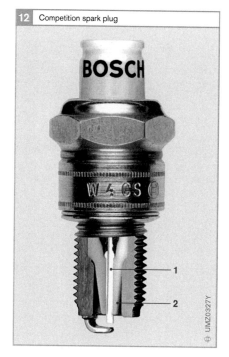

12 Competition spark plug

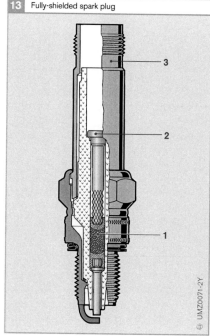

13 Fully-shielded spark plug

Spark-plug type designations

Spark-plugs types are identified by a type designation (Fig. 1). This type designation contains all the spark-plug's data – with the exception of the electrode gap. The electrode gap is specified on the packaging. The spark plug which is suitable for a given engine is specified or recommended by the engine manufacturer and by Bosch.

1   Key to type designations for Bosch spark plugs

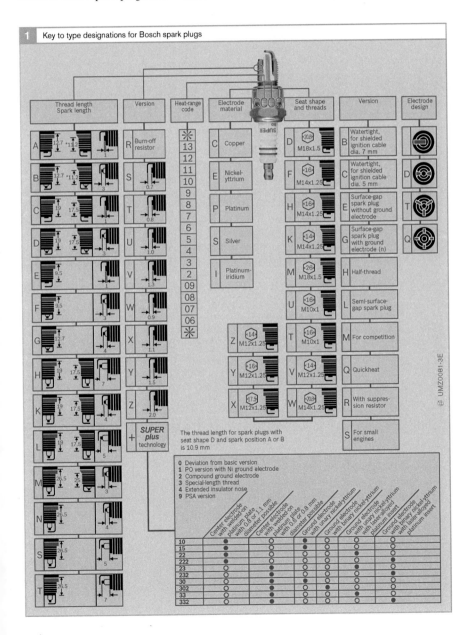

Thread length / Spark length	Version	Heat-range code	Electrode material	Seat shape and threads	Version	Electrode design
A 12.7 · 11.2	R Burn-off resistor	※ 13 12	C Copper	D M18x1.5	B Watertight, for shielded ignition cable dia. 7 mm	
B 12.7 · 11.2	S 0.7	11 10	E Nickel-yttrium	F M14x1.25	C Watertight, for shielded ignition cable dia. 5 mm	D
C 19 17.5	T 0.8	9 8 7	P Platinum	H M14x1.25	E Surface-gap spark plug without ground electrode	T
D 19 17.5	U 1.0	6 5 4	S Silver	K M14x1.25	G Surface-gap spark plug with ground electrode (n)	Q
E 9.5	V 1.3	3 2 09	I Platinum-iridium	M M18x1.5	H Half-thread	
F 9.5	W 0.9	08 07 06		U M10x1	L Semi-surface-gap spark plug	
G 12.7	X 1.1	※	Z M12x1.25	T M10x1	M For competition	
H 19 17.5	Y 1.5		Y M12x1.25	V M12x1.25	Q Quickheat	
K 19 17.5	Z 2.0		X M12x1.25	W M14x1.25	R With suppression resistor	
L 19 17.5	+ SUPER plus technology				S For small engines	
M 26.5 · 25						
N 26.5						
S 26.5						
T 26.5						

The thread length for spark plugs with seat shape D and spark position A or B is 10.9 mm

0 Deviation from basic version
1 PO version with Ni ground electrode
2 Compound ground electrode
3 Special-length thread
4 Extended insulator nose
9 PSA version

	Center electrode with welded-on platinum plate	Center electrode with 0.8 or 1.1 mm diameter possible	Center electrode with welded-on platinum plate with 0.6 or 0.8 mm diameter possible	Ground electrode with unary nickel-yttrium	Ground electrode with binary nickel-yttrium	Ground electrode with unary nickel-yttrium with laser-alloyed platinum insert	Ground electrode with binary nickel-yttrium with laser-alloyed platinum insert
10	●	○	●	○	●	○	
15	●	○	●	○	●	○	
22	●	○	●	○	●	○	
222	●	○	●	○			○
23	○	●	●	○	●	○	
232	○	●	●	○			●
30	○	●	●	○	●	○	
302	○	●	●	○			○
33	○	●	●	○	●	○	
332	○	●	●	○			●

UMZ0081-3E

Manufacture of spark plugs

Each day roughly one million spark plugs emerge from our Bamberg plant, the only Bosch facility manufacturing these products within Europe. Spark plugs conforming to the universal Bosch quality standards are also produced for local markets and original-equimpent customers in plants in India, Brazil, China, and Russia. Bosch has now produced a total of more than seven billion spark plugs.

The individual components joined to form the finished spark plug in final assembly are created in three parallel manufacturing processes.

Insulator

The basic material used in the high-quality ceramic insulator is aluminum oxide. Aggregate materials and binders are added to this aluminum oxide, which is then ground to a fine consistency. The granulate is poured into molds and processed at high pressure. This gives the raw castings their internal shape. The outer contours are ground to produce the soft core, which already displays a strong similarity to the later plug core. The next work step involves mechanically anchoring a platinum pin only a few millimeters in length in the soft core. The ceramic elements pass through a sintering furnace, where they obtain their final shape at a temperature of approximately 1600 °C, and the platinum pins are secured in the ceramic element. The soft core must be manufactured to compensate for the contraction that occurs in the sintering process, which is approximately 20 %.

Once the insulators have been fired, the labeling is applied to the insulator nose, which is then coated with a lead-free glaze.

Plug core

Electrical contacting with the platinum pin is effected by means of a contact pin, which is flattened at the rear end. This blade ensures subsequent secure anchoring in the plug core. *Paste* is filled into the hole once the center electrode has been inserted into the insulator. The paste consists of glass particles, to which conductive particles are added to produce a conductive connection to the terminal stud after sealing in. The individual components can also be varied to manipulate the paste's resistance. Resistance values of up to 10 k| can be achieved.

The *terminal stud* is manufactured from wire and formed by flattening and edge knurling. It receives a protective nickel surface and is inserted in the plug core. The plug core then passes through an oven, where it is heated to over 850 °C. The paste becomes molten at these temperatures. It flows around the center electrode and the terminal stud can then be pressed into this molten mass. The core cools to form a gas-tight and electrically conductive connection between the center electrode and the terminal stud.

Shell

The shell is manufactured from steel by means of extrusion. A section several centimeters in length is cut from the wire and then cold-formed in several pressing operations until the spark-plug shell assumes its final contours. Only a limited number of machining operations (to produce shrinkage and threaded sections) is then required. After the ground electrodes (up to four, depending on spark-plug type) have been welded to the shell, the thread is rolled and the entire shell is nickel-plated for protection against corrosion.

Spark-plug assembly

During spark-plug assembly, a sealing ring and the plug core are installed in the spark-plug shell. The upper shell is crimped and beaded to position the plug core. A subsequent shrinking process (induction is used to heat parts of the spark-plug shell to over 900 °C) provides a gas-tight union between spark-plug shell and core. Then an outer sealing ring is mounted on flat-seat spark plugs in an operation that reshapes the material to form a captive seal washer. This ensures that the combustion chamber will be effectively sealed when the spark plug is subsequently installed in a cylinder head.

On some spark-plug versions, an SAE nut must then be installed on the terminal stud's M4 thread and staked several times to form a firm attachment.

The assembly process is completed once the electrode gap has been adjusted to the engine manufacturer's specifications. The spark plugs are then prepared for sale in market- and customer-specific packaging.

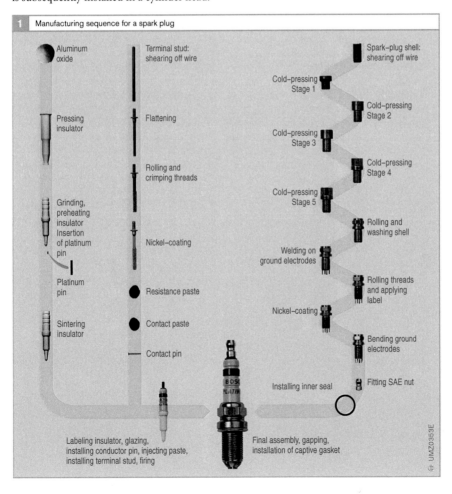

1   Manufacturing sequence for a spark plug

Aluminum oxide

Pressing insulator

Grinding, preheating insulator
Insertion of platinum pin

Platinum pin

Sintering insulator

Terminal stud: shearing off wire

Flattening

Rolling and crimping threads

Nickel-coating

Resistance paste

Contact paste

Contact pin

Labeling insulator, glazing, installing conductor pin, injecting paste, installing terminal stud, firing

Final assembly, gapping, installation of captive gasket

Spark-plug shell: shearing off wire

Cold-pressing Stage 1

Cold-pressing Stage 2

Cold-pressing Stage 3

Cold-pressing Stage 4

Cold-pressing Stage 5

Rolling and washing shell

Welding on ground electrodes

Rolling threads and applying label

Nickel-coating

Bending ground electrodes

Installing inner seal

Fitting SAE nut

UMZ0353E

Simulation-based spark-plug development

The Finite Element Method (FEM) is a mathematical approximation procedure for solving differential equations which describe the behavior and properties of physical systems. The process entails dividing structures into individual sectors, or finite elements.

In spark-plug design, FEM is employed to calculate temperature fields, electrical fields, and problems of structural mechanics. It makes it possible to determine the effects of changes to a spark plug's geometry and constituent materials, and variations in general environmental conditions, in advance, without extensive testing. The results provide the basis for precisely focused production of test samples which are then used for verification of the calculation results.

Temperature field

The maximum temperatures of the ceramic insulator and the center electrode in the combustion chamber are decisive factors for the spark plug's heat range. Figure 1a shows an axisymmetrical model of a spark plug along with a section of the cylinder head.

The temperature fields as indicated in the colored sections show that the highest temperatures occur at the nose of the ceramic insulator.

Electrical field

The high voltage applied at the moment of ignition is intended to generate flashover at the electrodes. Breakdown in the ceramic material or current tracking between the ceramic insulator and the spark-plug shell can lead to delayed combustion and ignition misses. Figure 1b shows an axisymmetrical model with the corresponding field-strength vectors between center electrode and shell. The electrical field penetrates the nonconductive ceramic material and the intermediate gas.

Structural mechanics

High pressures within the combustion chamber during combustion make a gas-tight union between the spark-plug shell and the insulator essential. Figure 1c shows an axisymmetrical model of a spark plug after the shell is crimped and heat-shrunk. The retention force and the mechanical stress in the spark-plug shell are measured.

Fig. 1
Axisymmetical models of a spark plug
a   Temperature distribution in ceramic insulator and in center electrode
b   Electric field strength adjacent to center electrode and shell
c   Retaining force and mechanical stress in spark-plug shell

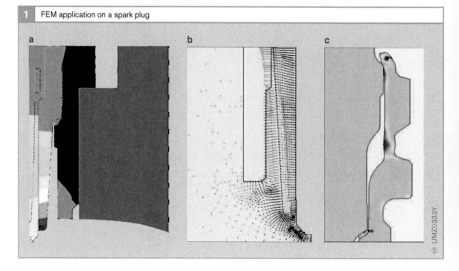

1   FEM application on a spark plug

a   b   c

Handling spark plugs

Spark-plug installation

Correct selection and installation will ensure that the spark plug continues to serve as a reliable component within the overall ignition system.

Readjusting the electrode gap is recommended only on spark plugs with front electrodes. Because this would involve actually changing the spark-plug concept, the gaps of the ground electrodes on surface-gap and surface-air-gap spark plugs should never be readjusted.

Removal

The first step is to screw out the spark plug by several thread turns. The spark-plug well is then cleaned using compressed air or a brush to prevent dirt particles from becoming lodged in the cylinder head threads or entering the combustion chamber. It is only after this operation that the spark plug should be completely unscrewed and removed.

To avoid damaging the threads in the cylinder head, respond to any tendency to seize in spark plugs by unscrewing them by only a small amount. Then apply oil or a solvent containing oil to the threads and screw the spark plug back in. Wait for the penetrating oil to work, then screw the plug back out all the way.

Installation

Please observe the following when installing the spark plug in the engine:
- The contact surfaces between spark plug and engine must be clean and free of all contamination.
- Bosch spark plugs are coated with anti-corrosion oil, thus eliminating the need for any other lubricant. Because the threads are nickel-plated, they will not seize in response to heat.

Wherever possible, spark plugs should be tightened down with a torque wrench. The torque applied to the spark plug's 6-point fitting is transferred to the seat and the socket's threads. Application of excessive torque or failure to keep the socket attachment correctly aligned within the spark-plug well can place stress on the shell and loosen the insulator. This destroys the spark plug's thermal-response properties and can lead to engine damage. This is one reason why torque should never be applied beyond the specified level. The specified tightening torques apply to new spark plugs, with a light coating of oil.

Under actual field conditions, spark plugs are often installed without a torque wrench. As a result, too much torque is usually used to install spark plugs. Bosch recommends the following procedure:

First: Screw the spark plug into the clean socket by hand until it is too tight to continue. Then apply the spark-plug wrench. At this point, we distinguish between:
- New spark plugs with flat seal seats, which are tightened by an angle of approximately 90° after initial resistance to turning
- Used spark plugs with flat seal seats, which are tightened by an angle of approximately 30°
- Spark plugs with conical seal seats, which are tightened by an angle of approximately 15°

Second: Do not allow the socket wrench to tilt to an angle relative to the plug while either tightening or loosening; this would apply excessive vertical or lateral force to the insulator, making the plug unsuitable for use.

Third: When socket wrench with a loose mandrel, ensure that the opening for the mandrel is above the top of the spark plug to allow the mandrel to be drawn through the socket wrench. If the opening is too low on the plug, resulting in the mandrel only engaging a short distance, spark plug damage can result.

Mistakes and their consequences

Only spark plugs specified by the engine manufacturer or as recommended by Bosch should be installed. Drivers should consult the professionals at a Bosch service center to avoid the possibility of incorrect spark-plug selection. Sales assistance and guidance are available from catalogues, sales displays with reference charts and application guides available on the premises.

Use of the wrong spark-plug type can lead to serious engine damage. The most frequently encountered mistakes are:
- Incorrect heat-range code number
- Incorrect thread length, or
- Modifications to the seal seat

Incorrect heat-range code number

It is essential to ensure that the spark plug's heat range corresponds to the engine manufacturer's specifications and/or Bosch recommendations. Use of spark plugs with a heat-range code number other than that specified for the specific engine can cause auto-ignition.

Incorrect thread length

The length of the threads on the spark plug must correspond precisely to the depth of the socket in the cylinder head. If the threads are too long, the spark plug will protrude too far into the combustion chamber. Possible consequences:
- Piston damage
- Carbon residue baked onto the spark-plug threads can make it impossible to remove the plug, or
- Overheated spark plugs

A threaded section that is too short will prevent the spark plug from reaching far enough into the combustion chamber. Possible consequences:
- Poor ignition and flame propagation to the mixture
- The spark plug fails to reach its burn-off (self-cleaning) temperature, and

- The lower threads in the cylinder head's socket become coated with baked-on carbon residue

Modifications to the seal seat

Never install a sealing ring, shim or washer on a spark plug featuring a conical, or tapered, seal seat. On spark plugs with a flat seal seat, use only the captive sealing ring already installed on the plug. Never remove this sealing ring, and do not replace it with another shim or washer of any kind.

The sealing ring prevents the spark plug from protruding too far into the combustion chamber. This reduces the efficiency of thermal transfer from the spark-plug shell to the cylinder head, while also preventing an effective seal at the mating surfaces.

Installation of a supplementary sealing ring prevents the spark plug from penetrating far enough into its socket, which also reduces thermal transfer between the spark plug-shell and the cylinder head.

Spark-plug profiles

Spark-plug profiles provide information on the performance of both engine and plugs. The appearance of the spark plug's electrodes and insulator – the spark-plug profile – provides indications as to how the spark plug is performing, as well as to the composition of the induction mixture and the combustion process within the engine (Figs. 1 to 3, following pages).

Assessing the spark-plug profiles is thus an important part of the engine-diagnosis procedure. It is essential to observe the following procedure in order to obtain accurate results: The vehicle must be driven before the spark-plug profiles can be assessed. If the engine is run for an extended period at idle, and especially after cold starts, carbon residue will form, preventing an accurate assessment of the spark plug's condition. The vehicle should first be driven a distance of 10 kilometers (6 miles) at various engine speeds and under moderate load. Avoid extended idling before switching off the engine.

1 Spark-plug profiles, Part 1

① **Normal.**
Insulator tip with color between grayish white-grayish yellow to russet. Engine satisfactory. Correct heat range. Mixture adjustment and ignition timing are good, no ignition miss, cold-starting device functioning properly. No residue from leaded fuel additives or engine-oil alloying constituents. No overheating.

② **Sooted.**
Insulator tip, electrodes and spark-plug shell covered with a felt-textured, matt-black coating of soot.
Cause: Incorrect mixture adjustment (carburetor, injection): mixture too rich, extremely dirty air filter, automatic choke or choke cable defective, vehicle used only for extremely short hauls, spark plug too cold, heat-range code number too low.
Effects: Ignition miss, poor cold starts.
Corrective action: Adjust mixture and starting device, check air filter.

③ **Oil-fouled.**
Insulator tip, electrodes and spark-plug shell covered with shiny, oily layer of soot or carbon.
Cause: Excessive oil in combustion chamber. Oil level too high, severe wear on piston rings, cylinders and valve guides.
Two-stroke engines: too much oil in fuel mixture.
Effects: Ignition miss, poor starting.
Corrective action: Overhaul engine, use correct oil/fuel mixture, replace spark plugs.

④ **Lead fouling.**
A brownish-yellow glaze, possibly with a greenish tint, forms on the insulator tip.
Cause: Fuel additives containing lead. The glaze forms when the engine is operated under high loads after extended part-load operation.
Effects: At higher loads, the coating becomes electrically conductive, leading to ignition miss.
Corrective action: New spark plugs, cleaning is pointless.

2 Spark-plug profiles, Part 2

⑤ **Severe lead fouling.**
Thick, brownish-yellow glaze with possible green tint forms on the insulator tip.
Cause: Fuel additives containing lead: the glaze forms during operation under heavy loads following an extended period of part-load operation.
Effects: At higher loads, the coating becomes electrically conductive, leading to ignition miss.
Corrective action: New spark plugs. Cleaning is pointless.

⑥ **Ash deposits.**
Serious ash residue from oil and fuel additives on the insulator tip, in the breathing space (annular gap) and on the ground electrode. Loose or cinder-flake deposits.
Cause: Substances from additives, especially those used for oil, can leave these ash deposits in the combustion chamber and on the spark plug.
Effect: Can produce auto-ignition with power loss as well as engine damage.
Corrective action: Restore engine to satisfactory operating condition. Replace spark plugs, change oil as indicated.

⑦ **Melted center electrode.**
Melted center electrode, insulator tip is soft, porous and spongy.
Cause: Thermal overloading due to auto-ignition.
Can stem from overadvanced ignition timing, residue in the combustion chamber, defective valves, faulty ignition distributor and low-quality fuel. May also possibly be caused by heat range that is too low.
Effects: Ignition miss, lost power (engine damage).
Corrective action: Check engine, ignition and mixture preparation. Install new spark plugs with correct heat range.

⑧ **Center electrode with severe heat erosion.**
Severe heat erosion on center electrode, simultaneous serious damage to ground electrode.
Cause: Thermal overloading due to auto-ignition.
Can stem from overadvanced ignition timing, residue in the combustion chamber, defective valves, faulty ignition distributor and low-quality fuel.
Effects: Ignition miss, power loss, possible engine damage. Insulator tip may rupture from overheated center electrode.
Corrective action: Check engine, ignition and mixture preparation. Replace spark plugs.

3 Spark-plug profiles, Part 3

⑨ **Melted electrodes.**
Electrodes melted to form a cauliflower pattern.
Possibly with deposits from other sources.
Cause: Thermal overloading due to auto-ignition.
Can stem from overadvanced ignition timing, residue in
the combustion chamber, defective valves, faulty ignition
distributor and low-quality fuel.
Effect: Power loss followed by complete engine failure
(engine damage).
Corrective action: Check engine, ignition and mixture
preparation. Replace spark plugs.

⑩ **Severely eroded center electrode.**
Cause: Failure to observe spark-plug replacement
intervals.
Effects: Ignition miss, especially during acceleration
(ignition voltage not adequate for bridging wider
electrode gap). Poor starting.
Corrective action: New spark plugs.

⑪ **Severely eroded ground electrode.**
Cause: Aggressive fuel and oil additives. Deposits or
other factors interfering with flow patterns in combustion
chamber. Engine knock. No thermal overloading.
Effects: Ignition miss, especially during acceleration
(ignition voltage not adequate to bridge across electrode gap). Poor starting.
Corrective action: New spark plugs.

⑫ **Insulator-tip breakage.**
Cause: Mechanical damage (e.g., impact, fall or pressure on the center electrode from incorrect handling).
In extreme cases, the insulator tip may be split by deposits between the center electrode and the insulator
tip, or by corrosion in the center electrode (especially
when replacement intervals are not observed).
Effect: Ignition miss. Flashover occurs in locations
with no reliable access to the fresh mixture.
Corrective action: New spark plugs.

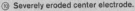

Electronic Control

"Motronic" is the name of an engine-management system that facilitates open- and closed-loop control of gasoline engines within a single ECU. The first Motronic system went into volume production at Bosch in 1979. Essentially, it comprised the functions of electronic fuel injection and electronic ignition. With the advances made in the field of microelectronics, it has been possible to continuously expand the capabilities of Motronic systems over the course of time. The range of functions has been continuously adapted in response to prevailing demands and the complexity of successive Motronic systems has consequently increased.

Although cost considerations limited application of early Motronic versions to luxury-class cars, progressively more stringent demands for clean emissions is gradually leading to widespread use of this system. Since the mid-1990s, all new engine projects in which Bosch has been involved use Motronic systems.

Open- and closed-loop electronic control

Motronic comprises all the components which control the gasoline engine (Fig. 1). The torque requested by the driver is adjusted by means of actuators or converters. The main individual components are
- The electrically actuated throttle valve (air system): This controls the air-mass flow to the cylinders and thus the cylinder charge.
- The fuel injectors (fuel system): These meter the correct amount of fuel for the cylinder charge.
- The ignition coils and spark plugs (ignition system): These provide for correctly timed ignition of the air/fuel mixture in the cylinder.

Modern-day engines are subject to exacting demands with regard to
- Exhaust-emission behavior
- Power output
- Fuel consumption
- Diagnostic capability, and
- Comfort/user-friendliness

Where necessary, additional components are installed on the engine for this purpose. All the manipulated variables are calculated in accordance with prespecified algorithms in the Motronic ECU. The actuating signals for the actuators are generated from these variables.

Acquisition of operating data
Sensor and setpoint generators
Motronic uses sensors and setpoint generators to collect the operating data required for open- and closed-loop control of the engine (Fig. 1).
Setpoint generators (e.g., switches) record settings made by the driver, such as
- The position of the ignition key in the ignition lock (terminal 15)
- The positions of A/C-control switches
- The cruise-control lever setting

Sensors detect physical and chemical variables, thus providing information about the engine's current operating state. Examples of such sensors are:
- Engine-speed sensor for detecting the crankshaft position and calculating the engine speed
- Phase sensor for detecting the phase angle (engine operating cycle) or the camshaft position
- Engine-temperature sensor and intake-air temperature sensor for calculating temperature-dependent correction variables
- Knock sensor for detecting engine knock
- Air-mass meter, and/or
- Intake-manifold pressure sensor for charge recording
- Lambda oxygen sensor for lambda closed-loop control

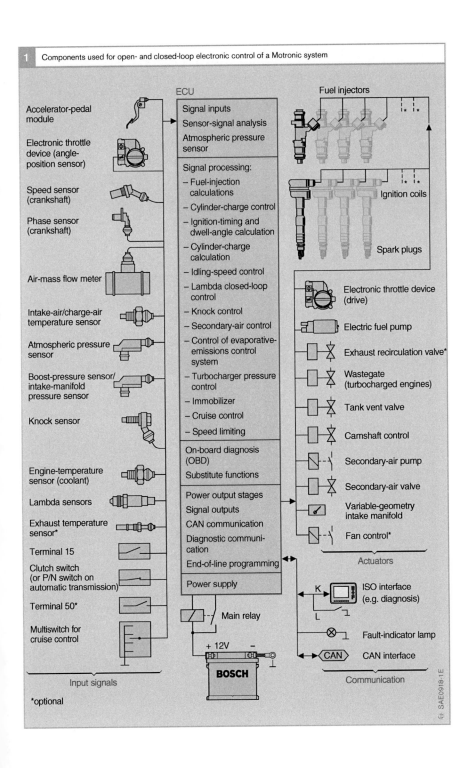

1   Components used for open- and closed-loop electronic control of a Motronic system

ECU

Fuel injectors

Accelerator-pedal
module

Electronic throttle
device (angle-
position sensor)

Speed sensor
(crankshaft)

Phase sensor
(crankshaft)

Air-mass flow meter

Intake-air/charge-air
temperature sensor

Atmospheric pressure
sensor

Boost-pressure sensor/
intake-manifold
pressure sensor

Knock sensor

Engine-temperature
sensor (coolant)

Lambda sensors

Exhaust temperature
sensor*

Terminal 15

Clutch switch
(or P/N switch on
automatic transmission)

Terminal 50*

Multiswitch for
cruise control

Input signals

*optional

Signal inputs
Sensor-signal analysis
Atmospheric pressure
sensor

Signal processing:
– Fuel-injection
  calculations
– Cylinder-charge control
– Ignition-timing and
  dwell-angle calculation
– Cylinder-charge
  calculation
– Idling-speed control
– Lambda closed-loop
  control
– Knock control
– Secondary-air control
– Control of evaporative-
  emissions control
  system
– Turbocharger pressure
  control
– Immobilizer
– Cruise control
– Speed limiting

On-board diagnosis
(OBD)
Substitute functions

Power output stages
Signal outputs
CAN communication
Diagnostic communi-
cation
End-of-line programming

Power supply

Main relay

+ 12V   –

BOSCH

Ignition coils

Spark plugs

Electronic throttle device
(drive)

Electric fuel pump

Exhaust recirculation valve*

Wastegate
(turbocharged engines)

Tank vent valve

Camshaft control

Secondary-air pump

Secondary-air valve

Variable-geometry
intake manifold

Fan control*

Actuators

K   ISO interface
    (e.g. diagnosis)
L

Fault-indicator lamp

CAN   CAN interface

Communication

SAE0918-1E

Signal processing in the ECU

The signals produced by the sensors can take the form of digital, pulse-type or analog voltage signals. Input circuits in the ECU, or in the sensors – which will increasingly be the case in the future – process these signals. These circuits transform voltages to the levels required for subsequent processing in the ECU's microprocessor.

Digital input signals are read in directly in the microcontroller and stored as digital information. Analog signals are converted into digital signals by an analog/digital converter.

Processing of operating data

From the input signals, the engine ECU detects the engine's current operating state of the engine and uses this information in conjunction with requests from auxiliary systems and from the driver (pedal-travel sensor and operating switches) to calculate the command signals for the actuators.

The tasks performed by the engine ECU are subdivided into functions. The algorithms are stored as software in the ECU's program memory.

ECU functions

Motronic has two basic functions: Firstly, metering the correct mass of fuel in accordance with the air mass drawn into the engine, and secondly, triggering the ignition spark at the best possible moment in time. In this way, fuel injection and ignition can be optimally matched.

The performance of the microcontrollers used in Motronic permits a wealth of further open- and closed-loop control functions to be integrated. Progressive tightening of emissions limits simultaneously spurs the demand for functions capable of improving the engine's exhaust behavior and exhaust-gas treatment. Functions that are capable of making a contribution here include:

- Idle-speed control
- Lambda closed-loop control
- Control of the evaporative-emissions control system (canister purge)
- Knock control
- Exhaust-gas recirculation for reducing NO_x emissions, and
- Control of the secondary-air system to ensure that the catalytic converter quickly reaches full operational readiness

Where there are increased demands on the drivetrain, the system can also be extended by the following additional functions:

- Control of the exhaust-gas turbocharger, and
- Variable-tract intake manifold in order to increase engine power and torque
- Camshaft control in order to reduce exhaust-gas emissions and fuel consumption, and to increase engine power
- Torque- and speed-limiting functions to protect engine and vehicle
- Control of gasoline direct injection in order to reduce exhaust-gas emissions and fuel consumption, and to increase engine power

Ever-increasing priority is being given in the design and development of motor vehicles to the driver's comfort and convenience. This also affects engine management. Examples of typical comfort and convenience functions are:

- Cruise control (vehicle-speed controller), and
- Adaptive Cruise Control (ACC)
- Torque adaptation during gearshifts on automatic transmissions, and
- Load-reversal damping (reducing abruptness of driver control commands)

Actuator triggering

The ECU functions are executed in accordance with the algorithms stored the program memory of the Motronic ECU. This produces variables (e.g., fuel mass to be injected), which are set by means of actuators (e.g., defined-time actuation of fuel injectors). The ECU generates the electrical actuating signals for the actuators.

Torque structure

The torque-based system structure was first introduced with ME7-Motronic. All performance requirements (Fig. 2) placed on the engine are consistently converted into a torque demand. The torque coordinator prioritizes the torque demands from internal and external loads/consumers and other requirements relating to engine efficiency. The resulting desired torque is allocated to the components of the air, fuel and ignition systems.

The charge component (air system) is implemented by varying the throttle-valve aperture and – in the case of turbocharged engines – by actuating the wastegate valve. The fuel component is essentially determined by means of the injected fuel, taking account of canister purge (evaporative-emissions control system).

The torque is adjusted via two channels. The air channel (main channel) involves calculating the required cylinder charge from the torque to be converted. From the required cylinder charge, the required throttle-valve aperture is calculated. The required injected-fuel mass is directly related to the cylinder charge due to the fixed lambda value specified. The air channel only permits gradual changes in torque (e.g. integral component of idle-speed control).

The crankshaft-synchronized channel uses the cylinder charge currently available to calculate the maximum possible torque for this operating point. If the desired torque is less than the maximum possible torque, a rapid reduction in torque (e.g., differential component of idle-speed control differential component, torque reduction during gear shifting, surge damping) can be achieved by retarding the ignition or blanking out one or more cylinders altogether (injection blank-out, e.g., TCS intervention or when the engine is overrunning).

On earlier M-Motronic systems without a torque structure, a reduction in torque (e.g., at the request of the automatic transmission when changing gear) is performed directly by the function concerned, for example, by retarding the ignition angle. There is no coordination between individual requests or of command implementation.

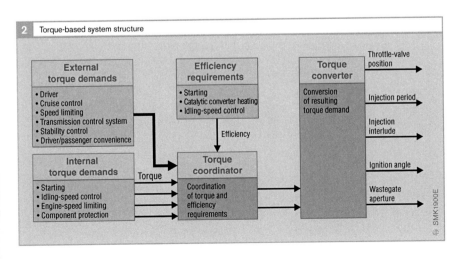

2 Torque-based system structure

Monitoring concept

It is imperative that, when the vehicle is in motion, it is never able to accelerate when the driver does not want it to. Consequently, the monitoring concept for the electronic engine-power control system must meet exacting requirements. To this end, the ECU includes a monitoring processor in addition to the main processor, and the two processors monitor one another.

Electronic diagnosis

The diagnosis functions integrated in the ECU monitor the Motronic system (ECU, sensors and actuators) for malfunctions and faults, store details of any faults detected in the data memory, and initiate substitute functions where necessary. A diagnosis lamp or a display within the instrument cluster alerts the driver to the faults.

System testers (e.g., KTS650) are connected via a diagnosis interface in the service garage/workshop. These allow the fault information stored in the ECU to be read out.

The diagnosis function was originally intended only to assist mechanics in conducting vehicle inspections and services in service garages/workshops. However, with the introduction of the Californian OBD (On-Board Diagnosis) emission-control legislation, diagnosis functions were now stipulated which check the entire engine system for exhaust-related faults and indicate these faults by way of a fault indicator lamp. Examples of such functions are catalytic-converter diagnosis, lambda-sensor diagnosis, and misfire detection. These requirements were later adopted in modified form in European legislation (EOBD).

Vehicle management

Motronic can communicate with the ECUs of other vehicle systems via bus systems, such as CAN (Controller Area Network). Figure 3 shows some examples. The ECUs can process the data from other systems in their control algorithms as input signals (e.g., Motronic reduces engine torque in response to a gearshift operation by the transmission to ensure a smoother gear change).

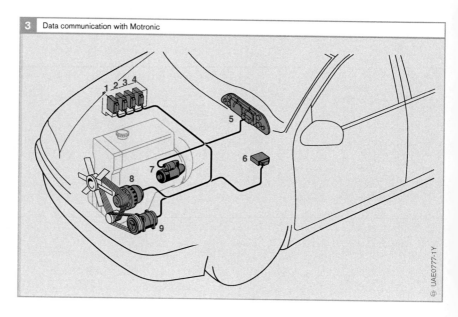

3   Data communication with Motronic

Fig. 3
1 Engine ECU (Motronic)
2 ESP ECU (Electronic Stability Program)
3 Transmission ECU
4 A/C ECU
5 Instrument cluster module with on-board computer
6 Immobilizer ECU
7 Starter
8 Alternator
9 A/C compressor

UAE0777-1Y

Simultaneously with the introduction of the Motronic systems on production vehicles, modified versions were also used on racing engines. Whereas the development objectives for production versions are aspects such as convenience, safety, reliability, emission limits and fuel consumption, the main focus in motor-racing applications is on maximum performance over a short period. The production costs with regard to choice of materials and dimensioning of components are a secondary consideration.

But the production and racing versions of the Motronic system are still based on identical principles because in both cases similar functions achieve contrasting aims. The excess-air factor and knock control systems are examples of this.

Environmental protection regulations are increasingly a consideration even in motorsport. The cars in the German Touring Car Championship, for example, are now fitted with catalytic converters. Noise and fuel-consumption levels have to be limited in more and more classes of racing. Consumption-reducing developments used on production cars quickly transfer to motor racing where shorter or less frequent refueling stops can make the difference between winning and losing. The 2001 Le Mans 24-hour race, for example, was won for the first time by a car with a Bosch gasoline direct-injection system.

The high revving speed of racing engines minimizes the time available during each operating cycle. The vast amount of process data requires high processor clock frequencies and the use of multiprocessor systems to a greater extent.

Not only the ECU but also the ignition and fuel-injection components have to operate at extremely high speeds. This requires ignition coils with fast charging times and fuel-system components that are capable of quicker throughput and higher pressures. Spark plugs with smaller thread diameters made of materials adapted to the operating temperatures encountered allow higher compression ratios.

During the race, data can be transmitted by radio from the car to the pits. Known as telemetry, this technology allows constant monitoring of operating parameters such as pressures and temperatures.

UAV0059Y

Motronic versions

Motronic comprises all the components which are needed to control a gasoline engine. The scope of the system is determined by the requirements with regard to engine power (e.g., exhaust-gas turbocharging), fuel consumption, and the stipulations of the relevant emission-control legislation. Californian emission-control and diagnosis legislation (Californian Air Resources Board, CARB) places particularly stringent requirements on the Motronic diagnosis system. Some emissions-related systems can only be diagnosed with the aid of additional components (e.g., evaporative-emissions control system).

In the course of the development history of Motronic systems, successive Motronic generations (e.g., M1, M3, ME7) have differed mainly in the design of the hardware. The basic distinguishing features are the microcontroller family, the peripheral modules and the output-stage modules (chipset). The hardware variations arising from the requirements of different vehicle manufacturers are distinguished by manufacturer-specific identification numbers (e.g., ME7.0).

In addition to the versions described in the following, there are also Motronic systems with integrated transmission management (e.g., MG and MEG-Motronic). However, these are not in extensive use, as the demands on hardware are considerable.

M-Motronic

M-Motronic is an engine-management system for manifold-injection gasoline engines. It is characterized by the fact that the air is supplied through a mechanically adjustable throttle valve.

The accelerator pedal is connected to the throttle valve by way of a linkage or a Bowden cable. The position of the accelerator pedal determines how far the throttle valve opens. This controls the air mass flowing through the intake manifold to the cylinders.

An idle actuator allows a defined air-mass flow to bypass the throttle valve. This provides a means with the additional air of holding the engine speed at a constant level, for example, when idling (idle-speed control). To do so, the engine ECU controls the opening cross-section of the bypass channel.

M-Motronic is no longer of significance to new developments in the European and North American markets, since it has been superseded by ME-Motronic.

ME-Motronic

ME-Motronic is characterized by electronic engine-power control. In this system, there is no longer a mechanical connection between the accelerator pedal and the throttle valve. The position of the accelerator pedal, i.e., the driver command, is detected by a potentiometer attached to the accelerator pedal (pedal-travel sensor in the accelerator-pedal module, Fig. 4 Pos. 23) and read in by the engine ECU (12) in the form of an analog voltage signal. In response, the ECU generates output signals that set the opening cross-section of the electrically actuated throttle valve (3) so that the engine produces the desired torque.

A system that regulates engine power in this way was first introduced by Bosch in 1986. In addition to the engine ECU, the original system also had a separate ECU for engine-power control.

The increasingly higher integration density of electronic systems allowed the combination of Motronic functions and engine-power control in a single ECU (1994). Nevertheless, functions remained divided between two microcontrollers. The next step was taken in 1998 with the launch of the new Motronic generation, the ME7, which executes all engine-management functions in a single microcontroller. This advance was made possible by the ever-increasing processing capacity of microcontrollers.

4 | Components used for open- and closed-loop electronic control of an ME-Motronic system (system diagram)

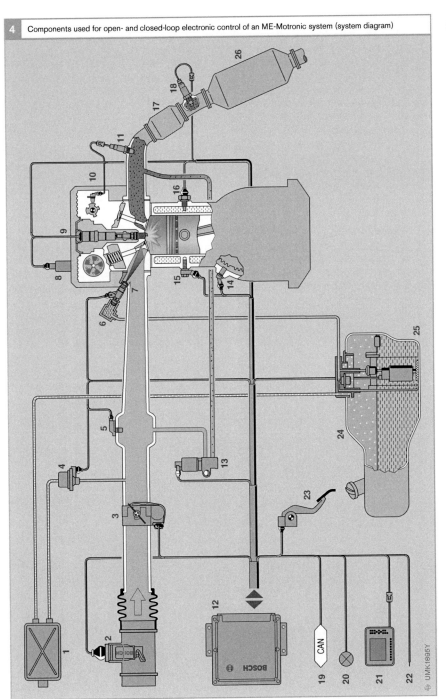

Fig. 1

1   Carbon canister
2   Hot-film air-mass
    meter with integrated
    temperature sensor
3   Throttle device (ETC)
4   Canister-purge valve
5   Intake-manifold
    pressure sensor
6   Fuel rail
7   Fuel injector
8   Actuators and
    sensors for camshaft
    control
9   Ignition coil and
    spark plug
10 Camshaft phase
    sensor
11 Lambda sensor
    upstream of primary
    catalytic converter
12 Engine ECU
13 Exhaust-gas
    recirculation valve
14 Speed sensor
15 Knock sensor
16 Engine-temperature
    sensor
17 Primary catalytic
    converter (three-way
    catalytic converter)
18 Lambda sensor
    downstream of
    primary catalytic
    converter
19 CAN interface
20 Fault indicator lamp
21 Diagnosis interface
22 Interface with
    immobilizer ECU
23 Accelerator-pedal
    module with pedal-
    travel sensor
24 Fuel tank
25 In-tank unit compris-
    ing electric fuel pump,
    fuel filter and fuel-
    pressure regulator
26 Main catalytic
    converter (three-way)

The on-board-diagnosis
system configuration
illustrated by the diagram
reflects the requirements
of EOBD

DI-Motronic

The introduction of direct injection in the gasoline engine necessitated a control concept which facilitates both homogeneous and stratified-charge operation.

In homogeneous mode, the fuel injector is actuated in such a way as to produce homogeneous mixture distribution in the combustion chamber. The fuel is injected during the induction stroke for this purpose. In stratified-charge mode, the process of retarding injection until the compression stroke, shortly before ignition, creates a locally limited mixture cloud in the area of the spark plug.

In addition to the systems which facilitate both stratified-charge and homogeneous operation, there are also purely homogeneous systems, in which the engine is operated homogeneously and stoichiometrically ($\lambda = 1$) over the entire operating range. These latter systems are increasingly being used in conjunction with supercharging.

The system diagram (Fig. 6) shows an example of a DI-Motronic system. The first series-production DI-Motronic was launched in the Volkswagen Lupo in 2000.

Operating-mode coordination and changeover

Further operating modes in addition to homogeneous and stratified-charge modes are also possible. Injection of a basic quantity of fuel during the induction stroke together with subsequent injection during the compression stroke result in a stratified charge at the spark plug, surrounded by a homogeneous lean mixture spread through the whole combustion chamber (homogeneous-lean). Further operating modes, for instance for heating the catalytic converter, are set by means of highly retarded injection points and moments of ignition.

DI-Motronic incorporates an operating-mode coordinator, which enables changeover to a different operating mode when engine requirements demand. The basis for selecting an operating mode is the operating-mode map, which plots operating mode against engine speed and torque. Deviating operating-mode requirements are evaluated in the priority list (Fig. 5). This produces the required operating mode. But before ignition and fuel injection can be changed over to the new operating mode, control functions for exhaust-gas recirculation, tank ventilation (canister purge), charge-flow control

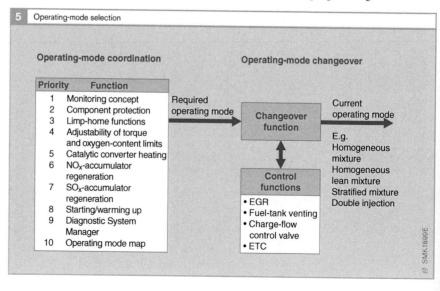

5   Operating-mode selection

Operating-mode coordination

Operating-mode changeover

Priority	Function
1	Monitoring concept
2	Component protection
3	Limp-home functions
4	Adjustability of torque and oxygen-content limits
5	Catalytic converter heating
6	NO_x-accumulator regeneration
7	SO_x-accumulator regeneration
8	Starting/warming up
9	Diagnostic System Manager
10	Operating mode map

Required operating mode →

Changeover function

Control functions
• EGR
• Fuel-tank venting
• Charge-flow control valve
• ETC

Current operating mode →

E.g.
Homogeneous mixture
Homogeneous lean mixture
Stratified mixture
Double injection

SMK1899E

6 Components used for open- and closed-loop electronic control of an DI-Motronic system (system diagram)

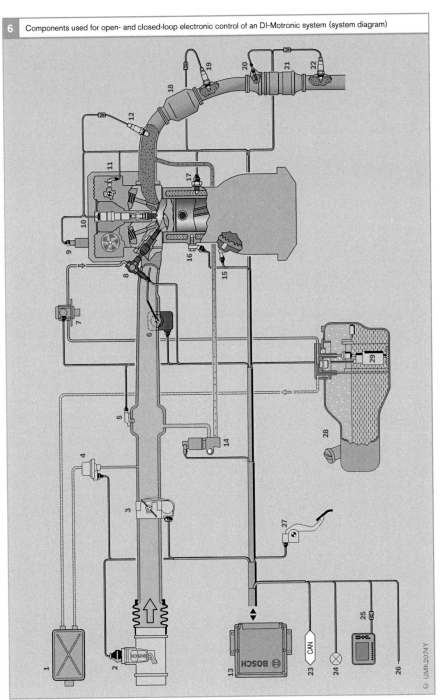

Fig. 6

1 Carbon canister
2 Hot-film air-mass
   meter
3 Throttle device (ETC)
4 Canister-purge valve
5 Intake-manifold
   pressure sensor
6 Charge-flow control
   valve
7 High-pressure pump
8 Fuel rail with high-
   pressure injector
9 Camshaft adjuster
10 Ignition coil with
   spark plug
11 Camshaft phase
   sensor
12 Lambda sensor
   (LSU)
13 Motronic ECU
14 Exhaust-gas
   recirculation valve
15 Speed sensor
16 Knock sensor
17 Engine-temperature
   sensor
18 Primary catalytic
   converter
19 Lambda sensor
20 Exhaust-gas
   temperature sensor
21 NO_X accumulator-
   type catalytic
   converter
22 Lambda sensor
23 CAN interface
24 Diagnosis lamp
25 Diagnosis interface
26 Interface to
   immobilizer ECU
27 Accelerator-pedal
   module
28 Fuel tank
29 Fuel-supply module
   with electric fuel
   pump

valve and throttle-valve setting are initiated if required. The system then waits for acknowledgement.

In stratified-charge mode at $\lambda > 1$, the throttle valve is fully open and the inducted air can enter the engine virtually unthrottled. The torque is proportional to the injected fuel mass.

During changeover to homogeneous mode, the air mass, which now determines the torque to a large extent, must be reduced very quickly and a desired lambda value set – for a stoichiometric mixture of $\lambda = 1$ (Fig. 7). The torque output by the engine now varies in relation to the accelerator-pedal position, although the driver is unaware of any change.

Brake-booster vacuum control

When the engine is operating with an unrestricted intake air flow, there is insufficient vacuum in the intake manifold to provide the vacuum required by the brake booster. A vacuum switch or pressure sensor is used to detect whether there is sufficient vacuum in the brake booster. If necessary, the engine has to be switched to a different operating mode in order to provide the required vacuum for the brake booster.

Bifuel-Motronic (natural gas/gasoline)

Bifuel-Motronic has been developed from ME-Motronic and therefore contains all the components for manifold injection familiar from ME-Motronic. Bifuel-Motronic also contains the components for the natural-gas system (Fig. 8).

Whereas in retrofit systems natural-gas operation is controlled by means of an external unit, with Bifuel-Motronic the CNG functionality is integrated in the engine-management system. The desired engine torque and the variables characterizing the operating state are generated only once in the Bifuel ECU. The physically based structure of the engine-management system makes it possible to easily integrate the parameters specific for gas operation.

Changeover

Depending on the engine design, it can be useful in the case of high load demands to switch automatically to the fuel type which provides the maximum engine power. Further automatic changeovers may also be useful in order, for example, to implement an optimized exhaust-gas strategy and to heat up the catalytic converter more quickly, or basically to effect fuel management. However, it is important in the case of automatic changeovers that these be implemented on a torque-neutral basis, i.e., they are not noticeable to the driver.

The 1-ECU concept enables fuel changeover to be performed in different ways. One option is direct changeover, comparable with a switch. Here, injected must not be interrupted, as this would increase the risk of misfiring during operation. However, the sudden injection of gas in comparison with gasoline operation results in a greater volume displacement to such an extent that the intake-manifold pressure increases and the cylinder charge decreases as a result of changeover by roughly 5 %. This displacement effect must be compensated for by a larger throttle-valve angle. In order to keep the engine torque constant during changeover under load conditions, it is necessary to effect an additional

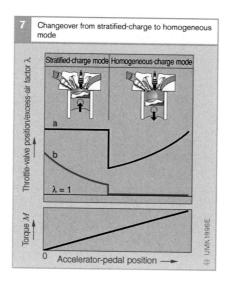

7   Changeover from stratified-charge to homogeneous mode

Stratified-charge mode | Homogeneous-charge mode

a

b

$\lambda = 1$

Throttle-valve position/excess-air factor λ

Torque M

0   Accelerator-pedal position →

UMK1898E

Fig. 7
a Throttle-valve position
b Excess-air factor λ

8   Components used for open- and closed-loop electronic control of a Bifuel-Motronic system (system diagram)

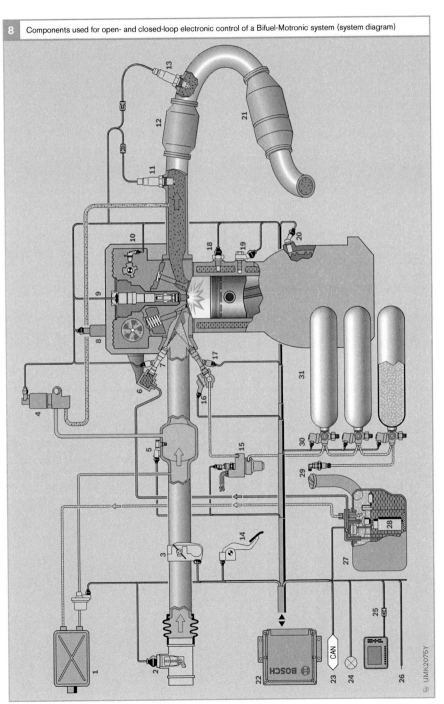

Fig. 8

1 Carbon canister with
   canister-purge valve
2 Hot-film air-mass
   meter
3 Throttle device (ETC)
4 Exhaust-gas
   recirculation valve
5 Intake-manifold
   pressure sensor
6 Fuel rail
7 Gasoline injector
8 Camshaft adjuster
9 Ignition coil with
   spark plug
10 Camshaft phase
   sensor
11 Lambda sensor
12 Primary catalytic
   converter
13 Lambda sensor
14 Accelerator-pedal
   module
15 Natural-gas pressure
   regulator
16 Natural-gas rail with
   natural-gas pressure
   and temperature
   sensor
17 Natural-gas injector
18 Engine-temperature
   sensor
19 Knock sensor
20 Speed sensor
21 Main catalytic
   converter
22 Bifuel-Motronic ECU
23 CAN interface
24 Diagnosis lamp
25 Diagnosis interface
26 Interface to
   immobilizer ECU
27 Fuel tank
28 Fuel-supply module
   with electric fuel
   pump
29 Filler neck for
   gasoline and
   natural gas
30 Tank shutoff valves
31 Natural-gas tank

intervention in the ignition angles, which facilitates a fast change of torque.

Another option for changeover is fading from gasoline to gas operation. In order to switch to gas operation, gasoline injection is reduced by a dividing factor and gas injection increased accordingly. In this way, jumps in the air charge are avoided. There is also the option of correcting an altered gas quality with lambda closed-loop control during changeover. With this method, it is possible to effect the changeover even at high load without a noticeable change of torque.

Retrofit systems often do not offer the option of switching between the gasoline and natural-gas operating modes under coordinated conditions. For this reason, many systems effect the changeover only during the overrun phases in order to avoid torque jumps.

European On-Board Diagnosis
Current EOBD legislation stipulates separate detection, handling and transmission of faults during operation with gasoline or CNG. This calls for the fault memory to be doubled in size. An alternative suggestion provides for a scenario where faults which are identified independently of the fuel (e.g., speed sensor faulty) are handled independently of the fuel. New fault paths are added for gas-specific faults so that these can also be stored and read out in the fault memory.

System structure

A few years ago, it was still possible to represent the functionality of Motronic systems with "simple" system and function descriptions. Now the open- and closed-loop control operations for gasoline engines have become so complex that a structured system description is necessary.

All torque demands placed on the engine are handled by Motronic as torque values and centrally coordinated. The requested torque is calculated and set by means of
- The electrically actuated throttle valve (air system)
- The ignition angle (ignition system)
- The fuel quantity in the case of gasoline direct injection (fuel system)
- The use of injection blank-outs, and
- Controlling the wastegate on exhaust-gas-turbocharged engines

Figure 9 shows the system structure used for new Motronic systems and their various subsystems.

In Figure 9, Motronic is referred to as the system. The different areas within the system are referred to as subsystems. Some subsystems are purely software constructs in the ECU (e.g., Torque Structure), while others also incorporate hardware components (e.g., Fuel System and Fuel Injectors). The various subsystems are interconnected by defined interfaces.

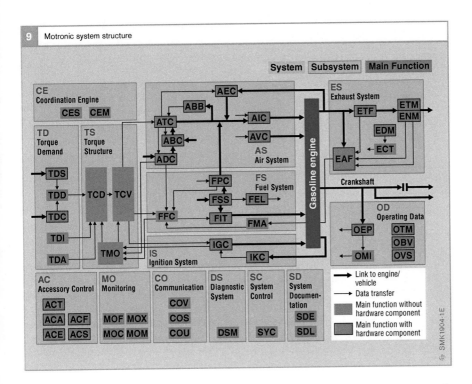

9   Motronic system structure

The system structure describes the Motronic engine-management system from the functional-sequence point of view. The system comprises the ECU (with hardware and software) and external components (actuators, sensors and mechanical components), which can be electrically connected to the ECU.

The system structure divides this mechatronic system hierarchically according to functional criteria into 14 subsystems (e.g., Air System, Fuel System), which in turn are subdivided into a total of 52 main functions (e.g., Boost-pressure Control, Closed-loop Lambda Control) (Fig. 9).

Since the introduction of Electronic Throttle Control (ETC) in the ME7, the torque demands on the engine have been centrally coordinated in the *Torque Demand* and *Torque Structure* subsystems. The control of cylinder charge by the electrically controlled throttle valve allows adjustment of the torque demand made by the driver via the accelerator pedal (driver command). At the same time, all other torque demands that arise from vehicle operation (e.g., when the A/C compressor is switched on) can be coordinated within the torque structure.

Subsystems and main functions

The description which follows provides a very general summary of the essential features of the main functions implemented in a Motronic system. A more detailed presentation is not possible within the scope of this publication.

System Documentation (SD)

System Documentation consists of technical documents which describe the customer project (e.g. description of ECUs, engine and vehicle data, and configuration descriptions).

System Control (SC)

The functions controlling the computer are combined in *System Control*.

The *System Control, SYC* main function defines the microcontroller states:
- Initialization (system run-up)
- Running state (normal operation) – this is the status in which the main functions are executed
- ECU run-on (e.g., for fan run-on, hardware test)

Coordination Engine (CE)

Both the engine status and the engine operating data are coordinated in *Coordination Engine*. This is done at a central point, because many further functionalities are affected in the overall engine-management, depending on this coordination.

The *Coordination Engine States, CES* main function contains both the various engine states, such as starting, running operation and switched-off engine, and coordination functions for injection activation (overrun fuel cutoff/restart) and for start/stop systems.

The operating modes for gasoline direct injection (DI-Motronic) are coordinated and changed over in the *Coordination Engine Operation, CEM* main function. In order to determine the required operating mode,

the requirements for various functionalities are coordinated on the basis of defined priorities in the operating-mode coordinator.

Torque Demand (TD)

All the torque demands on the engine are consistently coordinated on the torque level in the system structure of ME-Motronic and DI-Motronic. The *Torque Demand (TD)* subsystem detects all torque demands and makes them available to the *Torque Structure (TS)* subsystem as input variables.

The *Torque Demand Signal Conditioning, TDS* main function essentially consists of detecting the accelerator-pedal position. The pedal position is detected by two independent angle-position sensors and converted into a standardized accelerator-pedal angle. A number of plausibility checks are carried out to ensure that, in the event of a single fault, the standardized accelerator-pedal angle cannot adopt a greater value than the actual accelerator-pedal position.

The *Torque Demand Driver, TDD* main function calculates a setpoint value for the engine torque from the accelerator-pedal position. In addition, it defines the accelerator-pedal characteristic.

Torque Demand Cruise Control, TDC (vehicle-speed controller) holds the vehicle at a constant speed as long as the accelerator pedal is not depressed, assuming this is possible with the available engine torque. The most important shutdown conditions for this function include operating the "Off" button on the driver's control lever, operating the brakes or disengaging the clutch, and failure to reach the required minimum road speed.

Torque Demand Idle Speed Control, TDI regulates the speed of the engine at idle when the accelerator pedal is not depressed. The setpoint value for idle speed is defined to obtain even and smooth engine running at all times. Accordingly, the setpoint idle speed is set higher than the nominal idle speed under certain operating conditions

10 Excerpt from the structure diagram: *Torque Demand* and *Torque Structure* subsystems showing their main functions

nels, depending on the current operating mode.

Torque Conversion, TCV calculates from the desired-torque input variables the setpoint values for the relative air mass, the air/fuel ratio λ, the ignition angle, and injection blank-out (e.g., for overrun fuel cutoff). The setpoint air-mass value is calculated so that the setpoint for the air mass/torque is obtained at precisely the moment when the specified oxygen content and the specified ignition timing are applied.

Torque Modeling, TMO calculates a theoretically optimum indicated engine torque from the current values for cylinder charge, oxygen content (lambda), ignition timing, reduction stage, and engine speed. An indicated actual torque is determined with the aid of an efficiency chain. The efficiency chain consists of three different efficiency levels: the blank-out efficiency (proportional to the number of firing cylinders), the ignition-timing efficiency (resulting from the shift in the actual ignition angle relative to the optimum ignition angle), and the oxygen-content efficiency (obtained from plotting the efficiency characteristic against the air/fuel ratio).

(e.g., when the engine is cold). A higher idle speed may also be used to assist catalytic-converter heating, to increase the output of the A/C compressor, or if the battery charge level is low.

The *Torque Demand Auxiliary Functions, TDA* main function generates internal torque limitations and demands (e.g., engine-speed limitation, damping engine-bucking oscillations).

Torque Structure (TS)

The *Torque Structure* subsystem is where all torque demands are coordinated. The required torque is then set by the air, fuel and ignition systems.

Torque Coordination, TCD coordinates all torque demands. The various demands (e.g., from the driver, engine-speed limitation) are prioritized and converted into setpoint torque values for the various control chan-

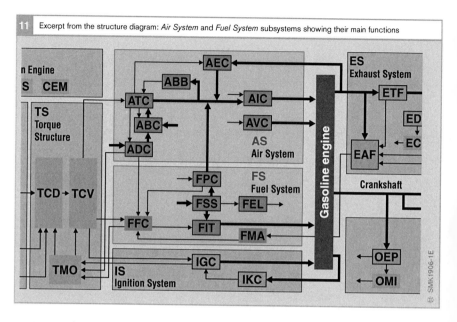

11   Excerpt from the structure diagram: *Air System* and *Fuel System* subsystems showing their main functions

Air System (AS)

The *Air System* subsystem is where the required cylinder charge for the torque to be implemented is set. In addition, the exhaust-gas recirculation, boost-pressure control, variable-tract intake-manifold geometry, charge-movement control and valve-timing functions are also part of the air system.

In *Air System Throttle Control, ATC*, the setpoint position for the throttle valve determining the air-mass flow entering the intake manifold is created from the setpoint air-mass flow.

Air System Determination of Charge, ADC determines the cylinder charge composed of fresh air and inert gas with the aid of the available load sensors. The air-mass flows are used to model the pressure conditions in the intake manifold (intake-manifold pressure model).

Air System Intake Manifold Control, AIC calculates the setpoint positions for the intake-manifold and charge-flow control valves.

The vacuum in the intake manifold allows exhaust-gas recirculation, which is calculated and adjusted in *Air System Exhaust Gas Recirculation, AEC*.

Air System Valve Control, AVC calculates the setpoint values for intake- and exhaust-valve positions and controls these settings. This influences the quantity of residual exhaust gas that is recirculated internally.

Air System Boost Control, ABC is responsible for calculating the charge-air pressure in exhaust-gas-turbocharged engines and controls the actuators for this system.

Engines with gasoline direct injection are run in stratified-charge mode with the throttle fully open at low loads. Consequently, the pressure in the intake manifold under such conditions is virtually atmospheric pressure. *Air System Brake Booster, ABB* ensures that there is sufficient vacuum in the brake booster by requesting a required amount of flow restriction.

Fuel System (FS)

The *Fuel System* (FS) subsystem calculates the output variables for the fuel-injection system relative to crankshaft position, i.e., the point(s) at which fuel is injected and the quantity of fuel injected.

Fuel System Feed Forward Control, FFC calculates the fuel mass from the setpoint cylinder charge, the setpoint oxygen content and additional corrections (e.g., transient compensation) or multiplicative corrections (e.g., corrections for engine start, warm-up and restart). Other corrections arise from the closed-loop lambda control, canister purge and mixture adaptation. In DI systems, specific values are calculated for the operating modes (e.g., fuel injection during the induction stroke or during the compression stroke, multiple injection).

Fuel System Injection Timing, FIT calculates the injection duration and the fuel-injection position. It ensures that the fuel injectors are open at the correct time relative to crankshaft rotation. The injection duration is calculated on the basis of previously calculated fuel mass and status variables (e.g., intake-manifold pressure, battery voltage, fuel-rail pressure, combustion-chamber pressure).

Fuel System Mixture Adaptation, FMA improves the pilot-control accuracy of the oxygen content by adjusting the longer-term lambda-controller errors relative to the neutral value. For smaller cylinder charges, the lambda-controller error is used to calculate an additive correction value. On systems with a hot-film air-mass meter, this normally reflects small amounts of intake-manifold leakage. On systems with an intake-manifold pressure sensor, the lambda controller corrects the pressure-sensor residual exhaust gas or offset error. For larger cylinder charges, a multiplicative correction factor is calculated. This essentially represents the hot-film air-mass meter gain error, fuel-rail pressure regulator inaccuracies (on DI systems) and fuel-injector characteristic-gradient errors.

Fuel Supply System, FSS has the function of delivering fuel from the fuel tank to the fuel rail at the required pressure and in the required quantity. In demand-controlled systems, the pressure can be regulated between 200 and 600 kPa. A pressure sensor provides feedback of the actual value.

In the case of gasoline direct injection, the fuel-supply system also includes a high-pressure circuit consisting of an HDP1-type high-pressure pump and a pressure-control valve (DSV), HDP2- and HDP5-type demand-controlled high-pressure pumps with fuel-supply control valve (MSV). This allows pressure in the high-pressure circuit to be varied between 3 and 11 MPa, depending on the engine operating point. The setpoint value is calculated depending on engine operating point and the actual pressure is detected by a high-pressure sensor.

Fuel System Purge Control, FPC controls regeneration during engine operation of the fuel that evaporates from the fuel tank and that is collected in the carbon canister of the evaporative-emissions control system. On the basis of the specified on/off ratio for operating the canister-purge valve and the pressure conditions, an actual value for the total mass flow through the valve is calculated. This is taken into account by the Air System Throttle Control (ATC) function. An actual fuel-content value is also calculated and is subtracted from the setpoint fuel mass.

Fuel System Evaporation Leakage Detection, FEL checks the gas-tightness of the fuel tank in accordance with the requirements of the Californian OBD II legislation. The design and method of operation of this diagnostic system are described in the chapter headed "Diagnosis/OBD functions".

Ignition System (IS)

The *Ignition System* subsystem calculates the output variables for ignition and actuates the ignition coils.

Ignition Control, IGC calculates the current setpoint ignition angle from the engine operating conditions, taking account of intervention by the torque structure. It then generates an ignition spark across the spark-plug electrodes at the required time. The resulting ignition angle is calculated from the basic ignition angle and the operating-point-dependent ignition-angle corrections and demands. When determining the engine-speed and load-dependent basic ignition angle, the effects of camshaft control, charge-flow control valve, cylinder-bank distribution, and special direct-injection operating modes are taken into account where applicable. In order to calculate the most advanced possible ignition angle, the basic ignition angle is corrected by the advance angles for engine warm-up, knock control and – where applicable – exhaust-gas recirculation.
The point at which the ignition driver stage needs to be triggered is calculated from the current ignition angle and the required charge time for the ignition coil. The driver stage is activated accordingly.

Ignition System Knock Control, IKC runs the engine at the knock limit for optimum efficiency, but prevents potentially damaging engine knock. The combustion process in all cylinders is monitored by means of knock sensors. The structure-borne noise signal detected by the sensors is compared with a reference level that is obtained for individual cylinders via a low-pass filter from previous combustion strokes. The reference level therefore represents the background engine noise when the engine is running free of en-

gine knock. The comparison analyzes how much louder current combustion is than the background level. Above a certain threshold, engine knock is assumed to occur. Both calculation of the reference level and detection of engine knock can take account of changes in operating conditions (engine speed, engine-speed dynamics, engine-load dynamics).

The knock-control function generates an ignition-timing adjustment for each individual cylinder. This is taken into account when calculating the current ignition angle (ignition retard). When engine knock is detected, this ignition-timing adjustment is increased by an applicable amount. The ignition-timing retard is then reduced in small increments if, over an applicable time period, engine knock does not occur.

If a hardware fault is detected, a safety function is activated (safety ignition-timing retard).

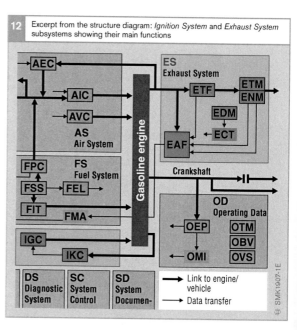

12 Excerpt from the structure diagram: *Ignition System* and *Exhaust System* subsystems showing their main functions

Exhaust System (ES)

The *Exhaust System* subsystem intervenes in the mixture-formation system, adjusts the excess-air factor, and controls the capacity utilization of the catalytic converters.

The prime functions of *Exhaust System Description and Modeling, EDM* are to model physical variables in the exhaust-gas system, analyze signals and diagnose the exhaust-gas temperature sensors (where present), and supply key exhaust-gas system data for tester output. The physical variables that are modeled are temperature (e.g., for component-protection purposes), pressure (primarily for residual-gas detection) and mass flow (for closed-loop lambda control and catalytic-converter diagnosis). In addition, the exhaust-gas excess-air factor is calculated (for NO_X accumulator-type catalytic-converter control and diagnosis).

The purpose of *Exhaust System Air Fuel Control, EAF* using the lambda sensor upstream of the front catalytic converter is to regulate the excess-air factor to a specified level. This minimizes harmful emissions, prevents engine-torque fluctuations and keeps the exhaust-gas composition on the right side of the lean-mixture limit. The input signals from the closed-loop lambda control system downstream of the main catalytic converter allows further minimization of emissions.

The *Exhaust System Three-Way Front Catalyst, ETF* main function uses the lambda sensor downstream of the front catalytic converter (if fitted). Its signal is a measure of the oxygen content in the exhaust gas and serves as the basis for reference-value regulation and catalytic-converter diagnosis. Reference-value regulation can substantially improve mixture control and permit optimum conversion response by the catalytic converter.

The *Exhaust System Three-Way Main Catalyst, ETM* main function basically operates in the same way as the ETF function described above. Reference-value regulation, however, may take different forms depend-ing on the system. A NO_X accumulator-type catalytic converter operated at $\lambda = 1$ displays optimum conversion response with a specific oxygen-accumulator content. The reference-value regulation function sets the capacity usage to this level. Deviations are corrected by compensation components.

The function of *Exhaust System NO_X Main Catalyst, ENM* is to ensure that NO_X emission requirements in particular are complied with when the engine is running on a lean mixture by means of adapted control of the mixture to the requirements of the NO_X accumulator-type catalytic converter.

Depending on the condition of the catalytic converter, the NO_X storage phase is terminated and the engine switched over to an operating mode ($\lambda < 1$) in which the NO_X accumulator is emptied and the stored NO_X emissions converted to N_2. Regeneration of the NO_X accumulator-type catalytic converter is terminated in response to the change of signal from the sensor downstream of the NO_X accumulator-type catalytic converter. In systems with a NO_X accumulator-type catalytic converter, changeover to a special mode allows desulfurization of the catalytic converter.

Exhaust System Control of Temperature, ECT controls the temperature of the exhaust-gas system. Its aim is to speed up the time it takes the catalytic converters to reach operating temperature after the engine is started (catalytic-converter heating), prevent the catalytic converters from cooling down during operation (catalytic-converter temperature retention), heat up the NO_X accumulator-type catalytic converter for desulfurization, and prevent thermal damage to exhaust-gas system components (component protection). A torque reserve for the TS (*Torque Structure*) subsystem is determined from the heat flow required for a temperature increase. The temperature increase is then achieved by retarding the ignition, for example. When the engine is idling, the heat flow can also be increased by raising the idle speed.

Operating Data (OD)

The *Operating Data* subsystem records
all important engine operating parameters,
checks their plausibility and provides substitute data where required.

*Operating Data Engine Position Manage-
ment, OEP* calculates the position of the
crankshaft and the camshaft using the processed input signals from the crankshaft and
camshaft sensors. It calculates the engine
speed from this information. The crankshaft
timing wheel (two missing teeth) and the
characteristics of the camshaft signal are
used to synchronize the engine and the
ECU and to monitor synchonization while
the engine is running.

The camshaft signal pattern and the engine shutoff position are analyzed in order
to optimize the start time. This allows rapid
synchronization.

*Operating Data Temperature Measure-
ment, OTM* processes the temperature readings provided by the temperature sensors,
performs plausibility checks on them, and
provides substitute data in the event of
faults. The ambient temperature and the
engine-oil temperature may also be detected
in addition to the temperature of the engine
and the intake air. The input voltage signals
are assigned to a temperature reading. This
is followed by calculation of a characteristic
curve.

Operating Data Battery Voltage, OBV
function is responsible for providing the
supply-voltage signals and performing diagnostic operations on them. The raw signal
is detected at terminal 15 and, if necessary,
the main relay.

Misfire Detection Irregular Running, OMI
monitors the engine for ignition and combustion misses (see the chapter entitled
"Diagnosis/OBD functions").

Operating Data Vehicle Speed, OVS is responsible for detecting, conditioning and
diagnosing the vehicle speed signal. This
variable is needed, among others, for cruise
control, for speed limiting (v_{max}) and in the
case of the hand switch for gear recognition.

Depending on the configuration, there is
the option of using the variables supplied
via the CAN by the instrument cluster or
the ABS/ESP ECU.

Communication (CO)

The *Communication* subsystem encompasses
all Motronic main functions that communicate with other systems.

Communication User Interface, COU provides the connection to diagnostic (e.g., engine analyzer) and calibration equipment.
Communication takes place via the K-line,
though the CAN interface can also be used
for this purpose. Different communication
protocols are available for various applications (e.g., KWP2000, McMess).

Communication Vehicle Interface, COV
looks after communication with other
ECUs, sensors and actuators.

Communication Security Access, COS
provides for communication with the immobilizer and – as an option – enables
access control for reprogramming the
Flash-EPROM.

Accessory Control (AC)

The *Accessory Control* subsystem controls
the auxiliary systems.

Accessory Control Air Condition, ACA controls operation of the A/C compressor and
analyzes the signals from the pressure sensor
in the air conditioner. The A/C compressor
is switched on when, for example, a request
is received from the driver or the A/C ECU
via a switch. The A/C ECU signals to the
Motronic that the A/C compressor needs to
be switched on. It is switched on shortly af-
terwards. When the engine is idling, the engine-management system has sufficient time
to develop the required torque reserves.

Various conditions can result in the air
conditioner being switched off (e.g., critical
pressure in the air conditioner, fault in the
pressure sensor, low ambient temperature).

Accessory Control Fan Control, ACF controls the radiator fan in response to demand and detects faults in the fan and the control system. Under certain circumstances, the fan may be required to run on when the engine is not running.

Accessory Control Thermal Management, ACT regulates the engine temperature according to operating conditions. The required engine temperature is determined depending on engine power, driving speed, engine operating state, and ambient temperature. This helps the engine to reach its operating temperature more quickly and is then adequately cooled. The coolant volumetric flow through the radiator is calculated and the map-controlled thermostat is operated accordingly based on the temperature setpoint.

Accessory Control Electrical Machines, ACE is responsible for controlling the "electrical machines", i.e. the starter motor and alternator.

The function *Accessory Control Steering, ACS* is to control the power-steering pump.

Monitoring (MO)

Function Monitoring, MOF monitors all Motronic elements that affect engine torque and speed. The core function is torque comparison. This compares the permissible torque calculated on the basis of driver request with the actual torque calculated from the engine data. If the actual torque is too large, suitable measures are initiated to ensure that a controllable status is re-established.

Monitoring Module, MOM combines all the monitoring functions that contribute to or perform reciprocal monitoring between the function processor and the monitoring module. The function processor and the monitoring module are components of the ECU. Reciprocal monitoring between them takes place by means of continuous query-and-response communication.

Microcontroller Monitoring, MOC combines all the monitoring functions that can detect a fault or malfunction in the processor and its peripherals. Examples include:
- Analog-digital converter test
- Memory test for RAM and ROM
- Program-run monitoring
- Command test

Extended Monitoring, MOX contains functions for expanded function monitoring. These function determine the maximal torque which the engine can plausibly output.

Diagnostic System (DS)

Component and system diagnosis are performed by the main functions of the subsystems. The *Diagnostic System* (DS) is responsible for coordinating the various diagnosis results.

The function of the *Diagnostic System Manager (DSM)* is to:
- Store details of faults and associated ambient conditions
- Switch on the malfunction indicator lamp
- Establish communication with the diagnostic tester
- Coordinate execution of the various diagnostic functions (taking account of priorities and bars) and verify faults

Sensors

Sensors register operating states (e.g. engine speed) and setpoint/desired values (e.g. accelerator-pedal position). They convert physical quantities (e.g. pressure) or chemical quantities (e.g. exhaust-gas concentration) into electric signals.

Automotive applications

Sensors and actuators represent the interfaces between the ECUs, as the processing units, and the vehicle with its complex drive, braking, chassis, and bodywork functions (for instance, the Engine Management, the Electronic Stability Program ESP, and the air conditioner). As a rule, a matching circuit in the sensor converts the signals so that they can be processed by the ECU.

The field of mechatronics, in which mechanical, electronic, and data-processing components are interlinked and cooperate closely with each other, is rapidly gaining in importance in the field of sensor engineering. These components are integrated in modules (e.g. in the crankshaft CSWS (Composite Seal with Sensor) module complete with rpm sensor).

Since their output signals directly affect not only the engine's power output, torque, and emissions, but also vehicle handling and safety, sensors, although they are becoming smaller and smaller, must also fulfill demands that they be faster and more precise. These stipulations can be complied with thanks to mechatronics.

Depending upon the level of integration, signal conditioning, analog/digital conversion, and self-calibration functions can all be integrated in the sensor (Fig. 1), and in future a small microcomputer for further signal processing will be added. The advantages are as follows:

- Lower levels of computing power are needed in the ECU
- A uniform, flexible, and bus-compatible interface becomes possible for all sensors
- Direct multiple use of a given sensor through the data bus
- Registration of even smaller measured quantities
- Simple sensor calibration

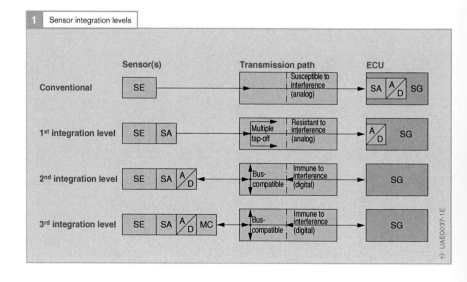

Fig. 1
SE Sensor(s)
SA Analog signal
    conditioning
A/D Analog-digital
    converter
SG Digital ECU
MC Microcomputer
    (evaluation
    electronics)

Temperature sensors

Applications

Engine-temperature sensor
This is installed in the coolant circuit
(Fig. 1). The engine management uses its
signal when calculating the engine temperature (measuring range −40...+130 °C).

Air-temperature sensor
This sensor is installed in the air-intake
tract. Together with the signal from the
boost-pressure sensor, its signal is applied in
calculating the intake-air mass. Apart from
this, desired values for the various control
loops (e.g. EGR, boost-pressure control)
can be adapted to the air temperature
(measuring range −40...+120 °C).

Engine-oil temperature sensor
The signal from this sensor is used in calculating the service interval (measuring range
−40...+170 °C).

Fuel-temperature sensor
Is incorporated in the low-pressure stage
of the diesel fuel circuit. The fuel temperature is used in calculating the precise injected fuel quantity (measuring range
−40...+120 °C).

Exhaust-gas temperature sensor
This sensor is mounted on the exhaust system at points which are particularly critical
regarding temperature. It is applied in the
closed-loop control of the systems used for
exhaust-gas treatment. A platinum measuring resistor is usually used (measuring range
−40...+1000 °C).

Design and method of operation

Depending upon the particular application,
a wide variety of temperature sensor designs
are available. A temperature-dependent
semiconductor measuring resistor is fitted
inside a housing. This resistor is usually
of the NTC (Negative Temperature Coefficient, Fig. 2) type. Less often a PTC (Positive
Temperature Coefficient) type is used. With
NTC, there is a sharp drop in resistance
when the temperature rises, and with PTC
there is a sharp increase.

The measuring resistor is part of a voltage-
divider circuit to which 5 V is applied.
The voltage measured across the measuring
resistor is therefore temperature-dependent.
It is inputted through an analog to digital
(A/D) converter and is a measure of the
temperature at the sensor. A characteristic
curve is stored in the engine-management
ECU which allocates a specific temperature
to every resistance or output-voltage.

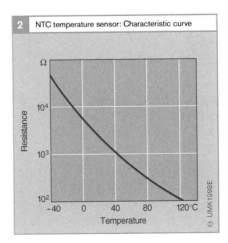

| 1 | Coolant temperature sensor |

1 cm

| 2 | NTC temperature sensor: Characteristic curve |

Resistance

10^4

10^3

10^2

−40   0   40   80   120 °C

Temperature

Fig. 1
1   Electrical
    connections
2   Housing
3   Gasket
4   Thread
5   Measuring resistor
6   Coolant

Engine-speed sensors

Application
Engine-speed sensors are used in Motronic systems for
- Measuring the engine speed, and
- Determining the crankshaft position (position of the engine pistons)

The rotational speed is calculated from the internal between the sensor's signals.

Inductive speed sensors
Design and method of operation
The sensor is mounted directly opposite a ferromagnetic trigger wheel (Fig. 1, Pos. 7) from which it is separated by a narrow air gap. It has a soft-iron core (pole pin, Pos. 4), which is enclosed by a winding (5). The pole pin is also connected to a permanent magnet (1), and a magnetic field extends through the pole pin and into the trigger wheel. The level of the magnetic flux through the coil depends on whether the sensor is opposite a trigger-wheel tooth or gap. Whereas the magnet's leakage flux is concentrated by a tooth, and leads to an increase in the working flux through the coil, it is weakened by a gap. When the trigger wheel rotates, these magnetic-flux changes induce a sinusoidal output voltage in the coil which is proportional to the rate of change of the flux and thus the engine speed (Fig. 2). The ampli-tude of the alternating voltage increases sharply along with increasing trigger-wheel speed (several mV... >100 V). At least about 30 rpm are needed to generate an adequate amplitude.

The number of teeth on the trigger wheel depends on the particular application. In Motronic systems, a 60-pitch trigger wheel is normally used, although 2 teeth are omitted (7) so that the trigger wheel has 60 – 2 = 58 teeth. The gap where the missing teeth would be situated is allocated to a defined crankshaft position and serves as a reference mark for synchronizing the ECU.

The geometries of the trigger-wheel teeth and the pole pin must be matched to each other. An evaluation circuit in the ECU converts the sinusoidal voltage, which is characterized by strongly varying amplitudes, into a constant-amplitude square-wave voltage for evaluation in the ECU microcontroller.

Active speed sensors
Active speed sensors operate according to the magnetostatic principle. The amplitude of the output signal is not dependent on the rotational speed. This makes it possible for very low speeds to be sensed (quasistatic speed sensing).

Fig. 1
1   Permanent magnet
2   Sensor housing
3   Crankcase
4   Pole pin
5   Winding
6   Air gap
7   Trigger wheel with reference mark

Fig. 2
1   Tooth
2   Tooth gap
3   Reference mark

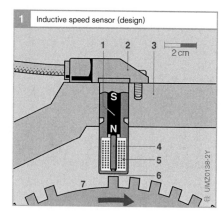

1   Inductive speed sensor (design)

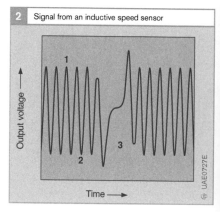

2   Signal from an inductive speed sensor

Differential Hall-effect sensor

A voltage U_H proportional to the magnetic field (Hall voltage) can be picked off horizontally to the current direction at a current-carrying plate which is permeated vertically by a magnetic induction B (Fig. 3). In a differential Hall-effect sensor, the magnetic field is generated by a permanent magnet (Fig. 4, Pos. 1). Two Hall-effect sensor elements (2 and 3) are situated between the magnet and the trigger wheel (4). The magnetic flux by which these are permeated depends on whether the sensor is opposite a tooth or a gap. By establishing the difference between the signals from the two sensors, it is possible to

- Reduce magnetic interference signals, and
- Obtain an improved signal-to-noise ratio

The edges of the sensor signal can be processed without digitization directly in the ECU.

Multipole wheels are used instead of the ferromagnetic trigger wheel. Here, a magnetizable plastic is attached to a non-magnetic metallic carrier and alternately magnetized. These north and south poles adopt the function formerly performed by the teeth of the trigger wheel.

AMR sensors

The electrical resistance of magnetoresistive material (AMR, Anisotropic Magneto Resistive) is anisotropic, i.e., it depends on the direction of the magnetic field to which it is exposed. This property is utilized in an AMR sensor. The sensor is located between a magnet and a trigger wheel. The field lines change direction when the trigger wheel rotates (Fig. 5). This generates a sinusoidal voltage, which is amplified in an evaluation circuit in the sensor and converted into a square-wave signal.

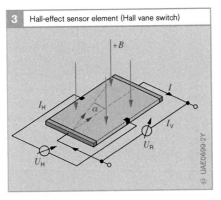

3 Hall-effect sensor element (Hall vane switch)

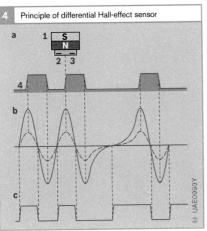

4 Principle of differential Hall-effect sensor

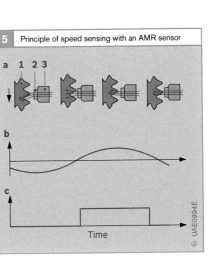

5 Principle of speed sensing with an AMR sensor

Fig. 3
I Plate current
I_H Hall current
I_V Supply current
U_H Hall voltage
U_R Longitudinal voltage
B Magnetic induction
α Deflection of the electrons by the magnetic field

Fig. 4
a   Arrangement
b   Signal of Hall-effect sensor
– high amplitude with small air gap
– low amplitude with large air gap
c   Output signal

1   Magnet
2   Hall-effect sensor 1
3   Hall-effect sensor 2
4   Trigger wheel

Fig. 5
a   Arrangement at different times
b   Signal from AMR sensor
c   Output signal

1   Trigger wheel
2   Sensor element
3   Magnet

Hall-effect phase sensors

Application

The engine's camshaft rotates at half the crankshaft speed. Taking a given piston on its way to TDC, the camshaft's rotational position is an indication as to whether the piston is in the compression or exhaust stroke.

The phase sensor on the camshaft provides the ECU with this information. This is required, for example, for ignition systems with single-spark ignition coils and for Sequential fuel injection (SEFI).

Design and method of operation

Hall-effect rod sensors

Hall-effect rod sensors (Fig. 1a) utilize the Hall effect: A rotor made ferromagnetic material (Pos. 7, trigger wheel with teeth, segments or aperture plate) rotates along with the camshaft. The Hall-effect IC (6) is located between the trigger wheel and a permanent magnet (5), which generates a magnetic field strength perpendicular to the Hall-effect element.

If one of the trigger-wheel teeth (Z) now passes the current-carrying sensor element (semiconductor plate), it changes the magnetic-field strength perpendicular to the Hall-effect element. This results in a voltage signal (Hall voltage) which is independent of the relative speed between sensor and trigger wheel. The evaluation electronics integrated in the sensor's Hall-effect IC conditions the signal and outputs it in the form of a square-wave signal (Fig. 1b).

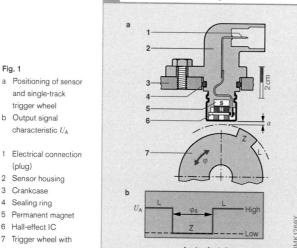

1 Hall-effect rod sensor (design)

Fig. 1

a Positioning of sensor and single-track trigger wheel

b Output signal characteristic U_A

1 Electrical connection (plug)
2 Sensor housing
3 Crankcase
4 Sealing ring
5 Permanent magnet
6 Hall-effect IC
7 Trigger wheel with tooth/segment (Z) and gap (L)

a Air gap
φ Angle of rotation

Fig. 2

TIM = Twist Intensive Mounting (i.e., the sensor can be rotated as desired about the sensor axis without any loss of accuracy. Important for minimizing type diversity).
TPO = True Power On (i.e., the sensor detects directly on switching on whether it is located above a tooth or a gap. Important for short synchronization times between crankshaft and camshaft signals).

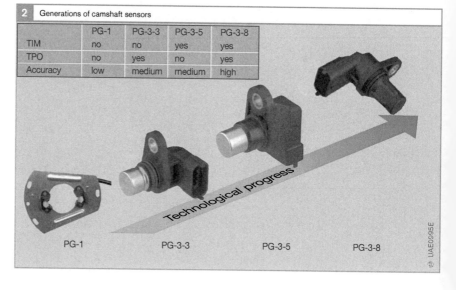

2 Generations of camshaft sensors

	PG-1	PG-3-3	PG-3-5	PG-3-8
TIM	no	no	yes	yes
TPO	no	yes	no	yes
Accuracy	low	medium	medium	high

Technological progress

PG-1   PG-3-3   PG-3-5   PG-3-8

▶ Miniaturization

Thanks to micromechanics it has become possible to locate sensor functions in the smallest possible space. Typically, the mechanical dimensions are in the micrometer range. Silicon, with its special characteristics, has proved to be a highly suitable material for the production of the very small, and often very intricate mechanical structures. With its elasticity and electrical properties, silicon is practically ideal for the production of sensors. Using processes derived from the field of semiconductor engineering, mechanical and electronic functions can be integrated with each other on a single chip or using other methods.

Bosch was the first to introduce a product with a micromechanical measuring element for automotive applications. This was an intake-pressure sensor for measuring load, and went into series production in 1994. Micromechanical acceleration and yaw-rate sensors are more recent developments in the field of miniaturisation, and are used in driving-safety systems for occupant protection and vehicle dynamics control (**E**lectronic **S**tability **P**rogram, ESP). The illustrations below show quite clearly just how small such components really are.

▼ Micromechanical acceleration sensor

Electric circuit

Bonding wire   Sensor chip

Evaluation
circuit

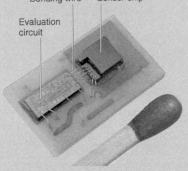

Comb-like structure compared to an insect's head

Suspension spring   Seismic mass with
movable electrodes

200 µm   Fixed electrodes   UAE0787E

▼ Micromechanical yaw-rate sensor

DRS-MM1 vehicle-dynamics control (ESP)

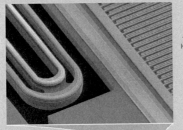

100 µm

DRS-MM2 roll-over sensing, navigation

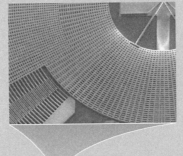

3.3 cm

UAE0788Y

Hot-film air-mass meter

Application

To provide precise pilot control of the air/fuel ratio, it is essential for the supplied air mass to be exactly determined in the respective operating state. The hot-film air-mass meter measures some of the actually inducted air-mass flow for this purpose. It takes into account the pulsations and reverse flows caused by the opening and closing of the engine's intake and exhaust valves. Intake-air temperature or air-pressure changes have no effect upon measuring accuracy.

HFM5 design

The housing of the HFM5 hot-film air-mass meter (Fig. 1, Pos. 5) extends into a measuring tube (2), which can have different diameters depending on the air mass required for the engine (370…970 kg/h).

The measuring tube normally contains a flow rectifier, which ensures that the flow in the measuring tube is uniform. The flow rectifier is either a combination of a plastic mesh with straightening action and a wire mesh, or is a wire mesh on its own (Fig. 3, Pos. 8). The measuring tube is installed in the intake tract downstream from the air filter. Plug-in versions are also available which are installed inside the air filter.

The most important components in the sensor are the measuring cell (Fig. 1, Pos. 4) in the air inlet (8) and the integrated evaluation electronics (3).

The sensor measuring cell consists of a semiconductor substrate. The sensitive surface is formed by a diaphragm which has been manufactured in micromechanical processes. This diaphragm incorporates temperature-sensitive resistors. The elements of the evaluation electronics (hybrid circuit) are installed on a ceramic substrate. This principle permits very compact design. The evaluation electronics is connected to the ECU by means of electrical connections (1).

The partial-flow measuring passage (6) is shaped so that the air flows past the measuring cell smoothly (without whirl effects) and back into the measuring tube via the air outlet (7). The length and location of the inlet and outlet of the partial-flow measuring passage have been chosen to provide good sensor performance even in the event of sharply pulsating flows.

Method of operation

The HFM5 hot-film air-mass meter is a thermal sensor which operates according tot he following principle: A centrally situated heating resistor on the measuring cell (Fig. 3, Pos. 3) heats a sensor diaphragm (5) and maintains it at a constant temperature. The temperature drops sharply on each side of this controlled heating zone (4).

Fig. 1

1   Electrical connections (plug)
2   Measuring-tube or air-filter housing wall
3   Evaluation electronics (hybrid circuit)
4   Measuring cell
5   Sensor housing
6   Partial-flow measuring passage
7   Outlet, partial air flow Q_M
8   Inlet, partial air flow Q_M

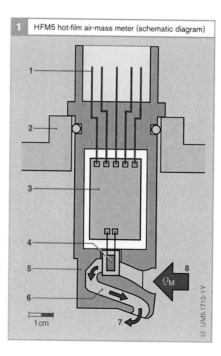

1   HFM5 hot-film air-mass meter (schematic diagram)

1 cm

The temperature distribution on the diaphragm is registered by two temperature-dependent resistors which are mounted upstream and downstream of the heating resistor so as to be symmetrical to it (measuring points M1, M2). Without the flow of incoming air, the temperature profile (1) is the same on each side of the heating zone ($T_1 = T_2$).

As soon as air flows over the measuring cell, the uniform temperature profile at the diaphragm changes (2). On the inlet side, the temperature characteristic is steeper since the incoming air flowing past this area cools it off. On the opposite side, the temperature characteristic only changes slightly, because the incoming air flowing past has been heated by the heater element. The change in temperature distribution leads to a temperature differential (ΔT) between the measuring points M1 and M2.

The heat dissipated to the air, and therefore the temperature characteristic at the measuring cell is dependent on the air mass flowing past. The temperature differential is (irrespective of the absolute temperature of the air flow past) a measure of the air-flow mass. It is also direction-dependent so that the air-mass sensor can record both the amount and the direction of an air-mass flow. Due to its very thin micromechanical diaphragm, the sensor has a highly dynamic response (<15 ms), a point which is of particular importance when the incoming air is pulsating heavily.

The evaluation electronics integrated in the sensor converts the resistance differential at the measuring points M1 and M2 into an analog voltage signal of between 0 and 5 V. Using the sensor characteristic (Fig. 2) stored in the ECU, the measured voltage is converted into a value representing the air-mass flow (kg/h).

The shape of the characteristic curve is such that the diagnosis facility incorporated in the ECU can detect such malfunctions as an open-circuit line. An additional temperature sensor for evaluation functions can be integrated in the HFM5. It is not required for measuring the air mass.

Incorrect air-mass readings will be registered if the sensor diaphragm is contaminated with dust, dirty water or oil. For the

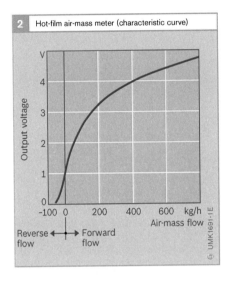

2 Hot-film air-mass meter (characteristic curve)

Reverse ←→ Forward
flow   flow

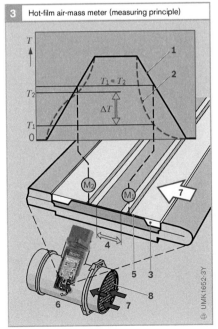

3 Hot-film air-mass meter (measuring principle)

Fig. 3
1 Temperature profile without air flow
2 Temperature profile with air flow
3 Measuring cell
4 Heating zone
5 Sensor diaphragm
6 Measuring tube with air-mass sensor
7 Intake-air flow
8 Wire mesh

M_1, M_2 Measuring points
T_1, T_2 Temperature values at measuring points M_1 and M_2
ΔT Temperature differential

purpose of increasing the robustness of the HFM5, a protective device has been developed which, in conjunction with a deflector mesh, keeps dirty water and dust away from the sensor element (HFM5-CI; with C-shaped bypass and inner tube (I), which together with the deflector mesh protects the sensor).

HFM6 hot-film air-mass meter

The HFM6 uses the same sensor element as the HFM5 and has the same basic design. It differs in two crucial points:
- The integrated evaluation electronics operates digitally in order to obtain greater measuring accuracy
- The design of the partial-flow measuring passage is altered to provide protection against contamination directly upstream of the sensor element (similar to the deflector mesh in the HFM5-CI)

Digital electronics

A voltage signal is generated with a bridge circuit from the resistance values at the measuring points M1 and M2 (Fig. 3); this voltage signal serves as the measure of the air mass. The signal is converted into digital form for further processing.

The HFM6 also takes into account the temperature of the intake air when determining the air mass. This increases significantly the accuracy of the air-mass measurement.

The intake-air temperature is measured by a temperature-dependent resistor, which is integrated in the closed control loop for monitoring the heating-zone temperature. The voltage drop at this resistor produces with the aid of an analog-digital converter a digital signal representing the intake-air temperature. The signals for air mass and intake-air temperature are used to address a program map in which the correction values for the air-mass signal are stored.

Improved protection against contamination

The partial-flow measuring passage is divided into two sections in order to provide better protection against contamination (Fig. 4). The passage which passes the sensor element has a sharp edge (1), around which air must flow. Heavy particulates and dirty-water droplets are unable to follow this diversion and are separated from the partial flow. These contaminants exit the sensor through a second passage (5). In this way, significantly fewer dirt particulates and droplets reach the sensor element (3) with the result that contamination is reduced and the service life of the air-mass sensor is significantly prolonged even when operated with contaminated air.

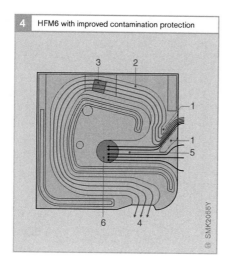

Fig. 4
1 Diverting edge
2 Partial-flow
   measuring passage
   (first passage)
3 Sensor element
4 Air outlet
5 Second passage
6 Particulate and
   water outlet

4 HFM6 with improved contamination protection

Piezoelectric knock sensors

Application
In terms of their principle of operation, knock sensors are basically vibration sensors and are suitable for detecting structure-borne acoustic oscillations. These occur as "knock", for instance, in gasoline engines when uncontrolled combustion takes place. They are converted by the knock sensor into electrical signals (Fig. 1) and transmitted to the Motronic ECU, which counteracts the engine knock by adjusting the ignition angle.

Design and method of operation
Due to its inertia, a mass (Fig. 2, Pos. 2) excited by a given oscillation or vibration exerts a compressive force on a toroidal piezoceramic element (1) at the same frequency as the excitation oscillation. These forces effect a charge transfer within the ceramic element. An electrical voltage is generated between the top and bottom of the ceramic element which is picked off via contact washers (5) and processed in the Motronic ECU.

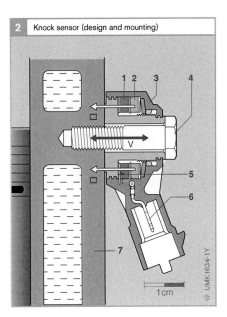

2 Knock sensor (design and mounting)

1 cm

Fig. 2
1 Piezoceramic element
2 Seismic mass with compressive forces F
3 Housing
4 Bolt
5 Contact surface
6 Electrical connection
7 Engine block

V Vibration

Mounting
In four-cylinder engines, one knock sensor is sufficient to record the knock signals for all the cylinders. Engines with more cylinders require two or more knock sensors. The knock-sensor installation point on the engine is selected so that knock can be reliably detected from each cylinder. The sensor is usually bolted to the side of the engine block. It must be possible for the generated signals (structure-borne-noise vibrations) to be introduced without resonance into the knock sensor from the measuring point on the engine block. A fixed bolted connection satisfying the following requirements is required for this purpose:
- The fastening bolts must be tightened to a defined torque
- The contact surface and the bore in the engine block must comply with prespecified quality requirements
- No washers of any type may be used for fastening purposes

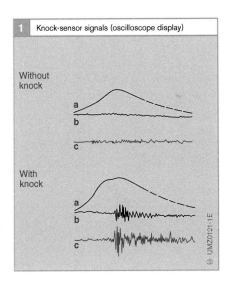

1 Knock-sensor signals (oscilloscope display)

Without knock

a
b

c

With knock

a
b

c

Fig. 1
a Cylinder-pressure curve
b Filtered pressure signal
c Knock-sensor signal

Micromechanical pressure sensors

Application

Pressure is a non-directional force acting in all directions which occurs in gases and liquids. Micromechanical pressure sensors detect the pressure of various media in the motor vehicle, e.g.:

- Intake-manifold pressure, e.g., for load sensing in engine-management systems
- Charge-air pressure for boost-pressure control
- Ambient pressure for taking into account air density, e.g., in boost-pressure control
- Oil pressure for taking into account engine load in the service display
- Fuel pressure for monitoring the level of fuel-filter contamination

Micromechanical pressure sensors determine the absolute pressure of liquids and gases by measuring the pressure differential in relation to a reference vacuum.

Version with the reference vacuum on the component side

Design

The measuring cell of a micromechanical pressure sensor consists of a silicon chip (Fig. 1a, Pos. 2), in which a thin diaphragm is micromechanically etched (1). Four deformation resistors (R1, R2) are diffused into the diaphragm. The electrical resistance in these resistors changes in response to mechanical elongation. The measuring cell is surrounded and sealed on its component side by a cap (Fig. 2, Pos. 6), which encloses the reference vacuum underneath.

A temperature sensor (Fig. 3, Pos. 1), the signals of which can be evaluated independently, can also be integrated in the pressure-sensor housing.

Method of operation

The diaphragm of the sensor cell is deflected to varying degrees depending on the external pressure acting on it (the center of the diaphragm is deflected by 10...1000 μm). The four deformation resistors on the diaphragm change their electrical resistance as a function of the mechanical stress resulting from the applied pressure (piezoresistive effect).

The four measuring resistors are arranged on the silicon chip so that when diaphragm deformation takes place, the resistance of two of them increases and that of the other two decreases. These measuring resistors are arranged in a Wheatstone bridge circuit (Fig. 1b) and a change in their resistance

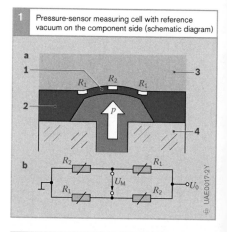

1   Pressure-sensor measuring cell with reference vacuum on the component side (schematic diagram)

a

b

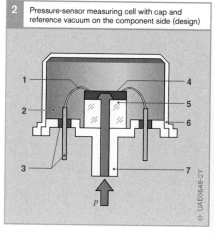

2   Pressure-sensor measuring cell with cap and reference vacuum on the component side (design)

Fig. 1
a   Sectional drawing
b   Bridge circuit

1   Diaphragm
2   Silicon chip
3   Reference vacuum
4   Glass (Pyrex)

p Measured pressure
U_0 Supply voltage
U_M Measured voltage
R_1 Deformation resistor (compressed)
R_2 Deformation resistor (elongated)

Fig. 2
1, 3   Electrical connections with glass-enclosed lead-in
2   Reference vacuum
4   Measuring cell (chip) with evaluation electronics
5   Glass base
6   Cap
7   Supply for measured pressure p

values leads to a change in the ratio of the voltages across them. This leads to a change in the measurement voltage U_M. This as yet unamplified voltage is therefore a measure of the pressure applied to the diaphragm.

The measurement voltage is higher with a bridge circuit than would be the case if an individual resistor were used. The Wheatstone bridge circuit thus permits a higher sensor sensitivity.

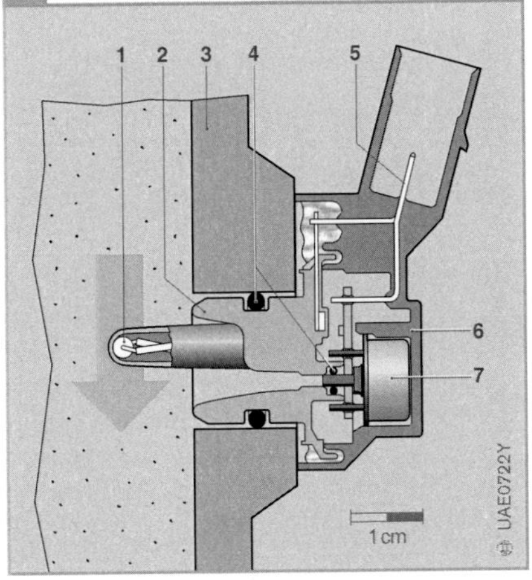

The signal-conditioning electronic circuitry is integrated on the chip. Its function is to amplify the bridge voltage, compensate for temperature influences, and linearize the pressure curve. The output voltage is in the range of 0...5 V and is supplied via electrical connections to the engine ECU (Fig. 3, Pos. 5). The ECU uses this output voltage to calculate the pressure.

Version with the reference vacuum in a special chamber

Design

A pressure sensor with the reference vacuum in a special chamber (Fig. 4) for use as an intake-manifold or boost-pressure sensor is easier to install than a sensor with the reference vacuum on the component side: A silicon chip (Fig. 5, Pos. 6) with etched diaphragm and four deformation resistors in a bridge circuit is located – like the pressure sensor with cap and reference vacuum on the component side – as a measuring cell on a glass base (3). In contrast to the sensor with the reference vacuum on the component side, there is no passage in the glass base through which the measured pressure can be applied to the sensor element. Instead, pressure is applied to the silicon chip from the side on which the evaluation electronics

---

3 Micromechanical pressure sensor with reference vacuum on the component side (design)

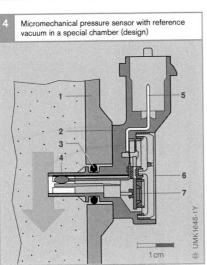

1 cm

---

4 Micromechanical pressure sensor with reference vacuum in a special chamber (design)

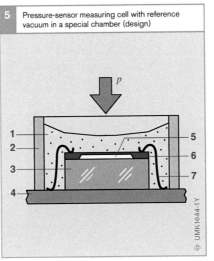

1 cm

Fig. 3
1 Temperature sensor (NTC)
2 Housing base
3 Manifold wall
4 Sealing rings
5 Electrical connection (plug)
6 Housing cover
7 Measuring cell

Fig. 4
1 Manifold wall
2 Housing
3 Sealing ring
4 Temperature sensor (NTC)
5 Electrical connection (plug)
6 Housing cover
7 Measuring cell

Fig. 5
1 Protective gel
2 Gel frame
3 Glass base
4 Ceramic hybrid
5 Chamber with reference vacuum
6 Measuring cell (chip) with evaluation electronics
7 Bonded connection

p Measured pressure

---

5 Pressure-sensor measuring cell with reference vacuum in a special chamber (design)

is situated. This means that a special gel must be used on this side of the sensor to protect it against environmental influences. The reference vacuum (5) is enclosed in the cavity (special chamber) between the silicon chip and the glass base. The complete measuring element is mounted on a ceramic hybrid (4), which incorporates the soldering surfaces for electrical contacting inside the sensor.

A temperature sensor can also be incorporated in the pressure-sensor housing. It protrudes into the air flow, and can therefore respond to temperature changes with a minimum of delay (Fig. 4, Pos. 4).

Method of operation
The method of operation, and with it the signal conditioning and signal amplification together with the characteristic curve, correspond to that used in the pressure sensor with cap and reference vacuum on the component side. The only difference is that the measuring cell's diaphragm is deformed in the opposite direction and therefore the deformation resistors are "bent" in the other direction.

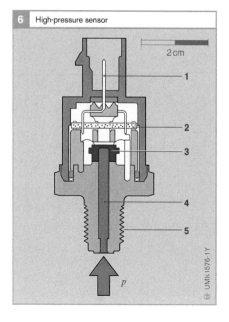

6   High-pressure sensor

2 cm

— 1

— 2

— 3

— 4

— 5

p

UMK1576-1Y

High-pressure sensors

Application
High-pressure sensors are used in a motor vehicle to measure fuel pressure and brake-fluid pressure, e.g:
- Rail-pressure sensor for gasoline direct injection (pressure up to 200 bar)
- Rail-pressure sensor for common-rail diesel-injection system (pressure up to 2000 bar)
- Brake-fluid pressure sensor in the hydraulic modulator of the Electronic Stability Program (pressure up to 350 bar)

Design and method of operation
High-pressure sensors operate according to the same principle as micromechanical pressure sensors. The core of the sensor is a steel diaphragm, onto which deformation resistors have been vapor-deposited in the form of a bridge circuit (Fig. 6, Pos. 3). The sensor's measuring range is dependent on the thickness of the diaphragm (thicker diaphragms for higher pressures, thinner diaphragms for lower pressures). As soon as the pressure to be measured is applied to one side of the diaphragm via the pressure port (4), the deformation resistors change their resistance values as a result of the deflection of the diaphragm. The output voltage generated by the bridge circuit is proportional to the applied pressure. This voltage is transmitted via connecting leads (bonding wires) to an evaluation circuit (2) in the sensor. The evaluation circuit amplifies the bridge signal to 0...5 V and transmits it to the ECU, which calculates the pressure with the aid of a characteristic curve.

▶ Micromechanics

Micromechanics is defined as the application of semiconductor techniques in the production of mechanical components from semiconductor materials (usually silicon). Not only silicon's semiconductor properties are used but also its mechanical characteristics. This enables sensor functions to be implemented in the smallest-possible space. The following techniques are used:

Bulk micromechanics

The silicon wafer material is processed at the required depth using anisotropic (alkaline) etching and, where needed, an electrochemical etching stop. From the rear, the material is removed from inside the silicon layer (Fig. 1, Pos. 2) at those points underneath an opening in the mask. Using this method, very small diaphragms can be produced (with typical thicknesses of between 5 and 50 μm, as well as openings (b), beams and webs (c) as are needed for instance for acceleration sensors.

Surface micromechanics

The substrate material here is a silicon wafer on whose surface very small mechanical structures are formed (Fig. 2). First of all, a "sacrificial layer" is applied and structured using semiconductor processes such as etching (A). An approx. 10 μm polysilicon layer is then deposited on top of this (B) and structured vertically using a mask and etching (C). In the final processing step, the "sacrificial" oxide layer underneath the polysilicon layer is removed by means of gaseous hydrogen fluoride (D). In this manner, the movable electrodes for acceleration sensors (Fig. 3) are exposed.

Wafer bonding

Anodic bonding and sealglass bonding are used to permanently join together (bonding) two wafers by the application of tension and heat or pressure and heat. This is needed for the hermetic sealing of reference vacuums for instance, and when protective caps must be applied to safeguard sensitive structures.

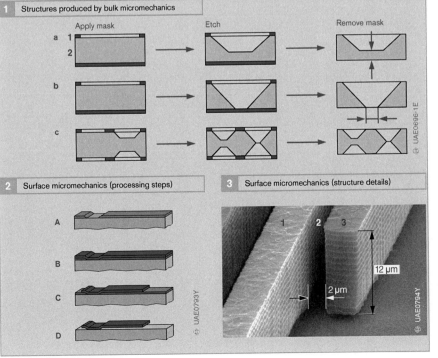

1 Structures produced by bulk micromechanics

Apply mask | Etch | Remove mask

a 1 2
b
c

2 Surface micromechanics (processing steps)

A
B
C
D

3 Surface micromechanics (structure details)

12 μm
2 μm

Fig. 1
a Diaphragms
b Openings
c Beams and webs

1 Etching mask
2 Silicon

Fig. 2
A Cutting and structuring the sacrificial layer
B Cutting the polysilicon
C Structuring the polysilicon
D Removing the sacrificial layer

Fig. 3
1 Fixed electrode
2 Gap
3 Spring electrodes

Two-step lambda oxygen sensors

Application

These sensors are used in gasoline engines equipped with two-step lambda control. They project between the engine exhaust and the catalytic converter into the exhaust pipe and simultaneously detect the exhaust-gas flow from all the cylinders. Because the lambda sensor is heated, it can be installed further away from the engine so that even extended periods of driving at full load present no problems. The LSF4 sensor is also suitable for use in exhaust systems with several sensors (e.g., OBD II).

Two-step lambda sensors compare the residual-oxygen content in the exhaust gas with the oxygen content in the reference atmosphere (surrounding air inside the sensor) and indicate whether a rich ($\lambda < 1$) or lean mixture ($\lambda > 1$) is present in the exhaust gas. The sudden jump in the characteristic curve of these sensors permits mixture control to $\lambda = 1$ (Fig. 1).

Method of operation

Two-step lambda oxygen sensors operated on the principle of a galvanic oxygen concentration cell with a solid electrolyte (Nernst principle). The ceramic element is conductive for oxygen ions from a temperature of approximately 350 °C (safe, reliable operation at > 350 °C). Due to the abrupt change in the residual-oxygen content on the exhaust-gas side in the range of $\lambda = 1$ (e.g., $9 \cdot 10^{-15}$ % vol. for $\lambda = 0.99$ and 0.2 % vol. for $\lambda = 1.01$), the different oxygen content on both sides of the sensor generates an electrical voltage between the two boundary layers. This means that the oxygen content in the exhaust gas can be used as a measure of the air/fuel ratio. The integrated heater ensures that the sensor functions even at extremely low exhaust-gas temperatures.

The voltage output by the sensor U_S is dependent on the oxygen content in the exhaust gas. In the case of a rich mixture ($\lambda < 1$), it reaches 800...1000 mV, and, in the case of a lean mixture ($\lambda > 1$), it reaches only about 100 mV. The transition from rich to lean occurs at $U_{reg} = 450...500$ mV.

The temperature of the ceramic element influences its ability to conduct the oxygen ions, and thus the shape of the output-voltage curve as a function of the excess-air factor λ (the values in Fig. 1 are therefore temperature-dependent). In addition, the response time for a voltage change when the mixture composition changes is also strongly dependent on temperature.

Whereas these response times at ceramic-element temperatures of below 350 °C are in the seconds range, the sensor responds at optimum operating temperatures of around 600 °C in less than 50 ms. When an engine is started, therefore, lambda closed-loop control is deactivated until the minimum operating temperature of about 350 °C is reached. During this period, the engine is open-loop-controlled.

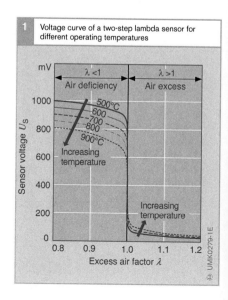

1   Voltage curve of a two-step lambda sensor for different operating temperatures

Fig. 1
a   Rich mixture
     (air deficiency)
b   Lean mixture
     (excess air)

Design

LSH25 finger-type sensor

Sensor ceramic element with protective tube

The solid electrolyte is a ceramic element which is impermeable to gas. It is composed of a mixed oxide of the elements zirconium and yttrium in the shape of a tube closed off at one end (finger). The surfaces have been provided on both sides with electrodes made from a microporous, thin noble-metal coating.

The platinum electrode on the exhaust side, which protrudes into the exhaust pipe, acts like a small catalytic converter: Exhaust gas which reaches this electrode is treated catalytically and brought to a stoichiometric balance ($\lambda = 1$). In addition, the side that is exposed to the exhaust gas has a porous, ceramic multiple layer (spinel layer) to protect it against contamination and erosive damage. The ceramic element is also

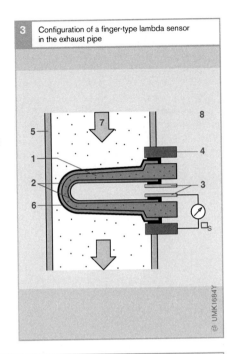

3   Configuration of a finger-type lambda sensor in the exhaust pipe

Fig. 3
1 Sensor ceramic element
2 Electrodes
3 Contacts
4 Housing contact
5 Exhaust pipe
6 Ceramic protective layer (porous)
7 Exhaust gas
8 Outside air

U_S Sensor voltage

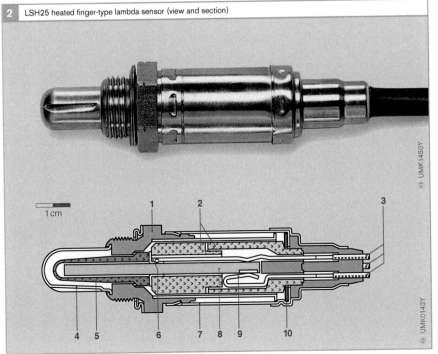

2   LSH25 heated finger-type lambda sensor (view and section)

1 cm

Fig. 2
1 Sensor housing
2 Ceramic support tube
3 Connecting cable
4 Protective tube with slots
5 Active sensor ceramic element
6 Contact element
7 Protective sleeve
8 Heater element
9 Clamp-type connections for the heater element
10 Disc spring

protected against mechanical impact and thermal shocks by a metal tube. Various slots in the protective tube are designed in such a way that they, on the one hand, provide particularly effective protection and thermal and chemical loads, and, on the other hand, prevent the ceramic element from cooling excessively when the exhaust gas is "cool".

The sensor's "open" inner chamber facing away from the exhaust gas is connected to the outside air, which acts as a reference gas (Fig. 3).

Sensor element with heater element and electrical connection
A ceramic support tube and a disc spring hold and seal the active, finger-shaped sensor ceramic element in the sensor housing. A contact element between the support tube and the active sensor ceramic element provides the contact between the inner electrode and the connecting cable.

The outer electrode is connected to the sensor housing by the metal sealing ring. A protective metal sleeve, which at the same time serves as the support for the disc spring, locates and fixes the sensor's complete inner structure. It also protects the sensor interior against contamination.

The connecting cable is crimped to the contact element which protrudes from the sensor, and is protected against moisture and mechanical damage by a temperature-resistant cap.

The finger-type sensor (Fig. 2) is also equipped with an electrical heater element. This ensures that the ceramic-element temperature remains sufficiently high, even at low engine load and thus low exhaust-gas temperature.

This external heating is so quick that the sensor reaches operating temperature 20...30 s after the engine has started and therefore lambda closed-loop control can come into operation. Finally, sensor heating provides for an optimal ceramic-element operating temperature above the operating limit of 350 °C and thus ensures low and stable exhaust-gas emissions.

Fig. 4
1 Porous protective layer
2 Outer electrode
3 Sensor foil
4 Inner electrode
5 Reference-air-passage foil
6 Insulation layer
7 Heater
8 Heater foil
9 Connection contacts

Fig. 5
1 Exhaust gas
2 Porous ceramic protective layer
3 Measuring cell with microporous noble-metal coating
4 Reference-air passage
5 Heater

U_A Output voltage

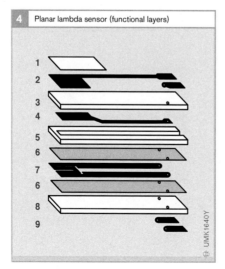

4 | Planar lambda sensor (functional layers)

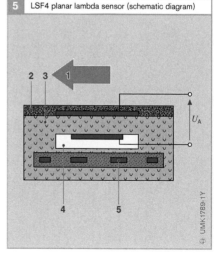

5 | LSF4 planar lambda sensor (schematic diagram)

LSF4 planar lambda sensor

In terms of its function, the planar lambda sensor corresponds to the heated finger-type sensors with a voltage-jump curve at $\lambda = 1$. However, on the planar sensor, the solid electrolyte is comprised of a number of individual laminated foils stacked one on top of the other (Fig. 4). The sensor is protected against thermal and mechanical influences by a double-walled protective tube.

The planar ceramic element (measuring cell and heater are integrated) is shaped like a long stretched-out wafer with rectangular cross-section.

The surfaces of the measuring cell are provided with a microporous noble-metal coating, which also has a porous ceramic coating on the exhaust-gas side to protect it against the erosive effects of the exhaust-residues. The heater is a wave-shaped element containing noble metal. It is integrated and insulated in the ceramic wafer and ensures that the sensor heats up quickly even in the event of low power input.

The reference-air passage inside the LSF4 lambda sensor – operating as a reference-gas sensor (Figs. 5 and 6) – has access to the ambient air. It can therefore compare the residual oxygen in the exhaust gas with the oxygen in the reference atmosphere, i.e., the ambient air inside the sensor. Thus, the planar-sensor voltage also demonstrates an abrupt change (Fig. 1) in the area of the stoichiometric composition of the air/fuel mixture ($\lambda = 1$).

6   LSF4 planar lambda sensor (view and section)

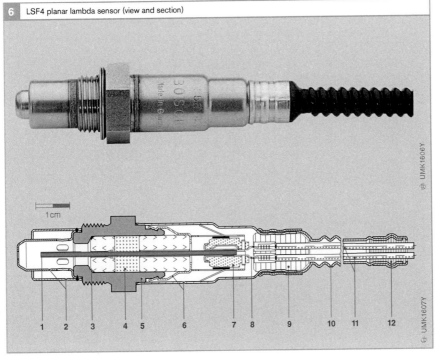

1 cm

UMK1606Y

UMK1607Y

Fig. 6
1 Planar measuring
   cell
2 Double protective
   tube
3 Sealing ring
4 Seal pack
5 Sensor housing
6 Protective sleeve
7 Contact holder
8 Contact clip
9 PTFE grommet
10 PTFE shaped sleeve
11 Five connecting
   leads
12 Seal

LSU4 planar broad-band lambda oxygen sensor

Application

As its name implies, the broad-band lambda sensor is used across a very extensive range to determine the oxygen concentration in the exhaust gas. The figures provided by the sensor are an indication of the air/fuel ratio in the combustion chamber. The excess-air factor λ describes this air/fuel ratio.

The sensor projects into the exhaust pipe and registers the exhaust-gas mass flow from all cylinders. It is capable of making precise measurements not only at the stoichiometric point at $\lambda = 1$, but also in the lean ($\lambda > 1$) and rich ($\lambda < 1$) ranges. In conjunction with control electronics, it delivers an unmistakable, continuous electrical signal in the range of $0.7 < \lambda < \infty$ (air with 21 % O_2) (Fig. 3). These characteristics enable the broad-band lambda sensor to be used not only in engine-management systems with two-step control ($\lambda = 1$), but also in control concepts with lean and rich air/fuel mixtures. This type of lambda sensor is therefore also suitable for lambda closed-loop control with lean-burn concepts on gasoline engines, as well as for diesel engines, gas-powered engines and gas-powered central

heaters and water heaters (hence the German designation LSU: L̲ambda-S̲onde-U̲niversal = universal lambda sensor).

In a number of systems, several lambda sensors are installed for even greater accuracy. Here, for instance, they are fitted upstream and downstream of the catalytic converter as well as in the individual exhaust tracts (cylinder banks).

Design

The LSU4 broad-band lambda sensor (Fig. 2) is a planar dual-cell limit-current sensor. It features a measuring cell (Fig. 1) made of zirconium-dioxide ceramic (ZrO_2), and is a combination of a Nernst concentration cell (sensor cell which functions in the same way as a two-step lambda sensor) and an oxygen pump cell for transporting the oxygen ions. The oxygen pump cell (Fig. 1, Pos. 8) is arranged in relation to the Nernst concentration cell (7) in such a way that there is a 10...50 µm diffusion gap (6). Here, there are two porous platinum electrodes: one pump electrode and one Nernst measuring electrode. The diffusion gap is connected to the exhaust gas by way of a gas-access passage (10). A porous diffusion barrier (11) serves to limit the flow of oxygen molecules from the exhaust gas.

On the one side, the Nernst concentration cell is connected to the surrounding atmo-

Fig. 1

1 Exhaust gas
2 Exhaust pipe
3 Heater
4 Control electronics
5 Reference cell with reference-air passage
6 Diffusion gap
7 Nernst concentration cell with Nernst measuring electrode (on diffusion-gap side) and reference electrode (on reference-cell side)
8 Oxygen pump cell with pump electrode
9 Porous protective layer
10 Gas-access passage
11 Porous diffusion barrier

I_P Pump current
U_P Pump voltage
U_H Heater voltage
U_{Ref} Reference voltage (450 mV, corresponds to $\lambda = 1$)
U_S Sensor voltage

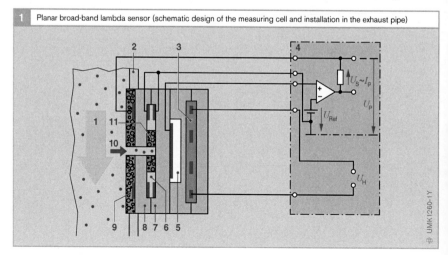

1   Planar broad-band lambda sensor (schematic design of the measuring cell and installation in the exhaust pipe)

sphere by a reference-air passage (5), and on the other, it is connected to the exhaust gas in the diffusion gap.

The sensor requires control-electronics circuitry to generate the sensor signal and to regulator the sensor temperature.

An integrated heater (3) heats the sensor so that it quickly reaches the operating temperature of 650...900 °C which is required for a signal that can be evaluated. This function drastically reduces the influence of the exhaust-gas temperature on the sensor signal.

Method of operation

The exhaust gas enters the actual measuring chamber (diffusion gap) of the Nernst concentration cell through the pump cell's small gas-access passage. In order that the excess-air factor λ can be adjusted in the diffusion gap, the Nernst concentration cell compares the gas in the diffusion gap with the ambient air in the reference-air passage.

The complete process proceeds as follows: By applying the pump voltage U_P across the pump cell's platinum electrodes, oxygen from the exhaust gas can be pumped through the diffusion barrier and into or out of the diffusion gap. With the aid of the Nernst concentration cell, an electronic circuit in the ECU controls the voltage (U_P) across the pump

cell in order that the composition of the gas in the diffusion gap remains constant at $\lambda = 1$. If the exhaust gas is lean, the pump cell pumps the oxygen to the outside (positive pump current). On the other hand, if the exhaust gas is rich, the oxygen (due to the decomposition of CO_2 and H_2O at the exhaust-gas electrode) is pumped from the surrounding exhaust gas and into the diffusion gap (negative pump current). At $\lambda = 1$, no oxygen needs to be transported, and the pump current is zero. The pump current is proportional to the oxygen concentration in the exhaust gas and is thus a (non-linear) measure of the excess-air factor λ (Fig. 3).

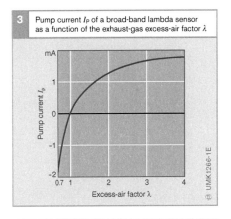

3 Pump current I_P of a broad-band lambda sensor as a function of the exhaust-gas excess-air factor λ

UMK1266-1E

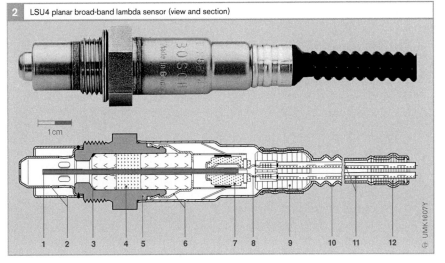

2 LSU4 planar broad-band lambda sensor (view and section)

1 cm

UMK1607Y

Fig. 2
1 Measuring cell (combination of Nernst concentration cell and oxygen pump cell)
2 Double protective tube
3 Sealing ring
4 Seal pack
5 Sensor housing
6 Protective sleeve
7 Contact holder
8 Contact clip
9 PTFE grommet
10 PTFE shaped sleeve
11 Five connecting leads
12 Seal

Electronic control unit (ECU)

Digital technology furnishes an extensive array of options for open and closed-loop control of automotive electronic systems. A large number of parameters can be included in the process to support optimal operation of various systems. After receiving the electric signals transmitted by the sensors, the ECU processes these data in order to generate control signals for the actuators. The software program for closed-loop control is stored in the ECU's memory. The program is executed by a microcontroller. The ECU and its components are referred to as hardware. The Motronic ECU contains all of the algorithms for open and closed-loop control needed to govern the engine-management processes (ignition, induction and mixture formation, etc.).

Operating conditions

The ECU operates in an extremely harsh and demanding environment.
It is exposed to
- Extreme temperatures
  (ranging from -40 to $+60...+125\,°C$)
  under normal operating conditions
- Abrupt temperature variations
- Exposure to fluids (oil, fuel, etc.)
- The effects of moisture and
- Mechanical stresses such as
  engine vibration

The engine-management ECU must continue to perform flawlessly in the face of fluctuations in electrical supply, during starts with a weak battery (cold starts, etc.) as well as at high voltages (surges in onboard electrical system).

Other requirements arise from the need for EMC (Electro-Magnetic Compatibility). The requirements for resistance to electromagnetic interference and for suppressing EMI emissions from the system itself are both very high.

Design

The printed-circuit board with the electrical components (Fig. 1) is installed in a housing of plastic or metal. A multipin plug connects the ECU to the sensors, actuators and electrical power supply. The high-power driver circuits that provide direct control of the actuators are specially integrated within the housing to ensure effective heat transfer to the housing and the surrounding air.

Most of the electronic components are SMDs (Surface-Mounted Devices). This concept provides extremely efficient use of space in low-weight packages. Only the power elements and the plugs are mounted using conventional insertion technology.

Hybrid versions combining compact dimensions with extreme resistance to thermal attack are available for mounting directly on the engine.

Data processing

Input signals
The sensors join the actuators as the peripheral components linking the vehicle and the central processing device, the engine-management ECU. The electrical signals from the sensors travel through the wiring harness and the plug to reach the control unit. These signals can be in various forms:

Analog input signals
Analog input signals can have any voltage level within a specific range. Samples of physical parameters monitored as analog data include induction air mass, battery voltage and intake-manifold pressure (including boost pressure) as well as the temperatures of the coolant and induction air. An analog/digital (A/D) converter within the control unit's microcontroller transforms the signal data into the digital form required by the microcontroller's central processing unit. The maximum resolution of these analog signals is

5 mV. This translates into roughly 1000 incremental graduations based on an overall monitoring range of 0...5 V.

Digital input signals

Digital input signals have only two conditions: high (logical 1) and low (logical 0). Samples of digital input signals are switch control signals (on/off) and digital sensor signals such as the rotational-speed pulses from Hall-effect and magnetoresistive sensors. The microcontroller can process these signals without prior conversion.

Pulse-shaped input signals

The pulse-shaped input signals with information on rpm and reference marks transmitted by inductive sensors are conditioned in special circuitry within the ECU. In this process interference pulses are suppressed while the actual signal pulses are converted to digital square-wave signals.

Signal conditioning

Protective circuits limit the voltages of incoming signals to levels suitable for conditioning. Most of the superimposed interference signals are removed from the useful signal by filters. When necessary, the useful signals are then amplified to the input voltage required by the microcontroller (0...5 V).

Some or all of this initial conditioning can be carried out in the sensor itself, depending upon its level of integration.

Signal processing

The ECU is the switching center governing all of the functions and sequences regulated by the engine-management system. The control algorithms are executed by the microcontroller. The input signals from sensors and interfaces linking other systems (from CAN bus, etc.) serve as the input parameters. The processor runs backup plausibility checks on these data. The ECU program supports calculation of the output signals used to control the actuators.

---

1   ECU structure, using ME Motronic as an example (housing cover cut open)

UAE0992Y

Microcontroller
The microcontroller is the central component of a control unit and controls its operative sequence (Fig. 2). Apart from the CPU (Central Processing Unit), the microcontroller contains not only the input and output channels, but also timer units, RAMs, ROMs, serial interfaces, and further peripheral assemblies, all of which are integrated on a single microchip. Quartz-controlled timing is used for the microcontroller.

Program and data memory
In order to carry out the computations, the microcontroller needs a program – the "software". This is in the form of binary numerical values arranged in data records and stored in a program memory.

These binary values are accessed by the CPU which interprets them as commands which it implements one after the other.

This program is stored in a Read Only Memory (ROM, EPROM, or Flash-EPROM) which also contains variant-specific data (individual data, characteristic curves, and maps). This is non-variable data which cannot be changed during vehicle operation. It is used to regulate the program's open and closed-loop control processes.

The program memory can be integrated in the microcontroller and, depending upon the particular application, extended in a separate component (e.g. an external EPROM or a Flash-EPROM).

ROM
Program memories can be in the form of a ROM (**Read Only Memory**). This is a memory whose contents have been defined permanently during manufacture and thereafter remain unalterable. The ROM installed in the microcontroller only has a restricted memory capacity, which means that an additional ROM is required in case of complicated applications.

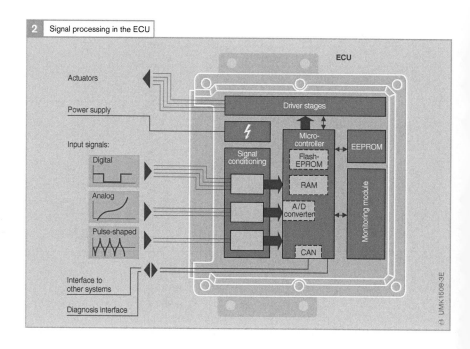

2   Signal processing in the ECU

EPROM

The data on an EPROM (Erasable Programmable **ROM**) can be erased by subjecting the device to UV light. Fresh data can then be entered using a programming unit. The EPROM is usually in the form of a separate component, and is accessed by the CPU through the Address/Data-Bus.

Flash-EPROM (FEPROM)

The contents of the Flash-EPROM can be electrically erased. In the process, the ECU is connected to the reprogramming unit through a serial interface.

If the microcontroller is also equipped with a ROM, this contains the programming routines for the Flash programming. Flash-EPROMs are available which, together with the microcontroller, are integrated on a single microchip.

Its decisive advantages have helped the Flash-EPROM to largely supersede the conventional EPROM.

Variable-data or main memory

Such a read/write memory is needed in order to store such variable data (variables) as the computational and signal values.

RAM

Instantaneous values are stored in the RAM (Random Access Memory) read/write memory. If complex applications are involved, the memory capacity of the RAM incorporated in the microcontroller is insufficient so that an additional RAM module becomes necessary. It is connected to the ECU through the Address/Data-Bus.

When the control unit is disconnected from the power supply, all data stored in the RAM is lost (volatile memory). However, the next time the engine is started the control unit has to have access to adaptation data (learned data relating to engine condition and operating status). That information must not be lost when the ignition is switched off. In order to prevent that happening, the RAM is permanently connected to the power supply (continuous power supply). If the battery is disconnected, however, the information will nevertheless be lost.

EEPROM (also known as the E²PROM)

Data that must not be lost even if the battery is disconnected (e.g. important adaptation data, codes for the immobilizer) must be permanently stored in a nonvolatile, non-erasable memory. The EEPROM is an electrically erasable EPROM in which (in contrast to the Flash-EPROM) every single memory location can be erased individually. Therefore, the EEPROM can be used as a non-volatile random-access memory.

Some control-unit variants also use separately erasable areas of the Flash-EPROM as nonvolatile memories.

ASIC

The ever-increasing complexity of ECU functions means that the computing powers of the standard microcontrollers available on the market no longer suffice. The solution here is to use so-called ASIC modules (Application Specific Integrated Circuit). These IC's are designed and produced in accordance with data from the ECU development departments and, as well as being equipped with an extra RAM for instance, and inputs and outputs, they can also generate and transmit pwm signals.

Monitoring module

The ECU is provided with a monitoring module. Using a "Question and Answer" cycle, the microcontroller and the monitoring module supervise each other, and as soon as a fault is detected one of them triggers appropriate back-up functions independent of the other.

Output signals

With its output signals, the microcontroller triggers driver stages which are usually powerful enough to operate the actuators directly. For particularly high power consumers (e.g. radiator fan) some driver stages can also operate relays.

The driver stages are proof against shorts to ground or battery voltage, as well as against destruction due to electrical or thermal overload. Such malfunctions, together with open-circuit lines or sensor faults are identified by the driver-stage IC as an error and reported to the microcontroller.

Switching signals

These are used to switch the actuators on and off (for instance, for the engine fan).

PWM signals

Digital output signals can be in the form of PWM (Pulse-Width Modulated) signals. Such "pulse-width modulated" signals are square-wave signals with a constant frequency and variable signal duration (Fig. 3). These signals can be used to move a variety of actuators (e.g. exhaust recirculation valve, turbocharger actuator) to any desired working position.

Communication within the ECU

In order to be able to support the microcontroller in its work, the peripheral components must communicate with it. This takes place using an address/data bus which, for instance, the microcomputer uses to issue the RAM address whose contents are to be accessed. The data bus is then used to transmit the relevant data. For former automotive applications, an 8-bit structure sufficed whereby the data bus comprised 8 lines which together can transmit 256 values simultaneously. The 16-bit address bus commonly used with such systems can access 65,536 addresses. Presently, more complex systems demand 16 bits, or even 32 bits, for the data bus. In order to save on pins at the components, the data and address buses can be combined in a multiplex system.

That is, data and addresses are dispatched through the same lines but offset from each other with respect to time.

Serial interfaces with only a single data line are used for data which need not be transmitted so quickly (e.g. data from the fault storage).

EoL programming

The extensive variety of vehicle variants with differing control programs and data records, makes it imperative to have a system which reduces the number of ECU types needed by a given manufacturer. To that end, the Flash-EPROM's entire memory area can be programmed at the end of the production line with the program and the variant-specific data record (this is referred to as End-of-Line, or EoL, programming).

A further means of reducing variant diversity is to have a number of data variants (e.g. gearbox variants) stored in the memory, which can then be selected by encoding at the end of the production line. This coding is stored in an EEPROM.

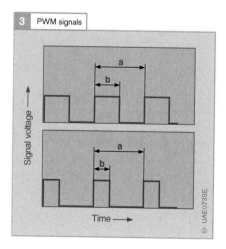

3 PWM signals

Signal voltage →

Time →

UAE0738E

Fig. 3

a Period duration
   (fixed or variable)

b Variable on-time

The performance of Electronic Control Units (ECUs) goes hand-in-hand with advances achieved in the field of microelectronics.
The first gasoline fuel-injection systems used analog technology and as such, they were not so versatile when it came to implementing control functions. These functions were constrained by the hardware.

Progress advanced in quantum leaps with the arrival of digital technology and the microcontroller. The entire engine management system was taken over by the universally applicable semiconductor microchip. In microcontroller-based systems, the actual control logic is accommodated on a programmable semiconductor memory chip.

From systems that initially simply controlled fuel injection, complex engine-management systems were then developed. They controlled not only fuel injection but also the ignition system including knock control, exhaust-gas recirculation and a whole variety of other systems. This continuous process of development is bound to continue in a similar vein over the next decade as well. The integration of functions and, above all, their complexity are constantly increasing. This pattern of development is only possible because the microcontrollers used are also undergoing a similar process of improvement.

For a long time microcontrollers of the Intel 8051 family were used until they were superseded at the end of the 1980s by the 80515 series which had extra input/output capabilities for timed signals and an integrated analog-digital converter. It was then possible to create relatively powerful systems. Figure 1 shows a comparison between the performance of a fuel-injection system (LH3.2) and an ignition system (EZ129K) – equipped with 80C515 controllers – and that of the succeeding Motronic systems. With a clock speed of 40 MHz, the ME7 has almost 40 times the processing power of the LH/EZ combination. With the benefit of a new generation of microcontrollers and a further increase in clock frequency on the ME9, this figure will increase to a factor of well over 50.

In the foreseeable future microcontrollers will process more than just digital control sequences. They will have integrated signal processors that will be able to process signals directly, such as signals from the engine-knock sensors, for example.

Advances in the development of semiconductor memory chips are also worthy of note. Complex control programs require an enormous amount of memory space. The capacity of memory chips at the start of the 1980s was still only 8 kilobytes. The ME7 now uses 1-megabyte chips and soon memory capacities of 2 megabytes will be required. Figure 1 shows this pattern of development and likely future trends.

Fig. 1
Chart illustrating
- The performance of engine-management systems
- Number of connector pins on the electronic control units
- Capacity of the program memory
- Capacity of the data memories (RAM)

By way of comparison, the performance of an engine-management system with the very latest technology far exceeds the capabilities of Apollo 13.

1 Development of electronic control units

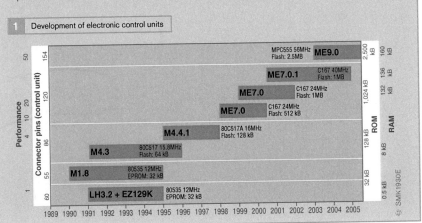

Exhaust emissions

The past few years have witnessed a drastic reduction in pollutant emissions from motor vehicles through the application of technical measures. In the case of passenger cars with gasoline engines, a significant role has been played by vehicles equipped with three-way catalytic converters.

Figure 1 illustrates the decrease in annual emissions from passenger cars. Only CO_2 experienced no significant decrease, and this is explained by the fact that CO_2 emissions are proportional to fuel consumption. It is therefore only possible to lower these emissions by reducing fuel consumption or by using lower-carbon fuels, such as natural gas.

The percentage contribution by motor vehicles to the total emissions generated by industry, traffic, households and power stations varies for the different substances and is as follows [1]

- 52 % for nitrous oxides
- 48 % for carbon monoxide
- 19 % for carbon dioxide
- 18 % for non-methane volatile hydrocarbons

Combustion of the air/fuel mixture

If a pure fuel were to be fully combusted under ideal conditions with sufficient oxygen, only water vapor (H_2O) and carbon dioxide (CO_2) would be created according to the following chemical reaction:

$$n_1 C_X H_Y + m_1 O_2 \rightarrow n_2 H_2O + m_2 CO_2$$

Because of the non-ideal combustion conditions in the combustion chamber (e.g., unvaporized fuel droplets) and the other constituents of fuel (e.g., sulfur), sometimes toxic by-products are also produced in addition to water and carbon dioxide in the combustion process.

The production of by-products is being increasingly reduced by procedures for optimizing combustion and improving fuel quality. The amount of CO_2 produced, however, is entirely dependent on the carbon content of fuel even under ideal conditions and can therefore not be influenced by combustion management.

[1] Provisional figures for 2001

Source:
German Federal
Environment Agency

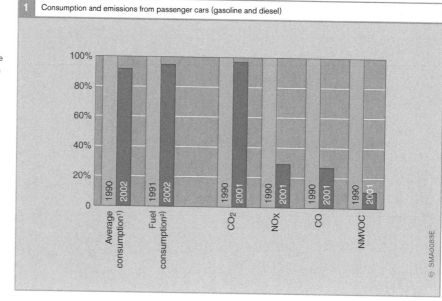

1   Consumption and emissions from passenger cars (gasoline and diesel)

Fig. 1
CO_2:   Carbon dioxide
NO_X: Nitrous oxides
CO: Carbon monoxide
NMVOC: Non-methane volatile hydrocarbons

[1] Average fuel consumption, passenger car (liter/distance)
[2] Absolute fuel consumption in passenger-car traffic

Source:
German Federal
Environment Agency

Main constituents of exhaust gas

Water (H$_2$O)
The hydrogen chemically bound within the fuel burns with the oxygen in the air to form water vapor, most of which condenses as it cools. This is the source of the exhaust plume visible on cold days. Water makes up about 13 % of the exhaust gas.

Carbon dioxide (CO$_2$)
During combustion, the carbon chemically bound within the fuel produces carbon dioxide (CO$_2$), which makes up approximately 14 % of the exhaust gas.

Carbon dioxide is a colorless, odorless, non-toxic gas and occurs naturally in the atmosphere. It is not classed as a pollutant with regard to motor-vehicle exhaust emissions. However, it is one of the substances responsible for the greenhouse effect and the global climate change that this causes. Since industrialization, the CO$_2$ content in the atmosphere has risen by roughly 30 % to today's figure of 367 ppm. Reducing CO$_2$ emissions by reducing fuel consumption is therefore seen more and more as a matter of urgency.

Nitrogen (N$_2$)
Nitrogen makes up 78 % of air and is therefore its primary constituent. It plays virtually no part in the combustion process and at roughly 71 % makes up the majority of the exhaust gas.

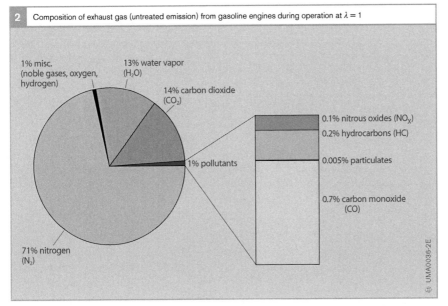

2   Composition of exhaust gas (untreated emission) from gasoline engines during operation at $\lambda = 1$

1% misc.
(noble gases, oxygen, hydrogen)

13% water vapor (H$_2$O)

14% carbon dioxide (CO$_2$)

1% pollutants

71% nitrogen (N$_2$)

0.1% nitrous oxides (NO$_X$)
0.2% hydrocarbons (HC)
0.005% particulates
0.7% carbon monoxide (CO)

UMA0036-2E

Fig. 2
Data in percent by volume

Actual concentrations of exhaust-gas constituents, especially pollutants, can vary in response to engine operating conditions, environmental factors (e.g., atmospheric humidity) and other parameters

Pollutants

During combustion, the air-fuel mixture generates a number of by-products. With the engine warmed to its normal operating temperature and running with a stoichiometric mixture composition ($\lambda = 1$), the proportion of these by-products in the engine's untreated emissions (exhaust gas after combustion, but before treatment) is about 1 % of the total exhaust-gas quantity.

The most significant of these combustion by-products are
- Carbon monoxide (CO)
- Hydrocarbons (HC), and
- Nitrous oxides (NO_x)

With the engine at normal operating temperature, catalytic converters can convert these pollutants at a rate of more than 99 % into harmless substances.

Carbon monoxide (CO)

Carbon monoxide results from incomplete combustion of rich air/fuel mixtures due to an air deficiency. Although carbon monoxide is also produced during operation with excess air, the concentrations are minimal, and stem from rich zones in the unhomogeneous air/fuel mixture. Fuel droplets that fail to vaporize form pockets of rich mixture that do not combust completely.

Carbon monoxide is a colorless and odorless gas. In humans, it inhibits the ability of the blood to absorb oxygen, thus leading to poisoning.

Hydrocarbons (HC)

Hydrocarbons are the chemical compounds of carbon (C) and hydrogen (H). HC emissions are caused by incomplete combustion of the air/fuel mixture where there is an oxygen deficiency. The combustion process also produces new hydrocarbon compounds not initially present in the original fuel (e.g., through the separation of extended molecular chains).

Aliphatic hydrocarbons (alkanes, alkenes, alkines, and their cyclic derivatives) are virtually odorless. Cyclic aromatic hydrocarbons (such as benzene, toluene, and polycyclic hydrocarbons) emit a discernible odor.

Some hydrocarbons are considered to be carcinogenic in the event of long-term exposure. Partially oxidized hydrocarbons (e.g., aldehydes, ketones) emit an unpleasant odor. The chemical products that result when these substances are exposed to sunlight are also considered to act as carcinogens in the event of extended exposure to specified concentrations.

Nitrous oxides (NO_x)

Nitrous oxide is the generic term for compounds of nitrogen and oxygen. Nitrous oxides are produced during all combustion processes with air as a result of secondary reactions with the nitrogen contained in the air. The main forms found in the exhaust gases from internal-combustion engines are nitrogen oxide (NO) and nitrogen dioxide (NO_2), with dinitrogen monoxide (N_2O) also present in small concentrations.

Nitrogen oxide (NO) is colorless and odorless and is slowly converted in air into nitrogen dioxide (NO_2). Pure NO_2 is a toxic, reddish-brown gas with pungent odor. NO_2 can induce irritation of the mucous membranes when present in the concentrations found in heavily polluted air.

Nitrous oxides play their part in damaging woods and forests (acid rain) and creating smog.

Sulfur dioxide (SO_2)

Sulfur compounds in exhaust gases – primarily sulfur dioxide – are produced by the sulfur content in fuels. SO_2 emissions are caused only to a small extent by motor vehicles and are not restricted by emission-control legislation.

Nevertheless, the production of sulfur compounds must be avoided to the greatest possible extent, since SO_2 sticks to catalytic converters (Three-Way Catalysts, TWC, NO_x

accumulator-type catalysts) and poisons them, i.e., reduces their reaction capability.

Like nitrous oxides, SO_2 contributes to the creation of acid rain, because it can be converted in the atmosphere or after settling into sulfuric or nitric acid.

Particulates

Solids are created in the form of particulates as a result of incomplete combustion. While exhaust composition varies as a function of combustion process and engine operating condition, these particulates basically consist of chains of carbon particles (soot) with an extremely extended specific surface ratio. Uncombusted and partly combusted hydrocarbons form deposits on the soot, where they are joined by aldehydes, with their overpowering odor. Aerosol components (minutely dispersed solids or fluids in gases) and sulfates bond to the soot. The sulfates result from the sulfur content in the fuel.

The problem of solids (particulates) in exhaust gas is primarily associated with diesel engines. Levels of particulate emissions from gasoline engines are negligible.

▶ Greenhouse effect

Short-wave solar radiation penetrates the Earth's atmosphere and continues to the ground, where it is absorbed. This process promotes warming in the ground, which then radiates long-wave heat, or infrared energy. A portion of this radiation is reflected by the atmosphere, causing the Earth to warm.

Without this natural greenhouse effect the Earth would be an inhospitable planet with an average temperature of −18°C. Greenhouse gases within the atmosphere (water vapor, carbon dioxide, methane, ozone, dinitrogen oxide, aerosols and particulate mist) raise average temperatures to approximately +15°C. Water vapor, in particular, retains substantial amounts of heat.

Carbon dioxide has risen substantially since the dawn of the industrial age more than 100 years ago an. The primary cause of this increase has been the burning of coal and petroleum products. In this process, the carbon bound in the fuels is released in the form of carbon dioxide.

The processes that influence the greenhouse effect within the Earth's atmosphere are extremely complex. While some scientists maintain that anthropogenic emissions (i.e., caused by humans) are the primary cause of climate change, this theory is challenged by other experts, who believe that the warming of the Earth's atmosphere is being caused by increased solar activity.

There is, however, a large degree of unanimity in calling for reductions in energy use to lower carbon-dioxide emissions and combat the greenhouse effect.

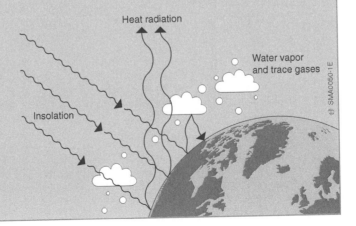

Factors affecting untreated emissions

The primary by-products when the air/fuel mixture is combusted are the pollutants NO_X, CO and HC. The quantities of these pollutants present in untreated exhaust gases (post-combustion gases prior to exhaust treatment) display major variations in response to different kinds of engine operation. The excess-air factor λ and the moment of ignition have a crucial influence on the formation of pollutants.

The catalytic-converter system converts pollutants to the greatest possible extent so that the emissions discharged by the vehicle to atmosphere are far lower than the untreated emissions. In order to minimize the discharged pollutants for tenable exhaust-gas treatment, it is essential however to keep untreated emissions as low as possible.

Influencing factors

Air/fuel ratio

Another primary factor defining the engine's toxic emissions is the air/fuel ratio (excess-air factor λ). To obtain maximum emissions reductions from three-way catalytic converters, manifold-injection engines run on a stoichiometric air/fuel mixture ($\lambda = 1$) in most operating ranges.

Engines with gasoline direct injection engines can be operated in stratified-charge or homogeneous mode, with selection varying according to the engine operating point. In homogeneous mode, the system injects fuel during the intake stroke to produce conditions comparable to those encountered with manifold injection. The system reverts to this mode of operation in response to demand for high torque and at high engine speeds. In this operating mode, the set excess-air factor is usually equal to or in the immediate vicinity of $\lambda = 1$.

The fuel is not distributed evenly throughout the entire combustion chamber during stratified-charge operation. The desired effect is achieved by waiting until the compression stroke to inject the fuel.

The mixture cloud formed at the center of the combustion chamber should be as homogeneous as possible, with an excess-air factor of $\lambda = 1$. Virtually pure air or an extremely lean mixture is present in the extremities of the combustion chamber. This results in an overall ratio of $\lambda > 1$ (lean) for the entire combustion chamber.

Mixture formation

In the interests of optimal combustion efficiency, the fuel destined for combustion should be thoroughly dispersed to form the most homogeneous mixture possible with the air. In manifold-injection engines, this refers to the overall combustion chamber, while in engines with gasoline direct injection this refers only to the stratified-charge cloud in the center of the combustion chamber.

Consistent distribution of uniform mixture to all cylinders is important for low pollutant emissions. Fuel-injection systems that employ their intake manifolds exclusively to transport air ensure consistent mixture distribution by discharging fuel into the intake port directly in front of the intake valve (manifold injection) or directly into the combustion chamber (gasoline direct injection). This type of consistency is less certain with systems relying on carburetors and single-point injection, as fuel tends to condense on the walls of the individual intake runners.

Engine speed

Higher engine speeds lead to greater friction losses within the engine as well as increased power consumption by auxiliary systems (e.g., water pump). Under these conditions, the power output per consumed unit of energy decreases. The engine's operating efficiency falls as engine speed rises.

Generating a given level of power at high engine speed equates with a higher level of fuel consumption than producing the same output at a lower engine speed. This leads to higher levels of pollutant emissions.

Engine load

The engine load or the generated engine torque has different effects on the pollutant components carbon monoxide (CO), unburnt hydrocarbons (HC) and nitrous oxides (NO$_x$). The various influences are described in more detail below.

Moment of ignition

The ignition of the air/fuel mixture, i.e., the period of time between flashover and the formation of a stable flame front, is of decisive significance to the combustion sequence. The character of the ignition process is shaped by the timing of the flashover, the ignition energy, and the composition of the mixture at spark plug. A large quantity of ignition energy translates into stable ignition with positive effects, both on the consistency of the consecutive combustion cycles and the composition of the exhaust gases.

Untreated HC emissions

The influence of torque

The temperature in the combustion chamber rises as torque increases. The depth of the zone in which the flame is extinguished close to the combustion-chamber wall as a result of low temperatures therefore shrinks as torque rises. Fewer unburnt hydrocarbons are then produced on account of more complete combustion.

In addition, the high exhaust-gas temperatures that accompany higher combustion-chamber temperatures under high-torque operation promote secondary reactions in the unburned hydrocarbons during the expansion and push-out phases to produce CO$_2$ and water. Because high-torque operation equates with higher temperatures in combustion chambers and exhaust gases, it leads to reductions in quantities of unburnt hydrocarbons relative to units of power generated.

The influence of the air/fuel ratio

The HC emissions from a gasoline engine increase as engine speeds rises, because the time available for preparing and combusting the mixture becomes shorter.

The influence of the air/fuel ratio

During operation with an air deficiency ($\lambda < 1$), incomplete combustion leads to the formation of unburnt hydrocarbons. Richer mixtures produce progressively greater HC concentrations (Fig. 1). In the rich range, therefore, HC emissions increase as the excess-air factor λ decreases.

HC emissions also increase in the lean range ($\lambda > 1$). Minimum HC generation coincides with the range $\lambda = 1.1...1.2$. The rise within the lean range is caused by incomplete combustion at the extremities of the combustion chamber. Extremely lean mixtures, where combustion lag can ultimately lead to ignition miss, aggravate this effect and produce a dramatic rise in HC emissions. This phenomenon is caused by unequal mixture distribution in the combustion chamber and thus poor ignition conditions in lean combustion-chamber zones.

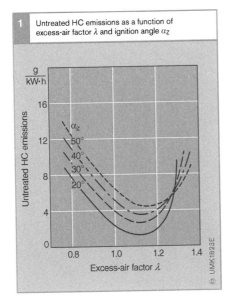

1 Untreated HC emissions as a function of excess-air factor λ and ignition angle α_z

Untreated HC emissions ($\frac{g}{kW \cdot h}$)

α_z: 50°, 40°, 30°, 20°

Excess-air factor λ

UMK1823E

The gasoline-engine's lean-burn limit is primarily dependent on the excess-air factor at the spark plug at the instant of ignition, and by the composite excess-air factor (air/fuel ration considered over the entire combustion chamber). The flow pattern of the charge in the combustion chamber can be manipulated to obtain a more homogeneous mixture to ensure more reliable ignition, while at the same time accelerating propagation of the flame front.

The stratified-charge method used in conjunction with gasoline direct injection presents a contrasting picture. Instead of focusing on obtaining a homogeneous air/fuel mixture throughout the combustion chamber, this concept creates a highly ignitable mixture only in the area immediately adjacent to the tip of the spark plug. This concept thus allows substantially higher composite excess-air factors than would be available using a homogeneous mixture. HC emissions in stratified-charge mode are essentially determined by the mixture formation process.

It is vital in the case of direct injection wherever possible to avoid depositing liquid fuel on the combustion-chamber walls and the piston, as the resulting wall-applied film usually fails to combust completely, leading to high HC emissions.

The influence of the moment of ignition
Increasing the ignition advance (high α_Z) produces a rise in emissions of unburnt hydrocarbons, as the resulting reduction in exhaust-gas temperature has a negative effect on secondary reactions in the expansion and exhaust phases (Fig. 1). It is only during operation with extremely lean mixtures that this response pattern is inverted. These types of lean mixtures result in such a low flame-front propagation rate that the combustion process will still be in progress when the exhaust valve opens if ignition is late. With late ignition, the engine reaches its lean-burn limit early, at an excess-air factor of λ.

Untreated CO emissions

The influence of torque
As with untreated HC emissions, the higher process temperatures that accompany high torque foster secondary reactions in CO during the expansion phase. The CO oxidizes to form CO_2.

The influence of engine speed
CO emissions also mirror the pattern of HC emissions in their response to variations in engine speed.

The influence of the air/fuel ratio
In the rich range, CO emissions display a virtually linear correlation with the excess-air factor (Fig. 2). This is the result of incomplete carbon oxidation during operation with an air deficiency.

In the lean range (excess surplus), CO emissions remain at extremely low levels, and the influence of changes in the excess-air factor is minimal. Under these conditions, the only source of CO generation is incomplete combustion of a poorly homogenized air/fuel mixture.

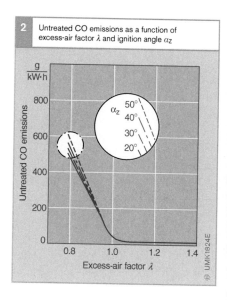

2   Untreated CO emissions as a function of excess-air factor λ and ignition angle α_Z

The influence of the moment of ignition
The moment of ignition has virtually no influence on CO emissions (Fig. 2), which are
almost entirely a function of the excess-air
factor λ.

Untreated NO_X emissions

The influence of torque
The higher combustion-chamber temperatures that accompany increased torque generation promote the formation of NO_X.
As torque output rises, untreated NO_X emissions display a disproportionate increase.

The influence of engine speed
As the response time available to form
NO_X is shorter at higher engine speeds,
NO_X emissions decrease as engine speed
increases. In addition, the residual-gas content in the combustion chamber must be
considered since it causes lower peak temperatures. Because this residual-gas content
tends to fall off as engine speed rises, this
effect counteracts the response pattern
described above.

The influence of the air/fuel ratio
The maximum level of NO_X emissions
lies with slight excess air in the range of
$\lambda = 1.05...1.1$. In the lean and rich ranges,
NO_X emissions drop, since the peak combustion temperatures decrease.
    A characteristic of stratified-charge opera-
tion in engines with gasoline direct injection
is a high excess-air factor. The NO_X emissions are low when compared with the operating point at $\lambda = 1$, since only some of the
gas takes part in the combustion process.

The influence of the moment of ignition
Throughout the range with excess-air factors
of λ, NO_X emissions rise as ignition advance
is increased (Fig. 3). The higher combustion
temperatures promoted by earlier ignition
timing not only shift the chemical equilibrium toward greater NO_X formation, but –
most significantly – they also accelerate the
speed at which this formation takes place.

Soot emissions
Gasoline engines produce only extremely
low soot emissions during operation on
mixtures in the vicinity of stoichiometric.
However, soot can be generated in engines
with gasoline direct injection during stratified-charge operation, when its formation
can be fostered by localized areas with extremely rich mixtures or even fuel droplets.
To ensure that adequate time remains available for efficient mixture formation, operation in stratified-charge mode must therefore be restricted to low and moderate
engine speeds.

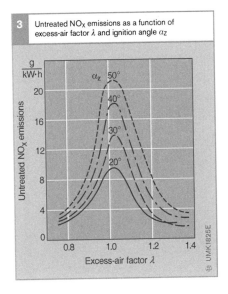

3 Untreated NO_X emissions as a function of
excess-air factor λ and ignition angle α_Z

Catalytic emission control

Emission-control legislation lays down limits governing pollutant emissions from motor vehicles. On-engine measures on their own are not enough to comply with these limits. In gasoline engines, catalytic exhaust-gas aftertreatment for the purpose of converting the pollutants has now taken center stage. To this end, the exhaust gas passes through one or more catalytic converters located in the exhaust-system branch before being discharged to atmosphere. The pollutants contained in the exhaust gas are converted on the surface of the catalytic converter(s) by chemical reactions into non-toxic substances.

Overview

Catalytic exhaust-gas aftertreatment with the aid of a three-way catalytic converter is currently the most effective form of emission control for gasoline engines. The three-way catalytic converter is an integral component of the exhaust-emission control systems of both manifold-injection engines and gasoline direct-injection engines (Fig. 1).

In the case of homogeneous mixture distribution with a stoichiometric air/fuel ratio ($\lambda = 1$), a three-way catalytic converter at normal operating temperature is able to convert the following pollutants virtually completely: carbon monoxide (CO), hydrocarbons (HCs) and nitrous oxides (NO_X). However, adhering exacting to a figure of $\lambda = 1$ requires mixture formation by means of electronically controlled gasoline injection; today, this system has completely replaced the carburetor, which was used primarily up until the introduction of the three-way catalytic converter. Precise lambda closed-loop control monitors the composition of the air/fuel mixture and regulates it at a value of $\lambda = 1$. Although these ideal conditions cannot always be maintained in all operating states, pollutant emissions can on average be reduced by more than 98 %.

Because the three-way catalytic converter is unable to convert the nitrous oxides in lean mode ($\lambda > 1$), an additional NO_X accumulator-type catalytic converter is used in engines with a lean operating mode. Another means of reducing NO_X at $\lambda > 1$ is Selective Catalytic Reduction (SCR). This process is already being used in diesel commercial vehicles, and development plans are currently underway to introduce it in diesel passenger

---

1   Exhaust-gas branch with lambda sensors and a three-way catalyst installed in the immediate vicinity of the engine

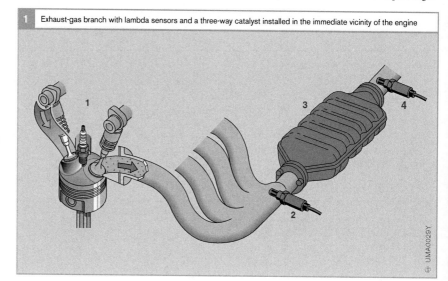

Fig. 1

1   Engine

2   Lambda sensor upstream of catalyst (two-step sensor or wide-band sensor depending on system)

3   Three-way catalyst

4   Two-step lambda sensor downstream of catalyst (only on systems with two-sensor lambda control)

cars. At this moment, however, it is not possible to foresee whether SCR will also be able to be used in gasoline engines.

The oxidation catalytic converter, which is used in diesel engines to oxidize HCs and CO, is not used separately in gasoline engines, since the three-way catalytic converter performs its function.

Development objectives

In view of the constantly tightening emission limits, reducing pollutant emissions remains an important objective in engine design and development. Whereas a catalytic converter which has reached normal operating temperature achieves very high conversion rates bordering on 100 %, considerably larger amounts of pollutants are emitted in the cold-start and warm-up phases: The actual amounts of pollutants emitted during the starting process and the subsequent post-start phase can make up to 90 % of the total emissions both in the European and American test cycles (NEDC or FTP 75). In order to reduce emissions, it is therefore essential both to ensure that the catalytic converter heats up quickly and to generate the fewest possible untreated emissions in the starting phase and while the catalyst is heating up. Early operational readiness on the part of the lambda oxygen sensor is also an important factor.

Catalytic-converter concepts

Catalytic converters can be divided into the following categories:
● Continuously operating catalysts, and
● Intermittently operating catalysts

Continuously operating catalytic converters convert pollutants uninterrupted and without actively intervening in the engine operating conditions. The following systems are classed as continuously operating: the three-way catalytic converter, the oxidation catalytic converter and the SCR catalytic converter (Selective Catalytic Reduction; currently used in diesel engines only).

Intermittently operating catalytic converters operate in different phases, which are initiated in each case by an active change in the boundary conditions by the catalytic-converter system. The NO_X accumulator-type catalytic converter operates intermittently: In the event of excess oxygen in the exhaust gas, NO_X is accumulated and the system is switched to rich mode (oxygen deficiency) for the subsequent regeneration phase.

Three-way catalytic converter

Method of operation

The three-way catalytic converter converts the following pollutant components into non-toxic components during the combustion of the air/fuel mixture: hydrocarbons (HCs), carbon monoxide (CO) and nitrous oxides (NO_X). The end products which result are water vapor (H_2O), carbon dioxide (CO_2) and nitrogen (N_2).

Conversion of pollutants
The conversion of pollutants can be divided into oxidation and reduction reactions. For example, oxidation of carbon monoxide and hydrocarbons takes place according to the following equations:

$$2\,CO + O_2 \quad\rightarrow 2\,CO_2 \qquad\qquad \text{(eq. 1)}$$

$$2\,C_2H_6 + 7\,O_2 \quad\rightarrow 4\,CO_2 + 6\,H_2O \quad \text{(eq. 2)}$$

Reduction of nitrous oxides takes place according to the following exemplary equations:

$$2\,NO + 2\,CO \quad\rightarrow N_2 + 2\,CO_2 \qquad \text{(eq. 3)}$$

$$2\,NO_2 + 2\,CO \quad\rightarrow N_2 + 2\,CO_2 + O_2 \quad \text{(eq. 4)}$$

The oxygen needed to oxidize HCs and CO is drawn either directly from the exhaust gas or from the nitrous oxides present in the exhaust gas, depending on the composition of the air/fuel mixture.

At $\lambda = 1$, a state of balance arises between the oxidation and reduction reactions. The residual-oxygen content in the exhaust gas at $\lambda = 1$ (approximately 0.5 %) and the oxygen bound in the nitrous oxide enable HCs and CO to oxidize completely; the nitrous oxides are simultaneously reduced. Thus, HCs and CO acts as reducing agents for the nitrous oxides. The catalytic converter can compensate for minor mixture fluctuations itself.

It has the ability to accumulate and release oxygen. Its substrate layer contains ceroxide, which can make oxygen available via the following balance reaction:

$$2\,Ce_2O_3 + O_2 \longleftrightarrow 4\,CeO_2 \qquad \text{(eq. 5)}$$

In the event of constant excess oxygen ($\lambda > 1$), HCs and CO are oxidized by the oxygen present in the exhaust gas. They are therefore not available for the reduction of the nitrous oxides. The raw NO_X emissions are therefore released untreated. In the event of a constant oxygen deficiency ($\lambda < 1$), the nitrous-oxide reduction reactions takes place with HCs and CO as the reducing agents. Excess hydrocarbons and carbon monoxide which cannot be converted for lack of oxygen are released untreated.

Conversion rate
The quantity of released pollutants is derived from the concentration of the pollutants in the untreated exhaust gas (Fig. 2a) and from the conversion rate, i.e., from the proportion that can be converted in the catalytic converter. Both variables are dependent on the set excess-air factor λ.

The highest possible conversion rate for all three pollutant components requires a mixture composition in the stoichiometric ratio of $\lambda = 1.0$. The window (lambda control range) in which the air/fuel ratio λ must be situated is therefore very small. Mixture formation must therefore be followed up in a lambda closed-loop control circuit.

The conversion rates for HCs and CO rise continuously as the excess-air factor increases, i.e., the emissions decrease (Fig. 2b). At $\lambda = 1$, there is only a very low level of pollutant components in the untreated exhaust gas. With high excess-air factors ($\lambda > 1$), the concentration of these pollutant components remains at this low level.

Conversion of the nitrous oxides (NO_X) is good in the rich range ($\lambda < 1$). Even a small increase in the oxygen content in the exhaust gas from $\lambda = 1$ impedes the reduction of nitrous oxides and causes a sharp increase in their concentration.

Fig. 2
a   Before catalytic
    aftertreatment
    (untreated exhaust
    gas)
b   After catalytic
    aftertreatment
c   Voltage curve of
    two-step lambda
    sensor

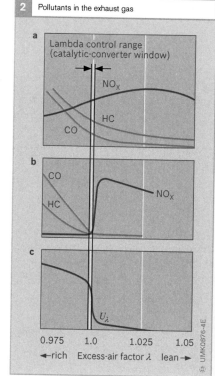

2   Pollutants in the exhaust gas

a
Lambda control range
(catalytic-converter window)
NO_X
HC
CO

b
CO
NO_X
HC

c
U_λ

0.975   1.0   1.025   1.05
←rich   Excess-air factor λ lean→

UMK0876-4E

Design

The catalytic converter (Fig. 3) essentially comprises a sheet-steel housing (6), a substrate (5), a substrate coating (washcoat), and the active catalytic noble-metal coating (4).

Substrates

Two substrate systems have come to the forefront:

Ceramic monoliths

These ceramic monoliths are ceramic bodies containing thousands of narrow passages through which the exhaust gas flows. The ceramic is a high-temperature-resistant magnesium-aluminum silicate. The monolith, which is highly sensitive to mechanical tension, is fastened inside a sheet-steel housing by means of mineral swell matting (2) which expands the first time it is heated up and firmly fixes the monolith in position. At the same time, the matting also ensures a 100 % gas seal. Ceramic monoliths are at present the most commonly used catalyst substrates.

Metallic monoliths

The metallic monolith is an alternative to the ceramic monolith. It is made of finely corrugated, thin metal foil approximately 0.03...0.05 mm thick which is wound and soldered in a high-temperature process. Thanks to the thin walls, it is possible to incorporate a greater number of passages per surface. This reduces the resistance experienced by the exhaust gas and brings advantages in the optimization of high-performance engines.

Coating

Ceramic and metallic monoliths require a substrate coating (washcoat) of aluminum oxide (Al_2O_3). This coating increases the effective surface of the catalytic converter by a factor of 7000. 1 l catalyst volume therefore has an area equating to a soccer field.

The effective catalytic coating contains the noble metals platinum and/or palladium and rhodium. Platinum and palladium accelerate the oxidation of hydrocarbons and of carbon monoxide. Rhodium accelerates the reduction of nitrous oxides (NO_X).

The amount of noble metal contained in a catalytic converter is roughly 1...5 g. This figure depends among other things on the engine displacement and on the exhaust-emission standard to be complied with.

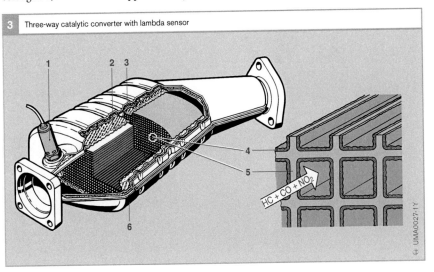

3 Three-way catalytic converter with lambda sensor

HC + CO + NO₂

Fig. 3
1 Lambda sensor
2 Swell matting
3 Thermally insulated double shell
4 Washcoat (Al_2O_3 substrate coating) with noble-metal coating
5 Substrate (monolith)
6 Housing

Operating conditions

Operating temperature

To enable the oxidation and reduction reactions for converting the pollutants to pass off, the reacting agents must be supplied with a specific amount of activation energy. This energy is provided by the heat from the heated-up catalytic converter.

The catalytic converter reduces the activation energy (Fig. 4) such that the light-off temperature (i.e., the temperature at which 50 % of the pollutants is converted) drops. The activation energy – and thus the light-off temperature – are greatly dependent on the respective reacting agents. Considering a three-way catalytic converter, no worthwhile conversion of pollutants takes place until the operating temperature exceeds 300 °C. Operation within a temperature range of 400...800 °C is ideal for high conversion levels and a long service life.

In the range of 800...1000 °C, thermal aging of the catalytic converter is significantly improved by sintering the noble metals and the Al$_2$O$_3$ substrate coat. This leads to a reduction of the active surface. The operating time in this temperature range also has a significant influence here. At temperatures in excess of 1000 °C, thermal aging of the catalytic converter increases sharply, causing the catalyst to be largely ineffective.

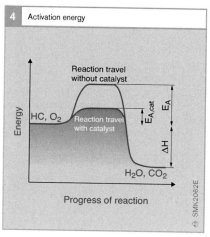

4   Activation energy

Reaction travel without catalyst

Energy

HC, O$_2$

Reaction travel with catalyst

$E_{A,cat}$ E_A

ΔH

H$_2$O, CO$_2$

Progress of reaction

SMK2062E

In the event of engine malfunction (e.g., ignition miss), the temperature in the catalytic converter can rise to up to 1400 °C if unburnt fuel ignites in the exhaust-system branch. Since such temperatures melt the substrate and completely destroy the catalyst, it is imperative that the ignition system be highly reliable so as to prevent this from occurring. Modern engine-management systems are able to detect ignition and combustion miss, and in such cases interrupt the fuel injection to the cylinder concerned so that unburnt mixture cannot enter the exhaust-system branch.

Catalytic-converter poisoning

In order to be able to use a three-way catalytic converter, it is essential for the engine to be run on unleaded fuel. Otherwise, lead compounds would be deposited in the active catalyst coating or clog the pores and thereby reduce their number.

Residues from the engine oil can also "poison" the catalyst, i.e., damage it so far that it becomes ineffective.

NO$_x$ accumulator-type catalytic converter

Function

During lean-burn operation, it is impossible for the three-way catalytic converter to convert the nitrous oxides (NO$_x$) which have been generated during combustion. CO and HCs are oxidized by the high residual-oxygen content in the exhaust gas and are therefore no longer available as reducing agents for the nitrous oxides. The NO$_x$ accumulator-type catalytic converter (NSC, NO$_x$ Storage Catalyst) converts the nitrous oxides in a different way.

Design and special coating

The NO$_x$ accumulator-type catalytic converter is similar in design to the conventional three-way converter. In addition to the platinum, palladium and rhodium coatings, the NO$_x$ converter is provided with

special additives which are capable of accumulating nitrous oxides. Typical accumulator
materials contain, for example, the oxides of
potassium, calcium, strontium, zirconium,
lanthanum, or barium.

Method of operation

At $\lambda = 1$, due to the noble-metal coating,
the NO$_X$ converter operates the same as
a three-way converter. It also converts the
non-reduced nitrous oxides in lean exhaust
gases. However, conversion is not a continuous process as it is with the hydrocarbons and
the carbon monoxide, but instead takes place
in three distinct phases:

1. NO$_X$ accumulation (storage)
2. NO$_X$ removal
3. Conversion

NO$_X$ accumulation (storage)

In lean-burn engine operation ($\lambda > 1$),
the nitrous oxides (NO$_X$) are catalytically oxidized on the surface of the platinum coating
to form nitrogen dioxide (NO$_2$). Then the
NO$_2$ reacts with the special oxides on the catalyst surface and with oxygen (O$_2$) to form
nitrates. Thus, for instance, NO$_2$ combines
chemically with barium oxide (BaO) to form
the chemical compound barium nitrate
Ba(NO$_3$)$_2$:

$$2\,BaO + 4\,NO_2 + O_2 \quad \rightarrow 2\,Ba(NO_3)_2$$

This enables the NO$_X$ converter to accumulate the nitrous oxides which have been
generated during engine operation with
excess air.

There are two methods in use to determine
when the NO$_X$ converter is saturated and the
accumulation phase has finished:

- The model-supported process calculates
  the quantity of stored NO$_X$ taking into
  account the catalyst temperature
- An NO$_X$ sensor (Fig. 5, Pos. 6) downstream
  of the NO$_X$ converter measures the NO$_X$
  concentration in the exhaust gas

NO$_X$ removal and conversion

As the amount of stored nitrous oxides
(charge) increases, so the ability to continue
to bind nitrous oxides decreases. This means
that regeneration must take place as soon as a
given level is exceeded, i.e., the accumulated
nitrous oxides must be removed and converted. To this end, the engine is switched
briefly to rich homogeneous mode ($\lambda < 0.8$).
The processes for removing the NO$_X$ and
converting it into nitrogen and carbon dioxide take place separately from each other. H$_2$,
HCs, and CO are used as reducing agents.
Reduction is slowest with HCs and most
rapid with H$_2$. Removal takes place as follows,

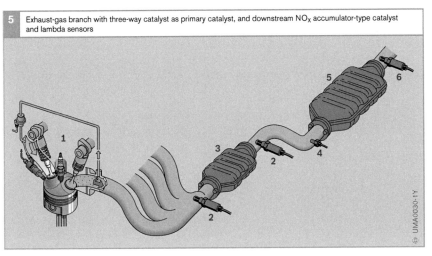

5 Exhaust-gas branch with three-way catalyst as primary catalyst, and downstream NO$_X$ accumulator-type catalyst and lambda sensors

Fig. 5
1   Engine with EGR
    system
2   Lambda sensor
3   Three-way catalyst
    (primary catalyst)
4   Temperature sensor
5   NO$_X$ accumulator-
    type catalyst (main
    catalyst)
6   Two-step lambda
    sensor, optionally
    available with
    integral NO$_X$ sensor

with CO being used as the reducing agent in the following description: The carbon monoxide reduces the nitrate (e.g., barium nitrate Ba(NO$_3$)$_2$) to an oxide (e.g., barium oxide BaO). Carbon dioxide and nitrogen monoxide are produced in the process:

$$Ba(NO_3)_2 + 3\,CO \quad \rightarrow 3\,CO_2 + BaO + 2\,NO$$

Subsequently, using carbon monoxide, the rhodium coating reduces the nitrous oxides to nitrogen and carbon dioxide:

$$2\,NO + 2\,CO \quad \rightarrow N_2 + 2\,CO_2$$

There are two different methods of detecting when the removal phase is complete.
● The model-supported process calculates the quantity of NO$_X$ still held by the NO$_X$ converter.
● A lambda sensor (6) downstream of the converter measures the oxygen concentration in the exhaust gas and outputs a voltage jump from "lean" to "rich" when removal has finished.

Operating temperature and installation point

The NO$_X$ converter's ability to accumulate/store NO$_X$ is highly dependent on temperature. Accumulation reaches its maximum in the range of 300...400 °C, which means that the favorable temperature range is much lower than that of the three-way catalytic converter. For this reason and because of the lower maximum permissible operating temperature of the NO$_X$ converter, two separate catalytic converters must be installed for catalytic emission control:
an upstream three-way catalyst as the primary converter (Fig. 5, Pos. 3) and a downstream NO$_X$ accumulator-type catalyst (5) as the main converter (underfloor converter).

Sulfur loading

The sulfur contained in gasoline presents the accumulator-type catalytic converter with a problem. The sulfur contained in lean exhaust gas reacts with the barium oxide (accumulator material) to form barium sulfate. The result is that, over time, the amount of accumulator material available for NO$_X$ accumulation diminishes. Barium sulfate is extremely resistant to high temperatures, and for this reason is only degraded to a slight degree during NO$_X$ regeneration.

When sour fuels (i.e., containing sulfur) are used, the catalytic converter must be constantly desulfurized. For this purpose, the converter is heated by specific heating measures to 600...650 °C and then exposed for a few minutes to alternately rich ($\lambda = 0.95$) and lean ($\lambda = 1.05$) exhaust gas. The barium sulfate reduces to barium oxide as a result.

Catalytic-converter configurations

Boundary conditions

The design of the exhaust system is defined by various boundary conditions: heat-up performance during cold starting, temperature loading at full load, space situation in the vehicle, and engine torque and power development.

The required operating temperature of the three-way catalytic converter limits the installation options. Upstream converters quickly reach operating temperature in the post-start phase, but can be exposed to extremely high thermal load at high engine load and speed. Downstream converters are less exposed to these temperature loads. But they do require more time in the heating-up phase to reach operating temperature, if this is not accelerated by an optimized strategy for heating the converter (e.g., secondary-air injection).

Strict emission-control regulations call for special concepts for heating the catalytic converter when the engine is started. The lower the heat flow which can be generated to heat the catalytic converter, and the lower the emission limits, the closer the converter should be installed to the engine – provided no additional measures for improving heating-up performance are taken. Air-gap-insulated manifolds are often used; these demonstrate lower heat losses up to the converter and thus have available a greater quantity of heat for heating the converter.

Primary and main catalytic converters

A widely used configuration of the three-way catalytic converter is the split arrangement with an upstream primary catalyst and an underfloor catalyst (main catalytic converter). Upstream catalytic converters (i.e., near the engine) require their coating to be optimized to provide for high-temperature stability. Underfloor converters, on the other hand, require optimization in the so-called "low light-off" direction (low start-up temperature) and good NO_X conversion characteristics. For faster heating and pollutant conversion, the primary catalyst is usually smaller and has a higher cell density and a greater noble-metal load.

Owing to their lower maximum permissible operating temperature, NO_X accumulator-type catalysts are always installed in the underfloor area.

As an alternative to the classic arrangement of splitting into two separate housings and installation positions, there are also two-stage catalytic-converter configurations (cascade catalytic converters), in which two catalyst substrates are accommodated in a common housing one after the other. The two substrates are separated from each other for thermal-isolation purposes by a small air gap. In a cascade catalytic converter, the thermal load of the second catalyst is, because of the spatial proximity, comparable with the first catalyst. Nevertheless, this arrangement permits independent optimization of the two catalysts

with regard to noble-metal load, cell density and wall thickness. The first catalyst general has a larger noble-metal load and a higher cell density for good light-off performance during cold starting. A lambda sensor can be installed between the two substrates for controlling and monitoring exhaust-gas treatment.

Even concepts with only one overall catalytic converter are used. It is possible with modern coating methods to create different noble-metal loads in the front and rear sections of the catalyst. This configuration has a smaller scope for design possibilities but is attractive in cost terms. Provided there is enough space available, the catalyst is located as close to the engine as possible. However, it can also be situated further away from the engine (i.e., downstream positioning) if an effective catalyst-heating process is used.

Multiflow configurations

The exhaust-system branches of the individual cylinders are brought together ahead of the catalyst at least partially by the exhaust manifold. Four-cylinder engines frequently use exhaust manifolds which bring together all four cylinders after a short distance. This makes it possible to use an upstream catalyst, which can be favorably positioned for heating performance (Fig. 6a).

For optimized-power engine configurations, 4-into-2 exhaust manifolds are the preferred choice in four-cylinder engines; in this arrangement, only two exhaust-gas branches are brought together in each case. Positioning one catalyst after the second junction into a single overall exhaust-gas branch is not good for the heating performance.
For this reason, the ideal solution is to install two upstream (primary) catalysts already after the first junction and if necessary a further (main) catalyst after the second join (Fig. 6b).

A similar situation arises in engines with more than four cylinders, especially in engines with more than one cylinder bank (V-engine). Primary and main catalysts can be used on each bank in accordance with the above descriptions. Two different arrangements may be used: The exhaust system is completely designed as a dual-flow arrangement (Fig. 6c), or a Y-shaped junction merging into an overall exhaust-gas branch is used. In the latter case, a joint main catalyst serving both cylinder banks can be used in a configuration with primary and main catalysts (Fig. 6d).

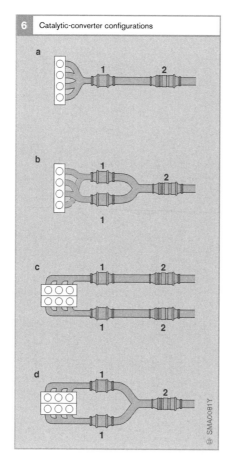

6   Catalytic-converter configurations

a

b

c

d

Catalytic-converter heating

The three-way catalytic converter must reach a minimum temperature of roughly 300 °C (light-off) before pollutants can be converted; this temperature threshold can be even higher on older catalysts. With the engine and exhaust system initially cold, the catalytic converter must be heated up as quickly as possible to operating temperature. This requires a short-term supply of heat, which can be provided by a variety concepts.

Purely on-engine measures
To ensure effective catalyst heating with on-engine measures, it is necessary to increase both the exhaust-gas temperature and the exhaust-gas mass flow. This is achieved using a variety of measures, which reduce engine efficiency and therefore create an increased exhaust-gas heat flow.

The heat-flow demand on the engine is dependent on the catalyst position and the layout of the exhaust system, because the exhaust gas cools on its way to the catalyst when the exhaust system is cold.

Ignition timing
The main measure for increasing the exhaust-gas heat flow is ignition-timing retardation. Combustion is initiated as late as possible and takes place during the expansion phase. The exhaust gas has a relatively high temperature at the end of the expansion phase.

Retarded combustion has an unfavorable effect on engine efficiency.

Idle speed
A supporting measure is to raise the idle speed and thereby increase the exhaust-gas mass flow. The increased engine speed permits a greater ignition-timing retardation; however, in order to ensure reliable ignition, the ignition angles are limited to roughly 10° to 15° after TDC. The heat output limited in this way is not always enough to achieve the current emission limits.

Exhaust-camshaft adjustment
A further contribution to increasing the heat flow can be achieved if necessary with exhaust-camshaft adjustment. The process of the exhaust valves opening as early as possible interrupts the retarded combustion early and the mechanical work generated is thus reduced further. The corresponding quantity of energy is available as a quantity of heat in the exhaust gas.

Stratified-charge/catalyst heating and homogeneous split
In the case of gasoline direct injection, there are further processes which can quickly heat the catalyst to operating temperature without the need for additional components.
In the "stratified-charge/catalyst heating" operating mode, during stratified-charge operation with high levels of excess air a further injection of fuel takes place at the end of combustion. This fuel burns partly in the exhaust manifold first and delivers an additional injection of heat to the catalyst. The "homogeneous-split" measure involves initially creating a homogeneous, lean basic mixture. A subsequent stratified-charge injection facilitates retarded moments of ignition and results in high exhaust-gas heat flows. These measures render the process of secondary-air injection unnecessary in engines with gasoline direct injection.

Secondary-air injection
Thermal afterburning of unburnt fuel constituents increases the temperature in the exhaust system. A rich ($\lambda = 0.9$) extending up to a very rich ($\lambda = 0.6$) basic mixture is set for this purpose. A secondary-air pump supplies oxygen to the exhaust system to produce a leaner exhaust-gas composition.

Where the basic mixture is very rich ($\lambda = 0.6$), the unburnt fuel constituents oxidize above a specific temperature threshold. To achieve this temperature, it is necessary on the one hand to raise the temperature level with retarded ignition angles and on the other hand to introduce the secondary air as closely as possible to the exhaust valves. The exothermic reaction in the exhaust system increases the heat flow to the catalyst and therefore shortens the heating period. Furthermore, by comparison with purely on-engine measures, HC and CO emissions are reduced before entering the catalyst.

Where the basic mixture is less rich ($\lambda = 0.9$), there is no significant reaction ahead of the catalyst. The unburnt fuel constituents oxidize in the catalyst and heat it up from the inside. For this purpose, however, the end face of the catalyst must first be brought up to a level above the light-off temperature by means of conventional measures (such as ignition-timing retardation).

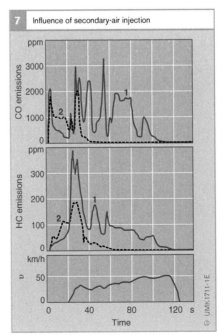

7   Influence of secondary-air injection

Fig. 7
1   Without secondary-air injection
2   With secondary-air injection

v Vehicle speed

In reality, it is always a combination of these two extreme cases that occurs. Figure 7 shows by way of example the curves for HC and CO emissions downstream of the catalyst in the first few seconds of an emission test (FTP 75), with and without secondary-air injection.

Secondary-air injection is performed with an electric secondary-air pump (Fig. 8, Pos. 1), which is switched by means of a relay (3) on account of the high power demand. Since the secondary-air valve (5) prevents backflow of exhaust gases into the pump, it must remain closed when the pump is deactivated. Either it is a passive non-return valve or it is actuated by purely electrical or (as shown here) pneumatic means with an electrically actuated control valve (6). When the control valve is actuated, the secondary-air valve opens in response to the intake-manifold vacuum. The secondary-air system is coordinated by the engine ECU (4).

Concepts for active heating

Measures for active catalytic-converter heating are processes which do not use the internal-combustion engine as the heat source or which do not use the engine as the sole heat source. The advantage of such processes lies in the fact that the heat can be supplied locally and does not have to pass through the exhaust system to the catalyst. Unlike conventional heating strategies with upstream catalysts, these concepts have failed to catch on.

Electrically heated catalytic converter

In the case of an electrically heated catalyst, the exhaust gas first flows through an approximately 20 mm thick catalytic substrate plate which can be heated with an electrical output of roughly 2 kW. A secondary-air system can be used to provide assistance. The additional exothermic effect (heat release) during the conversion of the exhaust-gas/secondary-air mixture in the heated plate accelerates the heating process further.

When compared with the heat currents of up to 20 kW which can be achieved with on-engine measures if necessary in conjunction with secondary-air injection, an electrical output of 2 kW seems relatively low. However, it is the temperature of the catalyst substrate and not the temperature of the exhaust gas which is crucial to catalyst operation. Direct electrical heating of substrate is highly effective and results in very good emission values.

In a conventional passenger car with a 12 V supply voltage, the high currents that occur for heating the catalyst place a significant strain on the car's electrical system. An amplified alternator and if necessary a second battery are therefore required. This system is better suited for use in electric hybrid vehicles, which have more powerful electrical systems anyway.

The electrically heated catalytic converter has previously been used in individual small-scale production projects.

8   Secondary-air system

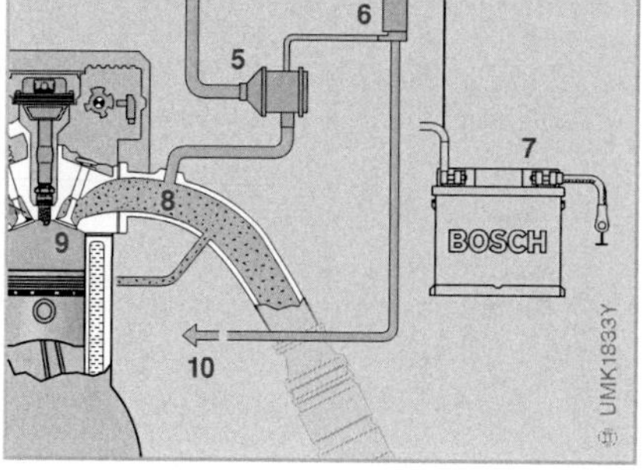

Fig. 8
1 Secondary-air pump
2 Induction air
3 Relay
4 Engine ECU
5 Secondary-air valve
6 Control valve
7 Battery
8 Point of introduction into exhaust pipe
9 Exhaust valve
10 To intake-manifold connection

Burner system

Another concept for locally heating the catalytic converter is a fuel-powered burner, the hot combustion exhaust gases of which are fed upstream of the catalyst to the engine exhaust system (Fig. 9). The burner system is supplied with air along the same lines as a secondary-air system and has its own ignition device and fuel-metering facility. In order to achieve low emission targets, good fuel atomization and mixture formation in the burner are crucial to low burner emissions, especially during starting.

Heating-output figures of 15 kW can be achieved with the burner system; however, in contrast to the electrically heated catalyst, the substrate is indirectly heated here.

Mixing the gas flows of the engine and burner (which differ greatly in temperature) can be problematic. However, as well as local heating, the burner system offers a further advantage: Unburnt constituents of the engine exhaust gas oxidize in the hot burner exhaust gas before entering the catalyst, and this results initially in a purely thermal reduction of the pollutants.

The burner system has only previously been used in prototypes and has not been used at all in volume-scale production.

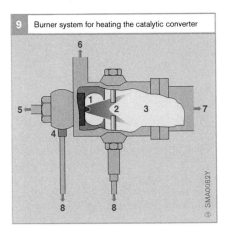

9 Burner system for heating the catalytic converter

Fig. 9
1 Burner nozzle
2 Ignition electrodes
3 Combustion chamber
4 Fuel shutoff valve
5 Fuel supply via fuel regulator
6 Air supply from secondary-air pump
7 To catalyst
8 To control system

SMA0082Y

▶ Catalysis

In order for chemical reactions to take place, bonds in the starting substances must be relaxed and then split. This requires a minimum amount of energy (activation energy) which must be supplied to the starting substances, for example in the form of heat. Many thermodynamically possible reactions do not take place at noticeable speed, since the activation energy is very high, and this energy barrier therefore is only overcome by a very small number of molecules.

A catalytic converter accelerates the reaction in two ways. Firstly, it reduces the activation energy of the reaction so that this energy barrier can be overcome by a greater number of molecules. Secondly, the adsorption of the reacting agents on the catalytic converter reduces the distance between them, and the probability that the reacting agents will interact with each other is thereby increased.

There are different mechanisms for the reaction on the catalytic-converter surface; for instance, the oxidation of CO on the platinum catalyst takes place according to the Langmuir-Hinshelwood mechanism, whereby both reacting agents (CO and O_2) are adsorbed on the catalyst surface. The bonds within the O_2 molecule are relaxed and the reactivity of the CO increases. Finally, the O_2 breaks down into atomic oxygen, which can now react with the adsorbed CO. The CO_2 created in this way is then desorbed again into the gaseous phase.

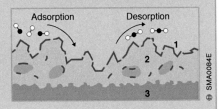

○ Oxygen
● Carbon

1 Catalytic layer
2 Substrate coating
3 Substrate

SMA0084E

Lambda control loop

Function

In order that the conversion rates of the three-way catalytic converter are as high as possible for the pollutant components HC, CO and NO$_X$, the reaction components must be present in the stoichiometric ratio. This requires a mixture composition of $\lambda = 1.0$, i.e., the stoichiometric air/fuel ratio must be adhered to very precisely. Mixture formation must therefore be followed up in a control loop, because sufficient accuracy cannot be achieved solely by controlling the metering of the fuel.

Method of operation

When the lambda control loop is used, deviations from a specific air/fuel ratio can be detected and corrected through the quantity of fuel injected. The residual-oxygen content in the exhaust gas, which is measured with lambda oxygen sensors, serves as the measure for the composition of the air/fuel mixture.

For the purpose of two-step control which can only maintain a value of $\lambda = 1$, a two-step lambda sensor (Fig. 10, Pos. 3a) is incorporated in the exhaust-gas branch upstream of the primary catalyst (4). However, the use of a wide-band lambda sensor also permits continuous lambda closed-loop control to λ values which deviate from the value 1.

Greater accuracy is achieved by a two-sensor control system in which a second lambda sensor (3b) is located downstream of the main catalyst (5).

Two-step control

Two-step lambda control maintains the air/fuel mixture at $\lambda = 1$. A two-step lambda sensor serving as a measuring sensor in the exhaust pipe constantly provides information on whether the mixture is richer or leaner than $\lambda = 1$. A high sensor voltage (e.g., 800 mV) indicates a rich mixture, while a low sensor voltage (e.g., 200 mV) indicates a leaner mixture.

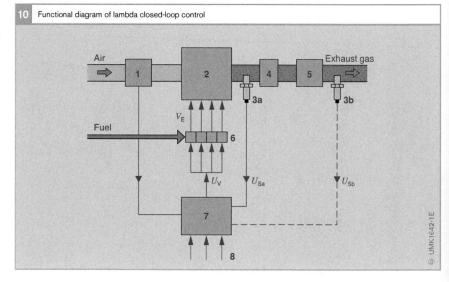

10   Functional diagram of lambda closed-loop control

On each transition from rich to lean and lean to rich, the sensor output signal demonstrates a voltage jump, which is evaluated by a control circuit. The manipulated variable changes its control direction in response to each voltage jump: The lambda controller lengthens or shortens the time that the fuel injector is activated and thereby increases or reduces the injected fuel quantity.

The manipulated variable (lengthening or shortening factor of the injection duration) is made up of a jump and a ramp (Fig. 11). In other words, when there is a jump in the sensor signal, the mixture is first altered immediately by a specific amount (jump) in order to bring about a mixture correction as quickly as possible. Then the manipulated variable follows an adaptation function (ramp) until another sensor-signal voltage jump occurs. The air/fuel mixture thus constantly changes its composition in a very narrow range around $\lambda = 1$.

The typical shift of the oxygen zero crossing (theoretically at $\lambda = 1.0$) and thus of the lambda-sensor jump, conditioned by the variation of the exhaust-gas composition, can be compensated for by shaping the manipulated-variable curve asymmetrically (rich/lean shift). The preferred option here is to maintain the ramp value for the sensor jump for a controlled dwell time t_V after the sensor jump (Fig. 11): During the shift to "rich", the manipulated variable remains for a dwell time t_V in the rich position, even though the sensor signal has already jumped in the "rich" direction. Only after the dwell time has elapsed do the jump and ramp follow the manipulated variable in the "lean" direction. If the sensor signal then jumps in the "lean" direction, the manipulated variable is controlled in the directly opposite direction (with jump and ramp) without remaining in the lean position.

During the shift to "lean", the behavior is reversed: If the sensor signal indicates a lean mixture, the manipulated variable remains for the dwell time t_V in the lean position and is only then controlled in the "rich" direction. On the other hand, countercontrol is effected immediately when the sensor signal jumps from "lean" to "rich".

Continuous-action lambda control
Continuous-action lambda control enables the composition of the air/fuel mixture to be controlled to values which deviate from the stoichiometric ratio. Thus, controlled enrichment ($\lambda < 1$), e.g., to protect components, and controlled leaning ($\lambda > 1$), e.g., for a leaner warm-up during catalyst heating, can be effected.

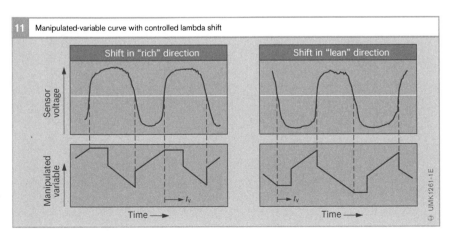

11   Manipulated-variable curve with controlled lambda shift

Shift in "rich" direction

Shift in "lean" direction

Sensor voltage

Manipulated variable

Time

Time

t_V

t_V

UMK1261-1E

Fig. 11
t_V Dwell time after sensor jump

The wide-band lambda sensor outputs a continuous voltage signal U_{Sa}. This means that not only the lambda range (rich or lean) but also the amount of the deviation from $\lambda = 1$ can be measured and directly evaluated. In this way, lambda control can react more quickly to a mixture deviation. The increased dynamic response results in an improved control response.

It is also possible with the wide-band lambda sensor (in contrast to control with a two-step lambda sensor) to effect control to mixture compositions which deviate from the stoichiometric ratio ($\lambda = 1$). The measurement range stretches to lambda values in the field of $\lambda = 0.7$ up to "pure air"; the range of active lambda control is limited, depending on the application. Continuous-action lambda control is therefore suitable for lean and rich operation.

Two-sensor control

When it is situated upstream of the catalyst, the lambda sensor is heavily stressed by high temperatures and untreated exhaust gas, and this leads to limitations in lambda-control accuracy. Locating a lambda sensor (Fig. 10, Pos. 3b) downstream of the catalyst means that these influences are considerably reduced. However, lambda control on its own with the sensor downstream of the catalyst demonstrates disadvantages in dynamic response on account of the gas travel times, and responds to mixture changes more slowly.

Greater accuracy is achieved with two-sensor control. Here, a slower correction control loop is superimposed on the two-step or continuous-action lambda control described by means of an additional two-step lambda sensor.

Lambda closed-loop control for gasoline direct injection

Lambda closed-loop control for gasoline direct injection does not differ for homogeneous operation ($\lambda = 1$) from the control strategies described above. Systems which additionally support lean engine operation ($\lambda > 1$) require control of the NO_X accumulator-type catalytic converter. The NO_X accumulator-type catalyst has a dual function. During lean-burn operation, NO_X accumulation and HC and CO oxidation must take place. In addition, at $\lambda = 1$, a stable three-way function is needed which provides for a minimum level of oxygen accumulation. The lambda sensor upstream of the catalyst monitors the stoichiometric composition of the mixture.

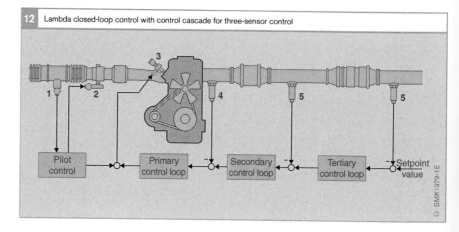

12   Lambda closed-loop control with control cascade for three-sensor control

Fig. 12
1   Air-mass sensor (HFM)
2   Throttle valve
3   Fuel injector
4   Wide-band lambda sensor
5   Two-step lambda sensor

SMK1979-1E

In addition to contributing to two-sensor control, the two-step sensor downstream of the NO$_X$ accumulator-type catalyst also monitors the combined O$_2$ and NO$_X$ accumulation performance (detection of the end of the NO$_X$ removal phase).

Three-sensor control
In the interests of both catalytic-converter diagnosis (separate monitoring of the primary and main catalysts) and exhaust-gas constancy, the use of a third sensor downstream of the main catalyst is recommended in SULEV vehicles (Super Ultra-Low-Emission Vehicle, a category defined in the Californian emission-control legislation) (Fig. 12). The two-sensor control system (single cascade) is expanded to include extremely slow control with the third sensor downstream of the main catalyst and thereby facilitates faster control with the second sensor.

Individual-cylinder control
Above all with upstream primary catalysts, it is not possible to ensure that the exhaust gases from the individual cylinders are sufficiently mixed together before they pass through the catalyst. The fact that exhaust gas passes through the catalyst segments in strands, depending on the deviation from $\lambda = 1$ of the cylinders, results in insufficient conversion. Lambda coordination of the individual cylinders can bring about significant reductions of exhaust emissions; this process involves adjusting the cylinders individually to $\lambda = 1$. Extremely high demands must be place on the dynamic response of the lambda sensor in order to obtain the lambda values for the individual cylinders from a measured lambda signal.

▶ **Ozone and smog**

Exposure to the sun's radiation splits nitrogen-dioxide molecules (NO$_2$). The products are nitrogen oxide (NO) and atomic oxygen (O), which combine with the ambient air's molecular oxygen (O$_2$) to form ozone (O$_3$). Ozone formation is also promoted by volatile organic compounds such as hydrocarbons. This is why higher ozone levels must be anticipated on hot, windless summer days when high levels of air pollution are present.

In naturally occurring concentrations, ozone is essential for human life. However, in higher concentrations it leads to coughing, irritation of the throat and sinuses, and burning eyes. It adversely affects lung function, reducing performance potential.

There is no direct contact or mutual movement between the undesirable ozone formed at ground level, and the stratospheric ozone that reduces the amount of ultraviolet radiation from the sun.

Smog can be created in winter in response to atmospheric layer inversions and low wind speeds. The temperature inversion in the air layers prevents the heavier, colder air containing the higher pollutant concentrations from rising and dispersing.

Smog leads to irritation of the mucous membranes, eyes, and respiratory system. It can also impair visibility. This last factor explains the origin of the term smog, which is a contraction of the words "smoke" and "fog".

Emission-control legislation

The US state of California has assumed a pioneering role in efforts to restrict by law pollutant emissions emanating from motor vehicles. The California legislature was spurred into action not least due to the fact that the city of Los Angeles is situated in a basin, preventing wind from dispersing exhaust gases and causing a blanket of haze to descend on and hang over the city. Given the high levels of pollutants in the air, this phenomenon led to the buildup of smog and as a result caused health problems in the city's population and massive restrictions on visibility.

Overview

California introduced the first emission-control legislation for gasoline engines in the mid-1960s. These regulations became progressively more stringent in the ensuing years. In the meantime, all industrialized countries have introduced emission-control laws which define limits for gasoline and diesel engines, as well as the test procedures employed to confirm compliance. In some countries, regulations governing exhaust emissions are supplemented by limits on evaporative losses from the fuel system.

The most important legal restrictions on exhaust emissions are (Fig. 1):
● CARB legislation
  (California Air Resources Board)
● EPA legislation
  (Environment Protection Agency), USA
● EU legislation (European Union)
● Japanese legislation

Classifications
Countries with legal limits on motor-vehicle emissions divide vehicles into various classes:
● Passenger cars: Emission testing is conducted on a chassis dynamometer.
● Light-duty trucks: The upper limit lies at a maximum permissible weight of between 3.5 and 6.35 metric tons, varying according to country. Testing is carried out on a chassis dynamometer (as with passenger cars).
● Heavy-duty trucks: Maximum permissible weight in excess of 3.5...6.35 metric tons. Testing is performed on an engine test bench, with no provision for in-vehicle testing.
● Off-highway (e.g., construction, agricultural, and forestry vehicles): Tested on an engine test bench, as for heavy-duty trucks.

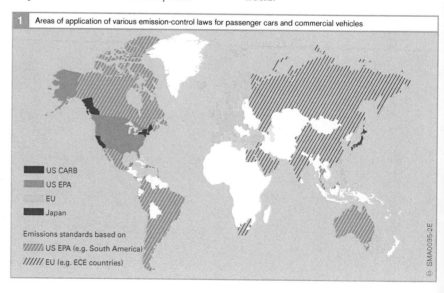

1   Areas of application of various emission-control laws for passenger cars and commercial vehicles

■ US CARB
■ US EPA
□ EU
■ Japan

Emissions standards based on
▨ US EPA (e.g. South America)
▨ EU (e.g. ECE countries)

SMA0095-2E

Test procedures

Japan and the European Union have followed the lead of the United States by defining test procedures for certifying compliance with emission limits. These procedures have been adopted in modified or unrevised form by other countries.

Legal requirements prescribe any of three different tests, depending on vehicle class and test objective:
- Type Approval (TA) to obtain General Certification
- Random testing of vehicles from serial production conducted by the approval authorities (Conformity of Production)
- In-field monitoring for testing certain exhaust-gas components in vehicle operation

Type approval

Type approvals are a precondition for granting General Certification for an engine or vehicle type. This process entails proving compliance with stipulated emissions limits in defined test cycles. Different countries have defined individual test cycles and emission limits.

Test cycles

Dynamic test cycles are specified for passenger cars and light-duty trucks. The country-specific differences between the two procedures are rooted in their respective origins:
- Test cycles designed to mirror conditions recorded in actual highway operation, e.g. Federal Test Procedure (FTP) test cycle in the USA, and
- Synthetically generated test cycles consisting of phases at constant cruising speed and acceleration rates, e.g., Modified New European Driving Cycle (MNEDC) in Europe

The mass of toxic emissions from each vehicle is determined by operating it in conformity with speed cycles precisely defined for the test cycle. During the test cycle, the exhaust gases are collected for subsequent analysis to determine the pollutant mass emitted during the driving cycle.

For heavy-duty trucks (on- and off-highway), steady-state exhaust-gas tests (e.g. 13-stage test in the EU), or dynamic tests (e.g. Transient Cycle in the USA) are carried out on the engine test bench.

All the test cycles are depicted at the end of this section.

Random testing

This testing is usually conducted by the vehicle manufacturer in the quality-control checks that accompany the production process. The same test procedures and limits are generally applied as for type approval. The authorities responsible for granting homologation approval can demand confirmation testing as often as deemed necessary. EU regulations and ECE directives (Economic Commission for Europe) take account of production tolerances by carrying out random testing on a minimum of 3 to a maximum of 32 vehicles. The most stringent requirements are encountered in the USA, and particularly in California, where the authorities require what is essentially comprehensive and total quality monitoring.

In-field monitoring

Random emission-control tests are conducted in driving mode on vehicles whose running performance and age are within specific limits. The emission-control test procedure is simplified compared with type approval.

CARB legislation (passenger cars/LDTs)

CARB, or California Air Resources Board emission limits for passenger cars and Light-Duty Trucks (LDTs) are defined in standards specifying exhaust-gas emissions:
- LEV I, and
- LEV II (Low Emission Vehicle)

Since model year 2004 the LEV II standard has governed all new vehicles up to a maximum permissible weight of 14,000 lb (lb: pound; 1 lb = 0.454 kg).

Phase-in

Following introduction of the LEV II standard, at least 25 % of new vehicle registrations must be certified to this standard. The phase-in rule stipulates that an additional 25 % of new vehicle registrations must then conform to the LEV II standard in each consecutive year. As of 2007 all new vehicle registrations must then be certified according to the LEV II standard.

Emission limits

The CARB legislation defines limits on
- Carbon monoxide (CO)
- Nitrous oxides (NO_X)
- Non-methane organic gases (NMOG)
- Formaldehyde (LEV II only)
- Particulate emissions (diesel engines: LEV I and LEV II; gasoline-engines: LEV II only)

Actual emission levels are determined using the FTP 75 driving cycle (Federal Test Procedure). Limits are defined in relation to distance and specified in grams per mile.

Within the period 2001 through 2004 the SFTP (Supplemental Federal Test Procedure) standard was introduced together with other test cycles. There are also further limits that require compliance in addition to FTP emission limits.

Emission categories

Automotive manufacturers are at liberty to deploy a variety of vehicle concepts within the permitted limits, providing they maintain a fleet average (see the section on "Fleet averages"). The concepts are allocated to the following exhaust-gas categories, depending on their emission values for NMOG, CO, NO_X, and particulate emissions (Fig. 2):

Fig. 2
[1] Limit in each case for "full useful life" (10 years/ 100,000 miles for LEV I or 120,000 miles for LEV II)
[2] Limit in each case for "intermediate useful life" (5 years/ 50,000 miles)
[3] Only limits for "full useful life" (see section entitled "Long-term compliance")

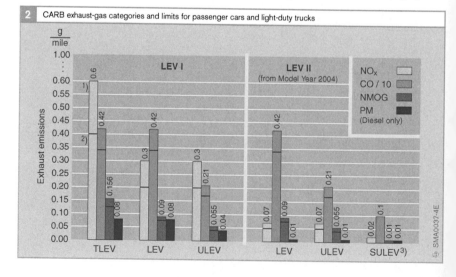

2 CARB exhaust-gas categories and limits for passenger cars and light-duty trucks

- TLEV (Transitional Low-Emission Vehicle; LEV I only)
- LEV (Low-Emission Vehicle, i.e., vehicles with low exhaust and evaporative emissions)
- ULEV (Ultra-Low-Emission Vehicle)
- SULEV (Super Ultra-Low-Emission Vehicle)

In addition to the categories of LEV I and LEV II, two categories define zero-emission and partial zero-emission vehicles:
- ZEV (Zero-Emission Vehicle, i.e., vehicles without exhaust and evaporative emissions), and
- PZEV (Partial ZEV, which is basically SULEV, but with more stringent limits on evaporative emissions and stricter long-term performance criteria)

Since 2004 new vehicle registrations have been governed by the LEV II exhaust-emission standard. At the same time, TLEV will be replaced by SULEV with its substantially lower limits. The LEV and ULEV classifications remain in place. The CO and NMOG

limits from LEV I remain unchanged, but the NO$_X$ limit is substantially lower for LEV II. The LEV II standard also includes new, supplementary limits governing formaldehyde.

Long-term compliance
To obtain approval for each vehicle model (type approval), the manufacturer must prove compliance with the official emission limits for pollutants. Compliance means that the limits may not be exceeded for the following mileages or periods of useful life:
- 50,000 miles or 5 years ("intermediate useful life"), and
- 100,000 miles (LEV I) / 120,000 miles (LEV II) or 10 years ("full useful life")

Manufacturers also have the option of certifying vehicles for 150,000 miles using the same limits that apply to 120,000 miles. The manufacturer then receives a bonus when the NMOG fleet average is defined (see the section entitled "Fleet averages").

The relevant figures for the PZEV emission-limit category are 150,000 miles or 15 years ("full useful life").

For this type of approval test, the manufacturer must furnish two vehicle fleets from series production:
- One fleet in which each vehicle must cover 4000 miles before testing.
- One fleet for endurance testing, in which deterioration factors for individual components are defined.

Endurance testing entails subjecting the vehicles to specific driving cycles over distances of 50,000 and 100,000/120,000 miles. Exhaust emissions are tested at intervals of 5000 miles. Service inspections and maintenance are restricted to the standard prescribed intervals.

Countries that base their regulations on the US test cycles allow application of defined deterioration factors to simplify the certification process.

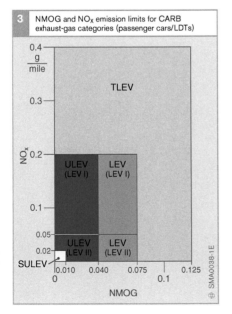

3 NMOG and NO$_X$ emission limits for CARB exhaust-gas categories (passenger cars/LDTs)

Fleet averages (NMOG)

Each vehicle manufacturer must ensure that exhaust emissions for its total vehicle fleet do not exceed a specified average. NMOG emissions serve as the reference category for assessing compliance with these averages. The fleet average is calculated from the average figures produced by all of the manufacturer's vehicles that comply with NMOG limits and are sold within one year. Different fleet averages apply to passenger cars and light-duty trucks and vans.

The compliance limits for the NMOG fleet average are lowered in each subsequent year (Fig. 4). To meet the lower limits, manufacturers must produce progressively cleaner vehicles in the more stringent emissions categories in each consecutive year.

The fleet averages apply irrespective of LEV I or LEV II standards.

Corporate average fuel economy

US legislation specifies the average amount of fuel an automotive manufacturer's vehicle fleet may consume per mile. The prescribed CAFE value (Corporate Average Fuel Economy) currently (2004) stands at 27.5 mpg for passenger cars. This corresponds to 8.55 liters per 100 kilometers in the metric system. At the present time it is not planned to reduce this limit.

The value for light-duty trucks is 20.7 miles per gallon or 11.4 liters per 100 kilometers. Between 2005 and 2007 "fuel economy" will be raised each year by 0.6 mpg. There are no regulations for heavy-duty trucks.

At the end of each year the average fuel economy for each manufacturer is calculated based on the numbers of vehicles sold. The manufacturer must remit a penalty fee of $5.50 per vehicle for each 0.1 mpg its fleet exceeds the target. Buyers will also have to pay a gas-guzzler tax on vehicles with especially high fuel consumption. Here, the limit is 22.5 miles per gallon (corresponding to 10.45 liters per 100 kilometers).

These penalties are intended to spur development of vehicles offering greater fuel economy.

The FTP 75 test cycle and the highway cycle are applied to measure fuel economy (see the section entitled "US test cycles").

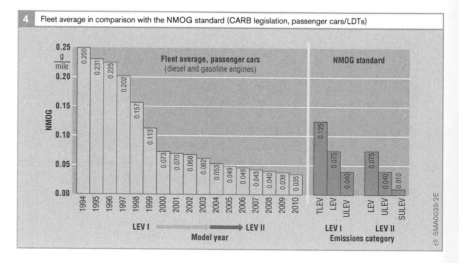

4   Fleet average in comparison with the NMOG standard (CARB legislation, passenger cars/LDTs)

Fig. 4
Fleet average for "intermediate useful life". Only "full useful life" exists for SULEV.

Zero-emission vehicles

As from 2005 10 % of new vehicle registrations will have to meet the requirements of the ZEV (Zero-Emission Vehicle) exhaust-gas category. These vehicles may emit no exhaust gas or evaporative emissions in operation. This category mainly refers to electric cars.

A 10 % percentage may partly be covered by vehicles of the PZEV (Partial Zero-Emission Vehicles) exhaust-gas category. These vehicles are not zero-emission, but they emit very few pollutants. They are weighted using a factor of 0.2 to 1, depending on the emission-limit standard. The minimum weighting factor of 0.2 is granted when the following demands are met:
- SULEV certification indicating long-term compliance extending over 150,000 miles or 15 years
- Warranty coverage extending over 150,000 miles or 15 years on all emission-related components
- No evaporative emissions from the fuel system (0 EVAP, Zero Evaporation), achieved by extensive encapsulation of the tank and fuel system

Special regulations apply to hybrid vehicles combining diesel engines and electric motors. These vehicles can also contribute to achieving compliance with the 10 % limit.

In-field monitoring
Unscheduled testing
Random emission testing is conducted on in-use vehicles using the FTP 75 test cycle and an evaporation test. Depending on the relevant exhaust-gas category, vehicles with mileage readings below 75,000/90,000/105,000 miles are tested.

Vehicle monitoring by the manufacturer
Official reporting of claims and damage to specific emissions-related components and systems has been mandatory for vehicle manufacturers since model year 1990. The reporting obligation remains in force

for a maximum period of 15 years, or 150,000 miles, depending on the warranty period applying to the component or assembly.

Reporting is divided into three stages with an incremental amount of detail:
- Emissions Warranty Information Report (EWIR)
- Field Information Report (FIR), and
- Emission Information Report (EIR)

Information concerning
- problem reports,
- malfunction statistics,
- defect analysis, and
- the effects on emissions

is then forwarded to the environment-protection authorities. The authorities use the FIR as the basis for issuing mandatory recall orders to the manufacturer.

EPA legislation (passenger cars/LDTs)

EPA (Environment Protection Agency) legislation applies to all of the states where the more stringent CARB stipulations from California are not in force. CARB regulations have already been adopted by some states, such as Maine, Massachusetts, and New York.

EPA regulations in force since 2004 conform to the Tier 2 standard.

Emission limits
The EPA legislation defines limits on the following pollutants:
- Carbon monoxide (CO)
- Nitrous oxides (NO_X)
- Non-Methane Organic Gases (NMOG)
- Formaldehyde (HCHO), and
- Solids (particulates)

Pollutant emissions are determined using the FTP 75 driving cycle. Limits are defined in relation to distance and specified in grams per mile.

The SFTP (Supplemental Federal Test Procedure) standard, comprising further test cycles, has been in force since 2001. The applicable limits require compliance in addition to FTP emission limits.

Since the introduction of Tier 2 standards, vehicles with diesel and gasoline engines have been subject to identical exhaust-emission limits.

Exhaust-gas categories

Tier 2 (Fig. 5) classifies limits for passenger cars in 10 bins (emission standards) and 11 bins for heavy-duty trucks. Bin 9 and Bin 10 in relation to passenger cars will be dropped in 2007.

The transition to Tier 2 has produced the following changes:
- Introduction of fleet averages for NO_x
- Formaldehydes (HCHO) are subject to a separate pollutant category
- Passenger cars and light-duty trucks up to 6000 lb (2.72 t) are combined in a single vehicle class

- MDPV (Medium-Duty Passenger Vehicle) forms a separate vehicle category; previously assigned to HDV (Heavy-Duty Vehicles)
- "Full useful life" is increased, depending on the emission standard (bin), to 120,000 miles or 150,000 miles

Phase-in

At least 25 % of all new passenger-car and LLDT (light LDTs) registrations are required to conform to Tier 2 standards when they take effect in 2004. The phase-in rule stipulates that an additional 25 % of vehicles will then be required to conform to the Tier 2 standards in each consecutive year. All vehicles are required to conform to Tier 2 standards by 2007. For the category HLDT/MDPV, the phase-in period ends in 2009.

Fleet averages

NO_x emissions are used to determine fleet averages for individual manufacturers under EPA legislation. CARB regulations, however, are based on NMOG emissions.

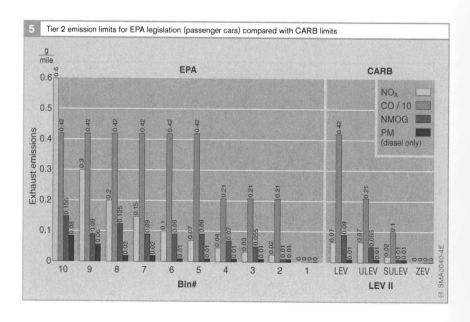

5 Tier 2 emission limits for EPA legislation (passenger cars) compared with CARB limits

Corporate average fuel economy

The regulations defining average fleet fuel consumption in the 49 states are the same as those applied in California. Again, the limit applicable to passenger cars is 27.5 miles per gallon (8.55 liters per 100 kilometers). Beyond this figure manufacturers are required to pay a penalty. The purchaser also pays a penalty tax on vehicles providing less than 22.5 miles per gallon.

In-field monitoring

Unscheduled testing

In analogy to CARB legislation, the EPA regulations require random exhaust-gas emission testing based on the FTP 75 test cycles for in-use vehicles. Testing is conducted on low-mileage (10,000 miles, roughly one year old) and higher-mileage vehicles (50,000 miles, in California however at least one vehicle per test group with 75,000 or 90,000 or 105,000 miles, depending on the emission standard; roughly four years old). The number of vehicles tested varies according to the number sold. At least one vehicle in each group is also tested for evaporative emissions.

Vehicle monitoring by the manufacturer

For vehicles after model year 1972, the manufacturer is obliged to make an official report concerning damage to specific emission-related components or systems if at least 25 identical emission-related parts in a model year are defective. The reporting obligation ends five years after the end of the model year. The report comprises a description of damage to the defective component, presentation of the impacts on exhaust-gas emissions, and suitable corrective action by the manufacturer. The environmental authorities use this information as the basis for determining whether to issue recall orders to the manufacturer.

EU legislation (passenger cars/LDTs)

The directives contained in European Union emission-control legislation are defined by the EU Commission. Emission-control legislation for passenger cars/light-duty trucks is Directive 70/220/EEC from 1970. For the first time it defined exhaust-emission limits, and the provisions have been updated ever since.

The emission limits for passenger cars and light-duty trucks (LDTs) are:
- Euro 1 (as from 1 July 1992)
- Euro 2 (as from 1 January 1996)
- Euro 3 (as from 1 January 2000)
- Euro 4 (as from 1 January 2005)

A new exhaust-emission standard is generally introduced in two stages. In the first stage, compliance with the newly defined emission limits is required in vehicle models submitted for initial homologation approval certification (TA, Type Approval). In the second stage – usually one year later – every new registration must comply with the new limits (First Registration, FR). The authorities can also inspect vehicles from series production to verify compliance with emissions limits (COP, Conformity of Production).

EU directives allow tax incentives for vehicles that comply with upcoming exhaust-gas emission standards before they actually become law. Depending on a vehicle's emission standard, there are a number of different motor-vehicle tax rates in Germany.

Emission limits

The EU standards define limits on the following pollutants:
- Carbon monoxide (CO)
- Hydrocarbons (HCs)
- Nitrous oxides (NO_X)
- Particulates, although these limits are initially restricted to diesel vehicles

The limits for hydrocarbons and nitrogen oxides for the Euro 1 and Euro 2 stages are combined into an aggregate value (HC+NO$_X$). Since Euro 3, there has been a special restriction for NO$_X$ emissions in addition to the aggregate value. The limits are defined based on mileage and are specified in grams per kilometer (g/km) (Fig. 6). Since Euro 3, emissions are measured on a chassis dynamometer using the MNEDC (Modified New European Driving Cycle).

The limits are different for vehicles with diesel or gasoline engines. In future, they will merge (probably by Euro 5).

The limits for the LDT category are not uniform. There are three classes (1 to 3) into which LDTs are subdivided, depending on the vehicle reference weight (unladen weight + 100 kg). The limits for Class 1 are the same as for cars.

Type approval

While type approval testing basically corresponds to the US procedures, deviations are encountered in the following areas: measurements of the pollutants HC, CO, NO$_X$ are supplemented by particulate and exhaust-gas opacity measurements on diesel vehicles. Test vehicles absolve an initial run-in period of 3000 kilometers before testing.

Deterioration factors used to assess test results are defined in the legislation for every pollutant component; manufacturers are also allowed to present documentation confirming lower factors obtained during specified endurance testing over 80,000 km (100,000 km starting with Euro 4).

Compliance with the specified limits must be maintained over a distance of 80,000 km (Euro 3), or 100,000 km (Euro 4), or after 5 years.

Type tests

There are six different type tests for type approval. The Type I, Type IV, Type V and Type VI tests applied to vehicles with gasoline engines.

The Type I test evaluates exhaust emissions after cold starting. Exhaust-gas opacity is also assessed on vehicles with diesel engines.

Type IV testing measures evaporative emissions from parked vehicles. These emissions consist primarily of the gases that evaporate from the fuel tank.

Type VI testing embraces hydrocarbon and carbon monoxide emissions immediately following cold starts at −7 °C. The first section of the MNEDC (urban portion) is employed for this test. This test assumed mandatory status in 2002.

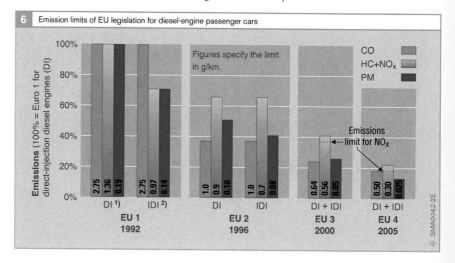

6   Emission limits of EU legislation for diesel-engine passenger cars

Figures specify the limit in g/km.

CO
HC+NO$_X$
PM

Emissions limit for NO$_X$

Emissions (100% = Euro 1 for direct-injection diesel engines (DI))

	DI [1]	IDI [2]	DI	IDI	DI + IDI	DI + IDI
CO	2.75	2.75	1.0	1.0	0.64	0.50
HC+NO$_X$	1.36	0.97	0.9	0.7	0.56	0.30
PM	0.19	0.14	0.10	0.08	0.05	0.025
	EU 1 1992		EU 2 1996		EU 3 2000	EU 4 2005

SMA0042-2E

Fig. 6
[1] For direct-injection engines
[2] For indirect-injection engines

The Type V test assesses the long-term durability of the emissions-reducing equipment. It may involve a specific test sequence, or it may be subjected to deterioration factors specified by the legislation.

CO_2 emissions

There are no legal emission limits for CO_2. However, the vehicle manufacturers (Association des Constructeurs Européen d'Automobiles, ACEA (Association of European Automobile Manufacturers)) have pledged to uphold a voluntary program. The objective is to achieve CO_2 emissions of max. 140 g/km for passenger cars by 2008 – this is equivalent to a fuel consumption of 5.8 l/100 km.

In Germany, vehicles with specially low CO_2 emissions (so-called 5-liter and 3-liter cars) will be tax-exempt until the end of 2005.

In-field monitoring

EU legislation also calls for conformity-verification testing on in-use vehicles as part of the Type I test cycle. The minimum number of vehicles of a vehicle type under test is three, while the maximum number varies according to the test procedure.

Vehicles under test must meet the following criteria:
- Mileages vary from 15,000 km to 80,000 km, and vehicle age from 6 months to 5 years (Euro 3). In Euro 4, the mileage specified ranges from 15,000 km to 100,000 km.
- Regular service inspections were carried out as specified by the manufacturer.
- The vehicle must show no indications of non-standard use (e.g. tampering, major repairs, etc.).

If emissions from an individual vehicle fail substantially to comply with the standards, the source of the high emissions must be determined. If more than one vehicle displays excessive emissions in random testing, no matter what the reason, the results of the random test must be classified as negative. If there are various reasons, the test schedule may be extended, providing the maximum sample size is not reached.

If the type-approval authorities detect that a vehicle type fails to meet the criteria, the vehicle manufacturer must devise suitable action to eliminate the defect. The action catalog must be applicable to all vehicles with the same defect. If necessary, a recall action must be started.

Periodic emission testing

In Germany, all passenger cars and light-duty trucks are required to undergo emission testing three years after their first registration, and then every two years. The procedure for emission testing in the garage/workshop comprises the following steps for a gasoline-engine vehicle:
- Visually inspecting the exhaust system
- Checking engine speed and engine temperature for the following CO and lambda-value measurement
- Checking the CO and lambda values within a defined engine-speed window (increased idle speed) at a minimum engine temperature
- For vehicles with On-Board Diagnosis (OBD): Reading out the fault memory; reading out the readiness codes to determine whether all the diagnosis functions have been carried out; if this is not the case, the lambda-sensor voltage is additionally check after the CO measurement at idle
- In vehicles with closed-loop-controlled three-way catalysts without OBD, the catalyst control loop is checked by feed-forwarding a disturbance value

Japanese legislation (passenger cars/LDTs)

The permitted emission values are also subject to gradual stages of severity in Japan. Further tightening of the regulations is planned for 2005.

Vehicles with a maximum permissible weight of up to 3.5 t are essentially divided into three categories: passenger cars (up to 10 seats), LDVs (Light-Duty Vehicles) up to 1.7 t, and MDVs (Medium-Duty Vehicles) up to 3.5 t. Compared with the other two vehicle categories, MDVs are governed by higher limits on CO and NO_X emissions (for gasoline-engine vehicles). For diesel engines, the vehicle categories are distinguished by NO_X and particulate limits.

Emission limits

Japanese legislation specifies limits on the following emissions:

● Carbon monoxide (CO)
● Nitrous oxides (NO_X)
● Hydrocarbons (NMHC)
● Particulates (diesel vehicles only)
● Smoke (diesel vehicles only)

1 Emission limits of Japanese legislation for gasoline-engine passenger cars		
CO	NO_X	NMHC
1.15 g/km	0.05 g/km	0.05 g/km

Pollutant emissions are determined from a combination of 11-mode and 10·15-mode test cycles (see section entitled "Japanese test cycle"). Cold-start emissions are thus also taken into account.

Corporate average fuel economy

In Japan, measures for reducing the CO_2 emissions of cars are planned. One proposal plans to fix the average fuel economy (CAFE value) for the entire passenger-car fleet. The proposal foresees a gradual phase-in of emission limits by vehicle weight.

US test cycles for passenger cars and LDTs

FTP 75 test cycle

The FTP 75 test cycle (Federal Test Procedure) consists of speed cycles that were actually recorded in commuter traffic in Los Angeles (Fig. 8a).

This test is also in force in some countries of South America besides the USA (including California).

Conditioning
The vehicle is subjected to an ambient temperature of 20...30 °C for a period of 6 to 36 hours.

Collecting pollutants
The vehicle is started and driven on the specific speed cycle. The emitted pollutants are collected in separate bags during various phases.

Phase ct (cold transient):
Collection of exhaust gas during the cold test phase.

Phase s (stabilized):
Start of stabilized phase 505 seconds after start. The exhaust gas is collected without interrupting the driving cycle. Upon termination of phase s, after a total of 1372 seconds, the engine is switched off for a period of 600 seconds.

Phase ht (hot transient):
The engine is restarted for the hot test. The speed cycle is identical to the cold transient phase (Phase ct).

Analysis
The bag samples from the first two phases are analyzed during the pause before the hot test. This is because samples may not remain in the bags for longer than 20 minutes.

The sample exhaust gases contained in the third bag are also analyzed on completion of the driving cycle. The total result includes emissions from the three phases rated at different weightings.

The pollutant masses of Phases ct and s are aggregated and assigned to the total distance of these two phases. The result is then weighted at a factor of 0.43.

The same process is applied to the aggregated pollutant masses from Phases ht and s, related to the total distance of these two phases, and weighted at a factor of 0.57. The test result for the individual pollutants (HC, CO, and NO$_X$) is obtained from the sum of the two previous results.

The emissions are specified as the pollutant emission per mile.

SFTP schedules

Tests according to the SFTP standard (Supplemental Federal Test Procedure) were phased in from 2001 to 2004. These are made up of the following driving cycles:
- FTP 75 cycle
- SC03 cycle (Fig. 8b), and
- US06 cycle (Fig. 8c)

The extended tests are intended to examine the following additional driving conditions:
- Aggressive driving
- Radical changes in vehicle speed
- Engine start and acceleration from a standing start
- Operation with frequent minor variations in speed
- Periods with vehicle parked
- Operation with air conditioner on

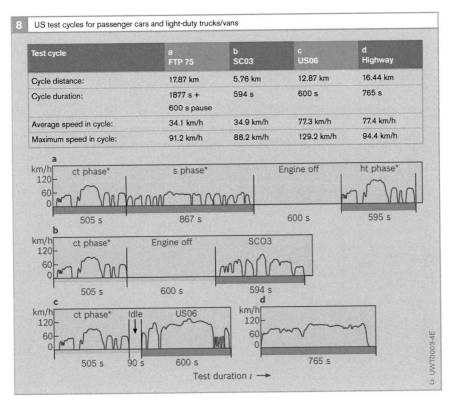

8   US test cycles for passenger cars and light-duty trucks/vans

Test cycle	a FTP 75	b SC03	c US06	d Highway
Cycle distance:	17.87 km	5.76 km	12.87 km	16.44 km
Cycle duration:	1877 s + 600 s pause	594 s	600 s	765 s
Average speed in cycle:	34.1 km/h	34.9 km/h	77.3 km/h	77.4 km/h
Maximum speed in cycle:	91.2 km/h	88.2 km/h	129.2 km/h	94.4 km/h

Fig. 8
* ct cold phase
  s stabilized phase
  ht hot test
  ■ Exhaust-gas collection phases
  □ Conditioning (other driving cycles are also possible)

Following preconditioning, the SC03 and US06 cycles proceed through the ct phase from FTP 75 without exhaust-gas collection. Other conditioning procedures may also be used.

The SC03 cycle is carried out at a temperature of 35 °C and 40 % relative humidity (vehicles with air conditioning only). The individual driving schedules are weighted as follows:
- Vehicles with air conditioning:
  35 % FTP 75 + 37 % SC03 + 28 % US06
- Vehicles without air conditioning:
  72 % FTP 75 + 28 % US06

The SFTP and FTP 75 test cycles must be successfully completed on an individual basis.

Cold-start enrichment, which is necessary when a vehicle is started at low temperatures, produces particularly high emissions. These cannot be measured in current emissions testing, which is conducted at ambient temperatures of 20...30 °C. A supplementary emissions test is performed at −7 °C to support enforcement of limits on these emissions. However, only carbon monoxide is subject to specified limits in this test.

Test cycles for determining corporate average fuel economy

Each vehicle manufacturer is required to provide data on corporage average fuel economy. Manufacturers that fail to comply with the emission limits are required to pay penalties.

Fuel consumption is determined from the exhaust-gas emissions produced during two test cycles: the FTP 75 test cycle (weighted at 55 %) and the highway test cycle (weighted at 45 %). An unmeasured highway test cycle (Fig. 8d) is conducted once after preconditioning (vehicle allowed to stand with engine off for 12 hours at 20...30 °C). The exhaust emissions from a second test run are then collected. The CO_2 emissions are used to calculate fuel consumption.

European test cycle for passenger cars and LDTs

MNEDC

The Modified New European Driving Cycle (MNEDC) has been in force since Euro 3. Contrary to the New European Driving Cycle (Euro 2), that only starts 40 seconds after the vehicle has started, the MNEDC also includes a cold-start phase.

Conditioning

The vehicle is allowed to start at an ambient temperature of 20...30 °C for a minimum period of 6 hours.

Since 2002 the starting temperature for Type VI testing has been lowered to −7 °C.

Collecting pollutants

The exhaust gas is collected in bags during two phases:
- Urban Driving Cycle (UDC) at a maximum of 50 km/h
- Extra-urban cycle at speeds up to 120 km/h

Analysis

The pollutant mass measured by analyzing the bag contents is referred to the distance covered.

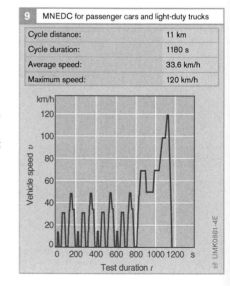

9   MNEDC for passenger cars and light-duty trucks

Cycle distance:	11 km
Cycle duration:	1180 s
Average speed:	33.6 km/h
Maximum speed:	120 km/h

Japanese test cycle for passenger cars and LDTs

The overall test is composed of two test cycles based on different, synthetically generated driving curves. Following a cold start, the 11-mode cycle is run four times, with evaluation of all four cycles. The 10·15-mode test cycle is conducted once with a hot start.

The preconditioning procedure for the hot test includes the prescribed exhaust-gas test at idle. The procedure is as follows: After the vehicle is allowed to warm up for approximately 15 minutes at 60 km/h, the concen-

trations of HC, CO, and CO_2 are measured in the exhaust pipe at idle. The 10·15-mode hot test commences after a second warm-up phase, consisting of 5 minutes at 60 km/h.

While the cold test examines pollutants in grams per test, the hot test's results are defined relative to distance, and are indicated in grams per kilometer.

The exhaust-gas regulations in Japan include limits on evaporative emissions, which are measured using the SHED method (see chapter entitled "Exhaust-gas measurement techniques").

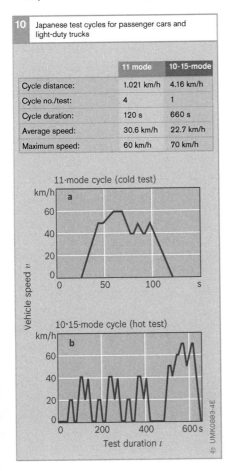

10	Japanese test cycles for passenger cars and light-duty trucks		
		11 mode	10·15-mode
Cycle distance:		1.021 km/h	4.16 km/h
Cycle no./test:		4	1
Cycle duration:		120 s	660 s
Average speed:		30.6 km/h	22.7 km/h
Maximum speed:		60 km/h	70 km/h

11-mode cycle (cold test)

10·15-mode cycle (hot test)

Vehicle speed v

Test duration t

UMK0883-4E

Exhaust-gas measuring techniques

During type-approval testing to obtain General Certification for passenger cars and light-duty trucks, the exhaust-gas test is conducted with the vehicle mounted on a chassis dynamometer. The test differs from exhaust-gas tests that are conducted using workshop/garage measuring devices for in-field monitoring.

Exhaust-gas test for type approval

The prescribed test cycles on the chassis dynamometer stipulate that practical on-road driving mode must be simulated as closely as possible. Testing on a chassis dynamometer offers many advantages compared with on-road testing:

- The results are easy to reproduce since the ambient conditions are constant
- The tests are comparable since a specified speed/time profile is driven irrespective of traffic flow
- The required measuring instruments are set up in a stationary environment

Besides type-approval testing, exhaust-gas measurements on the chassis dynamometer are conducted during the development of engine components.

Test setup

The test vehicle is placed on a chassis dynamometer with its drive wheels on the rollers (Fig. 1, Pos. 1). This means that the forces acting on the vehicle, i.e. the vehicle's moments of inertia, rolling resistance, and aerodynamic drag, must be simulated so that the test cycle on the test bench reproduces emissions comparable to those obtained during an on-road test. For this purpose, asynchronous machines, direct-current machines, or even electrodynamic retarders on older test benches, generate a suitable speed-dependent load that acts on the rollers. More modern machines use electric flywheel simulation to reproduce this inertia. Older test benches use real flywheels of different sizes attached by rapid couplings to the rollers to simulate vehicle mass. A fan mounted in front of the vehicle provides the necessary engine cooling.

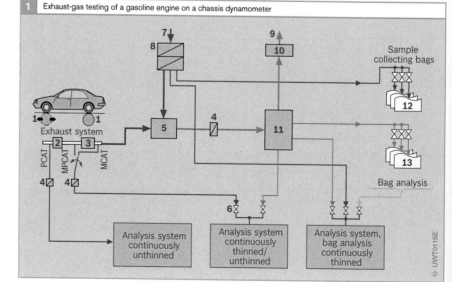

1   Exhaust-gas testing of a gasoline engine on a chassis dynamometer

Fig. 1
1 Roller with dynamometer
2 Primary catalyst
3 Main catalyst
4 Filter
5 Mix-T
6 Valve
7 Dilution air
8 Dilution-air conditioning
9 Exhaust air
10 Fan
11 CVS system
12 Dilution-air sample bag
13 Exhaust-gas sample bag

The test-vehicle exhaust pipe is generally a gas-tight attachment to the exhaust-gas collection system – the dilution system is described below. A proportion of the diluted exhaust gas is collected there. At the end of the test cycle, the gas is analyzed for pollutants limited by law (hydrocarbons, nitrous oxides, and carbon monoxide), and carbon dioxide (to determine fuel consumption).

In addition, and for development purposes, part of the exhaust-gas flow can be extracted continuously from sampling points along the vehicle's exhaust-gas system or dilution system to analyze pollutant concentrations.

The test cycle is repeated by a driver in the vehicle. During this cycle, the required and current vehicle speeds are displayed on a driver monitor. In some cases, an automated driving system replaces the driver to increase the reproducibility of test results by driving the test cycle with extreme precision.

CVS dilution method

The most commonly used method of collecting exhaust gases emitted from an engine is the Constant Volume Sampling (CVS) method. It was introduced for the first time in the USA in 1972 for passenger cars and light-duty trucks. In the meantime it has been updated in several stages. The CVS method is used in other countries, such as Japan. It has also been in use in Europe since 1982.

Objective

In the CVS method, the exhaust gas is only analyzed at the end of the test. It is necessary here to:
- Avoid the condensation of water vapor and the resulting nitrous-oxide losses
- Avoid secondary reactions in the collected exhaust gas

Principle of the CVS method

The exhaust gas emitted by the test vehicle is diluted with ambient air (7) at an average ratio of 1 : 5...1 : 10 and extracted using a special pump setup. This ensures that the total volumetric flow, comprising exhaust gas and dilution air, remains constant. The admixture of dilution air is therefore dependent on the momentary exhaust-gas volumetric flow. A representative sample is continuously extracted from the diluted exhaust-gas flow and is collected in one or more exhaust-gas sample bags (13). The sampling volumetric flow is constant during the bag-filling phase. The pollutant concentration in the exhaust-gas sample bags at the end of the test cycle corresponds to the average value of the concentrations in the diluted exhaust gas for the sample-bag filling period.

As the exhaust-gas sample bags are filled, a sample of dilution air is taken and collected in one or more air-sample bags in order to measure the pollutant concentration in the dilution air.

Filling the sample bags generally corresponds to the phases in which the test cycles are divided (e.g. the ht phase in the FTP 75 test cycle).

The pollutant mass emitted during the test is calculated from the total flow of diluted exhaust gas and the pollutant concentrations.

Dilution systems

There are two alternative methods to obtain a constant volumetric flow of diluted exhaust gas:
- Positive Displacement Pump (PDP) method: A rotary-piston blower (Roots blower) is used
- Critical Flow Venturi (CFV) method: A venturi tube and a standard blower are used in the critical state

Advances in the CVS method

Diluting the exhaust gas causes a reduction in pollutant concentrations as a factor of dilution. As pollutant emissions have dropped significantly in the past few years due to the growing severity of emission limits, the concentrations of some pollutants (in particular hydrogen compounds) in the diluted exhaust gas are equivalent to concentrations in diluted air in certain test phases (or are even lower). This poses a problem from the measuring-process aspect, as the difference between the two values is crucial for measuring exhaust-gas emissions. A further challenge is presented by the precision of analyzers used to measure small concentrations of pollutants.

To tackle these problems, more recent CVS dilution systems apply the following measures:
- Lower the dilution: This requires precautionary measures to guard against the condensation of water, e.g., heating parts of the dilution systems, in gasoline-engine vehicles also drying or heating the dilution air
- Reduce and stabilize pollutant concentrations in the dilution air, e.g., by using carbon canisters
- Optimize the measuring instruments (including dilution systems), e.g., by selecting or preconditioning the materials used and system setups; by using modified electronic components
- Optimize processes, e.g., by using special purge procedures

Bag mini diluter

As an alternative to advances in CVS technology described above, a new type of dilution system was developed in the US: the Bag Mini Diluter (BMD). Here, part of the exhaust gas flow is diluted at a constant ratio with dried, heated zero gas (e.g., cleaned air). During the test, part of the diluted exhaust-gas flow that is proportional to the exhaust-gas volumetric flow is filled in (exhaust-gas) sample bags and analyzed

at the end of the driving test. Diluting the exhaust gas with a pollution-free zero gas dispenses with air-sample bag analysis followed by taking the difference between the exhaust-gas and air-sample bag concentrations. However, a more complex procedure is required than for the CVS method, e.g., one requirement is to determine the (undiluted) exhaust-gas volumetric flow and the proportional sample-bag filling.

Exhaust-gas analyzers

Emission-control legislation defines internationally standardized test procedures for emission-limit pollutants in order to measure the concentrations in exhaust-gas and air-sample bags (Table 1).

1	Test procedure	
Component	**Procedure**	
CO, CO_2	Non-dispersive Infrared analyzer (NDIR)	
Nitrous oxides NO_X	Chemiluminescence detector (CLD) Note: NO_X is interpreted as the sum total of NO and NO_2	
Total hydrocarbon (THC)	Flame-ionization detector (FID)	
CH_4	Combination of gas-chromatographic procedure and flame-ionization detector (GC-FID)	
CH_3OH, CH_2O	Combination of impinger or cartridge procedure and chromatographic analysis techniques; necessary in the USA when certain fuels are used	
Particulates	Gravimetric procedure (weighing of particulate filters before and after the test drive), in Europe and Japan currently only required for diesel vehicles	

For development purposes, many test benches also include the continuous measurement of pollutant concentrations in the vehicle exhaust-gas system or dilution system. The reason is to capture data on emission-limit components, as well as other components not subject to legislation. Other test procedures than those mentioned in Table 1 required for this, e.g.:
- Paramagnetic method (for determining the O_2 concentration)
- Cutter FID: Combination of a flame-ionization detector with an absorber for nonmethane hydrocarbons (for determining the CH_4 concentration)
- Mass spectroscopy (multi-component analyzer)
- FTIR (Fourier Transform InfraRed) spectroscopy (multi components analyzer)
- Infrared-laser spectrometer (multi-component analyzer)

NDIR analyzer
The NDIR (Non-Dispersive InfraRed) analyzer uses the property of certain gases to absorb infrared radiation within a narrow characteristic wavelength band. The absorbed radiation is converted into vibrational and rotational energy of the absorbing molecules.

In the NDIR analyzer, the analysis gas flows through the absorption cell (vessel), where it is exposed to infrared radiation (Fig. 2, Pos. 2). The gas absorbs radiation energy within the characteristic wavelength band of the pollutant, whereby the radiation energy is proportional to the concentration of the pollutant under analysis. A reference cell (7) arranged in parallel to the absorption cell is filled with an inert gas (e.g., nitrogen, N_2).

The detector (9) is located at the opposite end of the cell to the infrared light source and measures the residual energy from infrared radiation in the measurement and reference cells. The detector comprises two chambers connected by a membrane and containing samples of the gas components under analysis. The reference-cell radiation characteristic for this component is absorbed in one chamber. The other absorbs radiation from the test-gas vessel. The difference between the radiation received and the radiation absorbed in the two detector chambers results in a pressure differential, and thus a deflection in the diaphragm between the measuring and reference detectors. This deflection is a measure of the pollutant concentration in the test-gas vessel.

A rotating chopper (8) interrupts infrared radiation cyclically, causing an alternating deflection of the diaphragm, and thus a modulation of the sensor signal.

NDIR analyzers have a strong cross-sensitivity[1]) to water vapor in the test gas, since H_2O molecules absorb infrared radiation across a broad wavelength band. This is the reason why NDIR analyzers are positioned downstream of a test-gas treatment system (e.g., a gas cooler) to dry the exhaust gas when they are used to make measurements on undiluted exhaust gas.

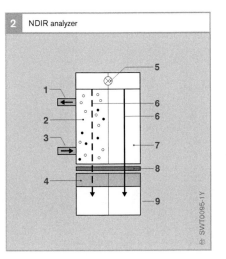

2   NDIR analyzer

1) The absorption of infrared radiation within a particular wavelength band is possible not only with the gas component measured, but also with water vapor

Fig. 2
1 Gas outlet
2 Absorption cell
3 Test-gas inlet
4 Optical filter
5 Infrared light source
6 Infrared radiation
7 Reference cell
8 Rotating chopper
9 Detector

Chemiluminescence detector (CLD)

Due to its measuring principle, the CLD is limited to determining NO concentrations. Before measuring the aggregate of NO_2 and NO concentrations, the test gas is first routed to a converter that reduces NO_2 into NO.

The test gas is mixed with ozone in a reaction chamber (Fig. 3) to determine the nitrogen monoxide concentration (NO). The nitrogen monoxide contained in the test gas oxidizes in this environment to form NO_2; some of the molecules produced are in a state of excitation. When these molecules return to their basic state, energy is released in the form of light (chemiluminescence). A detector, e.g. a photomultiplier, measures the emitted luminous energy; under specific conditions, it is proportional to the nitrogen-monoxide concentration in the test gas.

Flame-ionization detector (FID)

The hydrocarbons present in the test gas are burned off in a hydrogen flame (Fig. 4). This forms carbon radicals; some of the radicals are ionized temporarily. The radicals are discharged at a collector electrode; the current produced is measured and is proportional to the number of carbon atoms in the test gas.

GC-FID and Cutter FID

There are two equally common methods to measure methane concentration in the test gas. Each method consists of the combination of a CH_4-separating element and a flame-ionization detector. Either a gas-chromotography column (GC-FID), or a heated catalytic converter, oxidizes the non-CH_4 hydrocarbons (cutter FID) in order to separate methane. Unlike the cutter FID, the GC-FID can only determine CH_4 concentration discontinuously (typical interval between two measurements: 30...45 seconds).

Paramagnetic detector (PMD)

There are different designs of paramagnetic detectors (dependent on the manufacturer). The measuring principle is based on inhomogenous magnetic fields that exert forces on molecules with paramagnetic properties (such as oxygen), causing the molecules to move. The movement is proportional to the concentration of molecules in the test gas and is sensed by a special detector.

Fig. 3
1   Reaction chamber
2   Ozone inlet
3   Test-gas inlet
4   Gas outlet
5   Filter
6   Detector

Fig. 4
1   Gas outlet
2   Collector electrode
3   Amplifier
4   Combustion air
5   Test-gas inlet
6   Combustion gas
    (H_2/He)
7   Burner

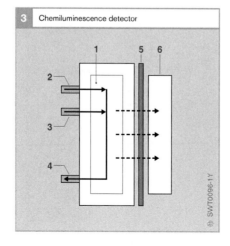

3 Chemiluminescence detector

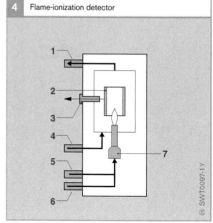

4 Flame-ionization detector

Evaporative-emissions test

A gasoline-engine motor vehicle emits hydrocarbons (HCs) through evaporation of fuel from the fuel tank and circuit. The amounts of fuel that evaporate vary according to the fuel temperature and the individual vehicle's design configuration. In some countries (e.g., USA and Europe), regulations are in place to limit these evaporative losses.

Test principle

These evaporative emissions are usually quantified with the aid of a hermetically sealed SHED (Sealed House for Evaporation Determination) chamber. For the test, HC concentrations are measured at the beginning and the end of the test, with the difference representing the evaporative losses.

Evaporative emissions are measured under some or all of the following conditions – depending upon individual country – and must comply with the stipulated limits:
- Evaporation emerging from the fuel system in the course of the day: "tank ventilation test" or "diurnal test" (EU and USA)
- Evaporation that emerges from the fuel system when the vehicle is parked with the engine warm following operation: "hot parking test" or "hot soak" (EU and USA)
- Evaporation during on-the-road operation: "running-loss test" (USA)

Evaporation is measured in several phases. Prior to testing, the vehicle undergoes preconditioning in a process including the carbon canister. With the tank filled to the stipulated level of 40 %, testing starts.

1st test: Hot-soak losses

Before testing to determine evaporative emissions in this phase, the vehicle is first warmed to normal operating temperature using the test cycle valid in the particular country. It is then parked in the SHED chamber. The increase in HC concentration within a period of 1 hour is measured during the vehicle's cooling period.

The vehicle's windows and trunk lid must remain open throughout the test. This makes it possible for the test to include evaporative losses from the vehicle's interior in its results.

2nd test: Tank-ventilation losses

In this test, a typical temperature profile for a hot summer day (maximum temperature for EU: 35 °C; EPA: 35.5 °C; CARB: 40.6 °C) is simulated within the hermetically sealed climate chamber. The hydrocarbons emitted by the vehicle under these conditions are then collected.

The USA requires testing in both 2-day diurnal (48-hour) and 3-day diurnal (72-hour) procedures. EU legislation prescribes a 24-hour test.

Running-loss test

The running-loss test is conducted prior to the hot-soak test. It is used to assess the hydrocarbon emissions generated during vehicle operation in the prescribed test cycles (1 FTP 72 cycle, 2 NYCC cycles, 1 FTP 72 cycle; refer to section entitled "US test cycles").

Other tests

Refueling test

The refueling test monitors evaporation of fuel vapors emitted during refueling by measuring HC emissions. In the US, this test is used in both CARB and EPA procedures.

Spit-back test

The spit-back test monitors the amount of fuel spray generated during each refueling process. The tank must be refueled to at least 85 % of its total volume. This test is only carried out if the refueling test has not be completed.

Diagnosis

The rise in the sheer amount of electronics in the automobile, the use of software to control the vehicle, and the increased complexity of modern fuel-injection systems place high demands on the diagnostic concept, monitoring during vehicle operation (on-board diagnosis), and workshop diagnosis (Fig. 1). Workshop diagnosis is based on a guided troubleshooting procedure that links the many possibilities of on-board and off-board test procedures and test equipment. As emission-control legislation becomes more and more stringent and continuous monitoring is now called for, lawmakers have now acknowledged on-board diagnosis as an aid to monitoring exhaust emissions, and have produced manufacturer-independent standardization. This additional system is termed the *OBD system* (On-Board-Diagnosis system).

Monitoring during vehicle operation (on-board diagnosis)

Overview

Integral diagnostic functions included within the ECU are a standard component in electronic engine-management systems. Besides a self-test of the ECU, input and output signals, and ECU intercommunication are monitored.

On-board diagnosis of an electronic system is the capability of an ECU to interpret and perform self-monitoring using "software intelligence", i.e., detect, store, and diagnostically interpret errors and faults. On-board diagnosis runs without the use of any additional equipment.

Monitoring algorithms check input and output signals during vehicle operation, and check the entire system and all its relevant functions for malfunctions and disturbances. Any errors or faults detected are stored in the ECU fault memory. Stored fault information can be read out via a serial interface.

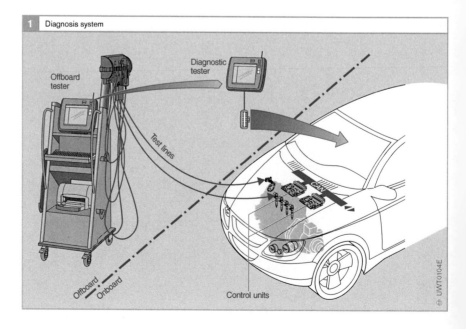

1   Diagnosis system

Offboard tester

Diagnostic tester

Test lines

Offboard   Onboard

Control units

UWT0104E

Monitoring of input signals

Sensors, plug connectors, and connecting lines (signal path) to the ECU (Fig. 2) are monitored by evaluating the input signal. This monitoring strategy is capable of detecting sensor errors, short circuits in the battery-power circuit U_{Batt} and vehicle-ground circuit, and line breaks. The following methods are used:

- Monitoring sensor supply voltage (if applicable)
- Checking the detected value for the permissible value range (e.g., $0.5...4.5\,V$)
- If additional information is available, a plausibility check is conducted using the detected value (e.g., comparison of crankshaft speed and camshaft speed)
- Critical sensors (e.g., pedal-travel sensor) are fitted in redundant configuration, which means that their signals can be directly compared with each other

Monitoring of output signals

Actuators triggered by an ECU via output stages (Fig. 2) are monitored. The monitoring functions detect line breaks and short circuits in addition to actuator faults. The following methods are used:

- Monitoring the progress of output signals through the output driver circuit. The electric circuit is monitored for short circuits to battery voltage U_{Batt}, to vehicle ground, and for open circuit.
- Impacts on the system by the actuator are detected directly or indirectly by a function or plausibility monitor. System actuators, e.g., exhaust-gas recirculation valves, throttle valves, or swirl flaps, are monitored indirectly via control loops (e.g. continuous control variance), and also partly by means of position sensors (e.g. position of the swirl flap).

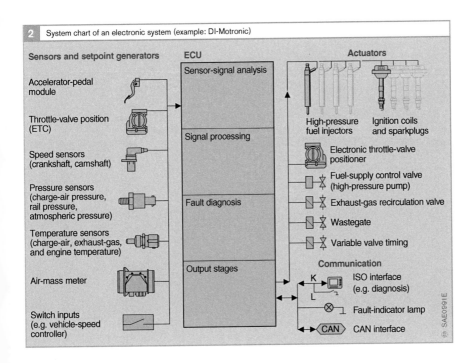

2   System chart of an electronic system (example: DI-Motronic)

Sensors and setpoint generators	ECU	Actuators

Sensor-signal analysis

Accelerator-pedal module

Throttle-valve position (ETC)

Signal processing

Speed sensors (crankshaft, camshaft)

Pressure sensors (charge-air pressure, rail pressure, atmospheric pressure)

Fault diagnosis

Temperature sensors (charge-air, exhaust-gas, and engine temperature)

Output stages

Air-mass meter

Switch inputs (e.g. vehicle-speed controller)

High-pressure fuel injectors   Ignition coils and sparkplugs

Electronic throttle-valve positioner

Fuel-supply control valve (high-pressure pump)

Exhaust-gas recirculation valve

Wastegate

Variable valve timing

Communication

K   ISO interface (e.g. diagnosis)

L

Fault-indicator lamp

CAN   CAN interface

SAE0991E

Monitoring of internal ECU functions

Monitoring functions are implemented in ECU hardware (e.g. "intelligent" output-stage modules) and software to ensure that the ECU functions correctly at all times. The monitoring functions check each of the ECU components (e.g., microcontroller, Flash-EPROM, RAM). Many tests are conducted immediately after startup. Other monitoring functions are performed during normal operation and repeated at regular intervals in order to detect component failure during operation. Test runs that require intensive CPU capacity, or that cannot be performed during vehicle operation for other reasons, are conducted in after-run more when the engine is switched off. This method ensures that the other functions are not interfered with. In the common-rail system for diesel engines, functions such as the injector switchoff paths are tested during engine runup or after-run. With a spark-ignition engine, functions such as the Flash-EPROM are tested in engine after-run.

Monitoring of ECU communication

Communication with the other ECUs usually takes place over the CAN bus. The CAN protocol contains control mechanisms to detect malfunctions. As a result, transmission errors are even detectable at CAN-module level. The ECU also runs a variety of other test routines. Since the majority of CAN messages are sent at regular intervals by the individual ECUs, the failure of a CAN controller in an ECU is detectable by testing at regular intervals. In addition, when redundant information is available in the ECU, the received signals are checked in the same way as all input signals.

Fault handling

Fault detection

A signal path is categorized as finally defective if a fault occurs over a definite period of time. Until the defect is categorized, the system uses the last valid value detected. When the defect is categorized, a standby function is triggered (e.g., engine-temperature substitute value $T = 90\,°C$).

Most faults can be rectified or detected as intact during vehicle operation, provided the signal path remains intact for a definite period of time.

Fault storage

Each fault is stored as a fault code in the non-volatile area of the data memory. The fault code also describes the fault type (e.g. short circuit, line break, plausibility, value range exceeded). Each fault-code input is accompanied by additional information, e.g. the operating and environmental conditions (freeze frame) at the time of fault occurrence (e.g. engine speed, engine temperature).

Limp-home function

If a fault is detected, limp-home strategies can be triggered in addition to substitute values (e.g., engine output power or speed limited). These strategies serve to
- Maintain driving safety
- Avoid consequential damage, or
- Minimize exhaust emissions

On-board-diagnosis system for passenger cars and light-duty trucks

The engine system and its components must be constantly monitored in order to comply with exhaust-emission limits specified by law in everyday driving situations. For this reason, regulations have come into force to monitor exhaust systems and components, e.g., in California. This has standardized and expanded manufacturer-specific on-board diagnosis with respect to the monitoring of emission-related components and systems.

Legislation

OBD I (CARB)
1988 marked the coming into force in California of OBD I, the first stage of CARB legislation (California Air Resources Board). The first OBD stage stipulates the following requirements:
- Monitoring emission-related electrical components (short circuits, line breaks) and storage of faults in the ECU fault memory
- A Malfunction Indicator Lamp (MIL) that alerts the driver to the malfunction
- The defective component must be displayed by means of on-board equipment (e.g. flashing code using a diagnosis lamp)

OBD II (CARB)
The second stage of the diagnosis legislation (OBD II) came into force in California in 1994. OBD II became mandatory for diesel-engine cars with effect from 1996. In addition to the scope of OBD I, system functionality was now monitored (e.g., plausibility check of sensor signals).

OBD II stipulates that all emission-related systems and components must be monitored if they cause an increase in toxic exhaust emissions in the event of a malfunction (by exceeding the OBD limits). Moreover, all components must be monitored if they are used to monitor emission-related components or if they can affect the diagnosis results.

Normally, the diagnostic functions for all components and systems under surveillance must run at least once during the exhaust-gas test cycle (e.g. FTP 72). A further stipulation is that all diagnostic functions must run with sufficient frequency during daily driving mode. For many monitoring functions, the law defines a monitoring frequency (In Use Monitor Performance Ratio) in daily operation starting model year 2005.

Since the introduction of OBD II, the law has been revised in several stages (updates). The last update came into force in model year 2004. Further updates have been announced.

OBD (EPA)
Since 1994 the laws of the EPA (Environmental Protection Agency) have been in force in the remaining US states. While the EPA stipulations governing diagnostic capabilities are essentially the same as the CARB rules (OBD II), the OBD regulations for CARB and EPA apply to all passenger cars with up to 12 seats and to light-duty trucks weighing up to 14,000 lbs (6.35 t).

EOBD (EU)
OBD adapted to European conditions is termed EOBD and is based on EPA-OBD.

EOBD has been valid for all passenger cars and light-duty trucks equipped with gasoline engines and weighing up to 3.5 t with up to 9 seats since January 2000. Since January 2003 the EOBD also applies to passenger cars and light-duty trucks with diesel engines.

Other countries
A number of other countries have already adopted or are planning to adopt EU or US-OBD legislation.

Requirements of the OBD system

The ECU must use suitable measures to monitor all on-board systems and components whose malfunction may cause a deterioration in exhaust-gas test specifications stipulated by law. The Malfunction Indicator Lamp (MIL) must alert the driver to a malfunction if the malfunction could cause an overshoot in OBD emission limits.

Emission limits

US OBD II (CARB and EPA) prescribes thresholds that are defined relative to emission limits. Accordingly, there are different permissible OBD emission limits for the various exhaust-gas categories that are applied during vehicle certification (e.g. TIER, LEV, ULEV). Absolute limits apply in Europe (Table 1).

Malfunction indicator lamp (MIL)

The Malfunction Indicator Lamp (MIL) alerts the driver that a component has malfunctioned. When a malfunction is detected, CARB and EPA stipulate that the MIL must light up no later than after two driving cycles of its occurrence. Within the scope of EOBD, the MIL must light up no later than in the third driving cycle with a detected malfunction.

If the malfunction disappears (e.g. loose contact), the malfunction remains entered in the fault memory for 40 trips (warm-up cycles). The MIL goes out after three fault-free driving cycles.

Communication with scan tool

OBD legislation prescribes standardization of the fault-memory information and access to the information (connector, communication interface) compliant with ISO 15031 and the corresponding SAE standards (Society of Automotive Engineers). This permits the readout of the fault memory using standardized, commercially available testers (scan tools, see Fig. 3).

The following communication protocols are approved for communication with an OBD scan tool:

ISO 15765-4	OBD II, EOBD
SAE J1850	OBD II up to MY 2007, EOBD
ISO 9141-2	OBD II up to MY 2007, EOBD
ISO 14230-4	OBD II up to MY 2007, EOBD

Scan tool

These serial interfaces operate at a bit rate (baud rate) of 5 to 10.4 kbaud. They are single-wire interfaces with a common transmitting and receiving line. There are also two-wire interfaces with separate data line (K-line) and initiate line (L-line).

Workshop tester

Several ECUs (such as Motronic, ESP, or EDC, and transmission-shift control, etc.) can be combined on one diagnosis connector.

1	OBD limits for passenger cars and light-duty trucks	
	Gasoline passenger cars	**Diesel passenger cars**

CARB	– Relative emission limits
	– Mostly 1.5 times the limit of
	  a specific exhaust-gas category
		– Relative emission limits
		– Mostly 1.5 times the limit of
		  a specific exhaust-gas category

EPA
(US Federal)
	– Relative emission limits
	– Mostly 1.5 times the limit of
	  a specific exhaust-gas category
		– Relative emission limits
		– Mostly 1.5 times the limit of
		  a specific exhaust-gas category

EOBD	2000 CO: 3.2 g/km HC: 0.4 g/km NO_x: 0.6 g/km	2003 CO: 3.2 g/km HC: 0.4 g/km NO_x: 1.2 g/km PM: 0.18 g/km

Table 1

Communication between the tester and ECU is set up in three phases:
- Initiate the ECU
- Detect and generate baud rate
- Read key bytes which serve to identify the communications protocol (not for CAN)

Evaluation is performed subsequently. The following functions are possible:
- Identify the ECU
- Read the fault memory
- Erase the fault memory
- Read the actual values

In future, the CAN bus (ISO 15 765-4) will be used increasingly to handle communication between ECUs and the tester. Starting in 2008 diagnosis will only be permitted over this interface in the USA.

To make it easier to read out the ECU fault-memory information, a standardized diagnosis socket will be fitted at an easily accessible place in every car (easy to reach from the driver's seat). The socket is used to connect the scan tool (Fig. 4).

Reading the fault information
Any workshop can use the scan tool to read out emission-relevant fault information from the ECU (Fig. 5). In this way, workshops not franchized to a particular manufacturer are also able to carry out repairs. Manufacturers are obliged to make the required tools and information available (on the internet) in return for a reasonable payment, to allow this.

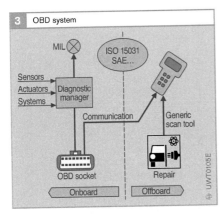

3 OBD system

MIL
ISO 15031 SAE...
Sensors
Actuators
Systems
Diagnostic manager
Communication
Generic scan tool
OBD socket
Repair
Onboard
Offboard

UWT0105E

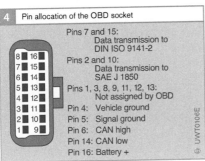

4 Pin allocation of the OBD socket

Pins 7 and 15:
    Data transmission to DIN ISO 9141-2
Pins 2 and 10:
    Data transmission to SAE J 1850
Pins 1, 3, 8, 9, 11, 12, 13:
    Not assigned by OBD
Pin 4: Vehicle ground
Pin 5: Signal ground
Pin 6: CAN high
Pin 14: CAN low
Pin 16: Battery +

UWT0106E

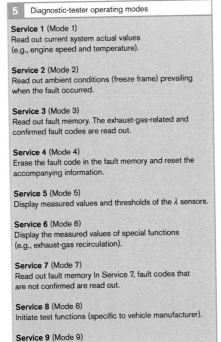

5 Diagnostic-tester operating modes

Service 1 (Mode 1)
Read out current system actual values (e.g., engine speed and temperature).

Service 2 (Mode 2)
Read out ambient conditions (freeze frame) prevailing when the fault occurred.

Service 3 (Mode 3)
Read out fault memory. The exhaust-gas-related and confirmed fault codes are read out.

Service 4 (Mode 4)
Erase the fault code in the fault memory and reset the accompanying information.

Service 5 (Mode 5)
Display measured values and thresholds of the λ sensors.

Service 6 (Mode 6)
Display the measured values of special functions (e.g., exhaust-gas recirculation).

Service 7 (Mode 7)
Read out fault memory In Service 7, fault codes that are not confirmed are read out.

Service 8 (Mode 8)
Initiate test functions (specific to vehicle manufacturer).

Service 9 (Mode 9)
Read out vehicle information.

Recalls

If vehicles fail to comply with OBD requirements by law, the authorities may demand the vehicle manufacturer to start a recall at his own cost.

Functional requirements

Overview

Just as for on-board diagnosis, all ECU input and output signals, as well as the components themselves, must be monitored.

Legislation demands the monitoring of electrical functions (short circuit, line breaks), a plausibility check for sensors, and a function monitoring for actuators.

The pollutant concentration expected as the result of a component failure (measured values), and the monitoring mode partly required by law determine the type of diagnosis. A simple functional test (black/white test) only checks system or component operability (e.g. swirl flap opens and closes). The extensive functional test provides more detailed information about system operability. As a result, the limits of adaptation must be monitored when monitoring adaptive fuel-injection functions (e.g., zero-delivery calibration for a diesel engine, lambda adaptation for a gasoline engine).

Diagnostic complexity has constantly increased as emission-control legislation has evolved.

Switch-on conditions

Diagnostic capabilities only run if the switch-on conditions are satisfied. These include, for instance:
- Torque thresholds
- Engine-temperature thresholds, and
- Engine-speed thresholds or limits

Inhibit conditions

Diagnostic capabilities and engine functions cannot always operate simultaneously. There are inhibit conditions that prohibit the performance of certain functions. In the diesel system, the hot-film air-mass meter (HFM) can only be monitored satisfactorily when the exhaust-gas recirculation valve is closed. For instance, tank ventilation (evaporative-emissions control system) in a gasoline system cannot function when catalytic-converter diagnosis is in operation.

Temporary disabling of diagnostic functions

Diagnostic capabilities may only be disabled under certain conditions in order to prevent false diagnosis. Examples include:
- Height too large
- Low ambient temperature on engine starting, or
- Low battery voltage

Readiness code

When the fault memory is checked, it is important to know that the diagnostic capabilities ran at least once. This can be checked by reading out the readiness code over the diagnosis interface. After the fault memory has been erased by service personnel, the readiness codes must be reset after the functions have been checked.

Diagnostic system management (DSM)

The diagnostic capabilities for all components and systems checked must normally run in driving mode, but at least once during the exhaust-gas test cycle (e.g. FTP 72, NEDC). Diagnostic System Management (DSM) can change the sequence for running the diagnostic capabilities dynamically, depending on driving conditions.

The DSM comprises the following three components (Fig. 6):

Diagnosis fault path management (DFPM)

The primary role of DFPM is to store fault states that are detected in the system. In addition to faults, it also stores other information, such as environmental conditions (freeze frame).

Diagnostic function scheduler (DSCHED)
DSCHED is responsible for coordinating assigned engine and diagnostic capabilities. It is supported in this process by information received from the DVAL and DFPM. In addition, it reports functions that require release by DSCHED to perform their readiness, after which the current system state is checked.

Diagnosis validator (DVAL)
The DVAL (only installed in gasoline systems to date) uses current fault-memory entries and additionally stored information to decide for each detected fault whether it is the actual cause, or a consequence of the fault. As a result, validation provides stored information to the diagnosis tester for use in reading out the fault memory.

In this way, diagnostic capabilities can be released in any sequence. All released diagnoses and their results are evaluated subsequently.

OBD functions
Overview
Whereas EOBD only contains detailed monitoring specifications for individual components, the requirements in CARB OBD II are much more detailed. The list below shows the current state of CARB requirements for gasoline-engine and diesel-engine passenger cars. Requirements that are also described in detail in the EOBD legislation are marked by (E):
- Catalytic converter (E), heated catalytic converter
- Combustion (ignition) miss (E, for diesel system not for EOBD)
- Evaporative-emission reduction system (tank-leakage diagnosis, with [E] at least electrical testing of the canister-purge valve)
- Secondary-air injection
- Fuel system
- Lambda sensors (E)
- Exhaust-gas recirculation
- Crankcase ventilation
- Engine cooling system
- Cold-start emission reduction system (presently only for gasoline system)
- Air conditioner (components)
- Variable valve timing (presently only in use with gasoline systems)
- Direct ozone reduction system (presently only in use with gasoline systems)
- Particulate filter (soot filter, presently only for diesel system) (E)
- Comprehensive components (E)
- Other emission-related components/systems (E)
- IMPPR (in use monitor performance ratio) for testing the pass frequency of diagnostic functions in everyday applications. Minimum values must be adhered to
- Protection against manipulation

"Other emission-related components/systems" refer to components and systems not mentioned in this list and that may exceed OBD emission limits, or block other diagnostic functions if they malfunction.

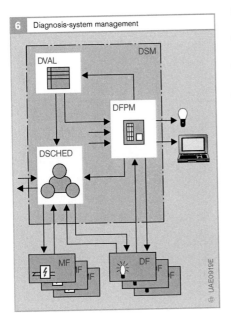

6   Diagnosis-system management

Catalytic-converter diagnosis

The function of the three-way catalytic converter is to convert to safer substances the harmful emissions of carbon monoxide (CO), nitrogen oxides (NO_X) and hydrocarbons (HC) that are produced by the combustion process. Aging or damage (thermal, poisoning) can diminish the converter's efficiency. Therefore, the effectiveness of the catalytic converter has to be monitored.

One measure of catalytic-converter efficiency is its oxygen-storage capacity. To date, it has been possible to demonstrate a correlation between oxygen-storage capacity and converter efficiency for all types of three-way catalytic-converter coating (washcoat with ceroxides as oxygen-storing component and noble metals as the actual catalysts).

Depending on the requirements regarding emission reduction, one or more main catalytic converters (usually underfloor catalytic converters) are used either alone or in combination with one or more primary catalytic converters close to the engine. Primary mixture control takes places using a lambda sensor upstream of the first catalytic converter in the exhaust system. Present-day concepts incorporate secondary lambda sensors downstream of the primary catalytic converter and/or the main converter. They serve the purpose firstly to fine-tune the primary lambda sensor(s) and secondly for the OBD functions. The basic principle

of catalytic-converter diagnosis is to compare the sensor signals upstream and downstream of the converter under review.

Diagnosis of primary catalytic converter

In systems in which the secondary lambda sensor is located directly after the primary catalytic converter, the primary catalytic converter can be separately monitored. Diagnosis is based on the following principle: The setpoint value for lambda closed-loop control is modulated at a specific frequency and amplitude (Fig. 7). The variations in the oxygen content of the exhaust gas resulting from these fluctuations in the control settings are attenuated in the catalytic converter by absorbing or releasing oxygen in the converter coating material, i.e. the rear sensor emits a signal with a very small amplitude (Fig. 9, top signal curve).

By contrast, a primary catalytic converter that has lost its oxygen-storage capacity as a result of aging or damage produces an oscillating two-point signal since almost no attenuation takes place (Fig. 9, bottom signal curve). Loss of oxygen-storage capacity can be calculated from the amplitude by filtering the signal in a special process. Conclusions can then be drawn as to converter efficiency.

Fig. 7

1   Exhaust-gas mass flow from engine
2   Broad-band lambda sensor
3   Primary catalyst
4   Two-step lambda sensor
5   Diagnosis lamp

U_S Sensor voltage
F_R Lambda control factor (control setpoint)

Fig. 8

1   Exhaust-gas mass flow from engine
2   Broad-band lambda sensor
3   Catalytic converter (primary and main catalysts)
4   Two-step lambda sensor
5   Diagnosis lamp

U_S Sensor voltage

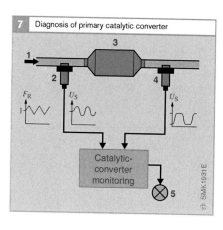

7   Diagnosis of primary catalytic converter

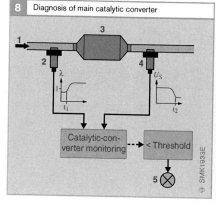

8   Diagnosis of main catalytic converter

Diagnosis of main catalytic converter

The entire arrangement of primary catalytic converter and main converter is monitored (Fig. 8) on systems in which the secondary lambda sensor is positioned downstream of the main catalytic converter. However, the oxygen-storage capacity of the main catalytic converter is far greater than a smaller primary catalytic converter. This means that the regulation of the control setpoint is still significantly damped even if the converter is damaged. For this reason, a change in the oxygen concentration downstream of the main catalytic converter is too small for passive assessment, as in the method described above. This necessitates a diagnosis process involving active intervention in the lambda closed-loop control system.

Diagnosis of the main catalytic converter is based on measuring oxygen storage directly on changeover from rich to lean mixture. A constant broad-band lambda sensor is fitted upstream of the catalytic converter and measures the oxygen content in the exhaust gas. Downstream of the catalytic converter is a two-step lambda sensor which detects the condition of the oxygen accumulator. The reading is taken under static engine-operating conditions in the part-load range.

In the first stage of the process, the oxygen accumulator is completely emptied while the engine is running with a rich mixture ($\lambda < 1$). The signal from the rear lambda sensor indicates this by a voltage of $>650\,mV$. In the next stage, while the engine is running with a lean mixture ($\lambda > 1$), the mass of oxygen absorbed to the point at which the accumulator reaches overflow is calculated with the aid of the air-mass flow and the primary lambda-sensor signal. The point of overflow is indicated by the signal voltage from the sensor downstream of the catalytic converter dropping to $<200\,mV$. The calculated integral of the oxygen mass indicates the oxygen-storage capacity. That figure must exceed a reference figure, otherwise a fault is recorded.

Theoretically, the analysis would also be possible by measuring the amount of oxygen released on changeover from lean to rich mixture. However, measuring the amount of oxygen absorbed on changeover from rich to lean mixture offers the following benefits:
- Less dependence on temperature, and
- Less dependence on sulfurization

This method allows more accurate measurement of oxygen-storage capacity.

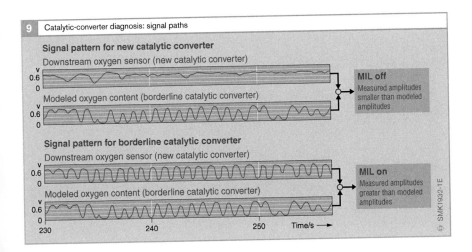

9   Catalytic-converter diagnosis: signal paths

Signal pattern for new catalytic converter

Downstream oxygen sensor (new catalytic converter)

V
0.6
0

Modeled oxygen content (borderline catalytic converter)

V
0.6
0

MIL off

Measured amplitudes smaller than modeled amplitudes

Signal pattern for borderline catalytic converter

Downstream oxygen sensor (new catalytic converter)

V
0.6
0

Modeled oxygen content (borderline catalytic converter)

V
0.6
0

MIL on

Measured amplitudes greater than modeled amplitudes

230   240   250   Time/s ⟶

SMK1932-1E

Diagnosis of NO$_X$ accumulator-type catalytic converter

As well as functioning as a three-way catalytic converter, the NO$_X$ accumulator-type catalytic converter required for gasoline direct-injection engines has the task of temporarily storing the nitrous oxides that cannot be converted when the engine is running on a lean mixture ($\lambda > 1$). It then converts them at a later stage when the engine is running with a homogeneously distributed air/fuel mixture of ($\lambda < 1$). The NO$_X$ storage capacity of this catalytic converter – indicated by the catalytic-converter quality factor – diminishes as a result of aging

or contamination (e.g., sulfur absorption). Therefore, its functional capacity has to be monitored. This can be achieved with the aid of lambda sensors fitted upstream and downstream of the catalytic converter, or an NO$_X$ sensor in place of the downstream lambda sensor.

In order to determine the catalytic-converter quality factor, the actual NO$_X$ accumulator capacity is compared with the NO$_X$ accumulator capacity of a new NO$_X$ accumulator-type catalytic converter (new catalytic-converter model, Fig. 10). The actual NO$_X$ storage capacity is equal to the metered consumption of reduction agents (HC and CO) during catalytic-converter regeneration. The quantity of reduction agents is determined by integrating the reduction-agent mass flow during the regeneration phase when $\lambda < 1$. The end of the regeneration phase is indicated by an abrupt change in the signal voltage detected by the secondary oxygen sensor.

Alternatively, the actual NO$_X$ accumulator capacity can be determined by means of an NO$_X$ sensor.

Fig. 10
1   Exhaust-gas mass flow from engine
2   Broad-band lambda sensor
3   NO$_X$ accumulator-type catalytic converter
4   Two-step lambda sensor/NO$_X$ sensor
5   Diagnosis lamp

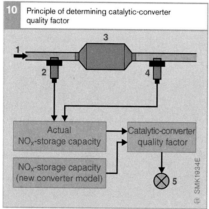

10 Principle of determining catalytic-converter quality factor

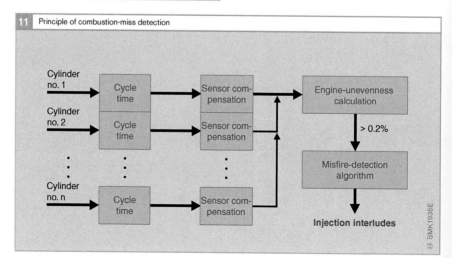

11 Principle of combustion-miss detection

Combustion-miss detection

Current legislation demands detection of combustion misses that may be caused by worn-out spark plugs, for example. A miss occurs when the spark plug fails to produce an ignition spark. The result is that combustion of the air/fuel mixture does not take place and unburnt fuel is emitted into the exhaust-gas system. Consequently, misses result in afterburning of unburnt fuel in the catalytic converter and cause a temperature rise. This can lead to more rapid aging or even complete destruction of the catalytic converter. In addition, misses increase exhaust-gas emissions, particularly of HC and CO. Detection of misses is therefore a necessity.

The miss detection function measures the time elapsed between one combustion stroke and the next – the cycle time – for each cylinder. This time is derived from the speed-sensor signal. The time it takes a certain number of teeth to pass the crankshaft sensor wheel is measured. If the engine has a combustion miss, it does not produce the amount of torque that would normally be expected. The result is that it rotates at a slower speed. A significant increase in the resulting cycle time is an indication of misfiring (Fig. 11).

At high engine speeds and low engine load, the increase in cycle time is only about 0.2 %. Consequently, precise monitoring of engine rotation and a complex calculation method are required in order to distinguish misfiring from other effects (e.g., judder caused by poor road surface).

Sensor-wheel compensation corrects for discrepancies arising from variations in manufacturing tolerances in the sensor wheel. This function is only active when the engine is overrunning since no acceleration torque is generated under these operating conditions. Sensor-wheel compensation supplies correction values for the cycle times.

If the combustion-miss rate exceeds a permissible level, fuel injection to the cylinder(s) concerned is deactivated in order to protect the catalytic converter.

Fuel-tank leakage diagnosis

It is not only exhaust-gas emissions that are harmful to the environment. Evaporative emissions escaping from the fuel system – especially the fuel tank – are equally undesirable and consequently subject to emission limits as well. In order to limit evaporative emissions, evaporated fuel is collected by the activated carbon canister in the evaporative-emissions control system. They are then released to the intake manifold via the canister-purge valve from where it joins the normal combustion process (Fig. 12). Monitoring the fuel-tank system is part of the on-board diagnostic functions.

Legislation for the European market is initially limited to straightforward checking of the electrical circuits for the fuel-tank pressure sensor and the canister-purge valve. In the US, on the other hand, detection of leaks from the fuel system is required. There are two different diagnosis methods for leakage detection. They can detect a major leak with a flow diameter of up to 1.0 mm and a minor leak with a flow diameter of up to 0.5 mm.

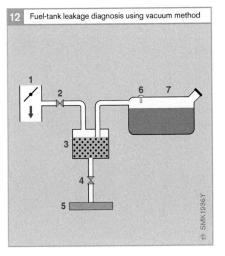

12   Fuel-tank leakage diagnosis using vacuum method

SMK1936Y

Fig. 12
1 Intake manifold with throttle valve
2 Canister-purge valve (regeneration valve)
3 Carbon canister
4 Shutoff valve
5 Air filter
6 Fuel-tank pressure sensor
7 Fuel tank

Vacuum-relief diagnosis method
When the vehicle is stationary and the engine idling, the canister-purge valve (Fig. 12, Pos. 2) is closed. As the shutoff valve (4) is open, air is drawn into the tank by the vacuum inside the tank. The pressure in the tank system then increases. If the pressure measured by the pressure sensor (6) does not reach atmospheric pressure within a certain period of time, it is assumed that the shutoff valve has failed to open sufficiently or at all and is therefore faulty.

If a shutoff-valve fault is not detected, the valve is closed. Fuel evaporation may now be expected to cause a pressure increase. The resulting pressure may not exceed specific upper and lower limits.

If the pressure measured is below the specified lower limit, the canister-purge valve is faulty. In other words, the lack of pressure is caused by a leaking canister-purge valve which is allowing vapor to be drawn out of the fuel tank by the vacuum in the intake manifold.

If the pressure measured is above the specified upper limit, this indicates that too much fuel is evaporating (e.g., due to high ambient temperature) for diagnosis to be carried out.

If the pressure produced by fuel evaporation is within the specified limits, then the pressure rise is stored as the compensation gradient for minor-leak detection.

Only after testing the shutoff and canister-purge valves can fuel-tank leakage diagnosis be continued.

Major-leak detection is performed first. When the engine is idling, the canister-purge valve (2) is opened. As a result, the vacuum in the intake manifold (1) "spreads" to the fuel-tank system. If the pressure change detected by the fuel-tank pressure sensor (6) is too small because air is able to enter due to a leak to balance out the induced pressure drop, a major leak is detected and the fault-diagnosis sequence is terminated.

The minor-leak diagnosis can start once the diagnosis system has failed to identify a major leak. This starts by closing the canister-purge valve (2) again. The pressure should then only increase by the fuel-evaporation rate (compensation gradient) previously stored, since the shutoff valve (4) is still closed. If the pressure increases at a higher rate, there must be a minor leak through which air is able to enter.

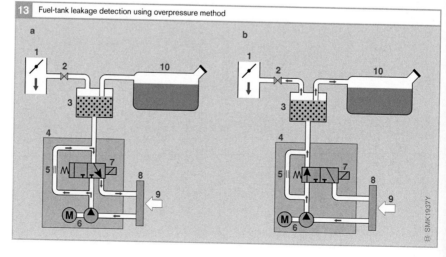

Fig. 13
a Reference-leak
  current measurement
b Minor- and major-
  leak test

1 Intake manifold with
  throttle valve
2 Canister-purge valve
  (regeneration valve)
3 Carbon canister
4 Diagnosis module
5 Reference leak
  0.5 mm
6 Vane pump
7 Changeover valve
8 Air filter
9 Intake air
10 Fuel tank

13 Fuel-tank leakage detection using overpressure method

Overpressure method

If the diagnosis-activation conditions are satisfied and the ignition is switched off, the overpressure test is started as part of the ECU run-on sequence.

For minor-leak detection, the electric vane pump (Fig. 13a, Pos. 6) integrated in the diagnosis module (4) pumps air through a "reference leak" (5) which has a diameter of 0.5 mm. As a result of the back pressure caused by the constriction at that point, the load on the pump increases, resulting in a drop in pump speed and an increase in electric current. That current level (Fig. 14) is measured and stored (4).

Next (Fig. 13b), the solenoid valve (7) is switched over and the pump pumps air into the fuel tank. If there are no leaks in the tank, the pressure increases and the pump current rises accordingly (Fig. 14) to a level higher than the reference current (3). If there is a minor leak, the pump current will reach the reference current but not exceed it (2). If the pump current fails to reach the reference current after an extended period, a major leak is present (1).

Diagnosis of secondary-air injection

Running the engine with a rich mixture ($\lambda < 1$) – which is necessary in cold weather, for example – leads to high concentrations of hydrocarbons and carbon monoxide in the exhaust gas. These emissions have to be re-oxidized in the exhaust system, i.e., passed through an afterburning process. Therefore, most vehicles have a secondary-air injection facility which blows the oxygen required for catalytic afterburning into the exhaust gas directly downstream of the exhaust manifold (Fig. 15).

If this system fails, there will be an increase in exhaust-gas emissions when cold-starting the engine or before the catalytic converter has reached operating temperature. For this reason, a diagnosis function is required.

Diagnosis of secondary-air injection is a function check which establishes whether the pump is working properly and whether there are problems with the air pipe to the exhaust-gas system. The check can be performed in two ways.

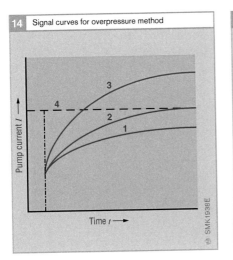

14 Signal curves for overpressure method

Pump current I →

Time t →

SMK1938E

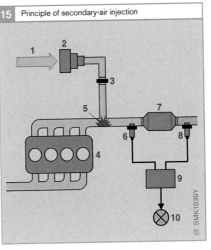

15 Principle of secondary-air injection

SMK1939Y

Fig. 14

1 Current curve for tank leak of $d > 0.5$ mm
2 Current curve for tank leak of $d \leq 0.5$ mm
3 Current curve for tank without leaks
4 Reference-current value

Fig. 15

1 Intake air
2 Secondary-air pump
3 Secondary-air valve
4 Engine
5 Point of injection into exhaust pipe
6 Upstream lambda sensor
7 Catalytic converter
8 Downstream lambda sensor
9 Engine ECU
10 Diagnosis lamp

The passive test takes place immediately after the engine is started and while the catalytic converter is being heated up. It involves the secondary-air system measuring the mass of secondary air injected. It uses an active lambda closed-loop control system and compares the secondary-air mass with a reference figure. If the calculated air mass differs from the reference figure, a fault is detected.

The active test involves activating secondary-air injection for diagnostic purposes only during the engine idle phase. The signals from the lambda sensors are used directly for calculating the secondary-air mass. As with the passive test, the calculated secondary-air mass flow rate is compared with a reference value.

Although less accurate, the passive test is necessary because it is precisely during the period when the engine has only been running for a short time that secondary-air injection is active and its correct functioning must be ensured.

Diagnosis of fuel system

Faults in the fuel system (e.g., defective fuel valve, leaks in the intake manifold) can prevent the optimum mixture form being formed. For this reason, the OBD needs to monitor the system. The ECU therefore processes data such as the intake air mass (air-mass-sensor signal), the throttle-valve position, the air/fuel ratio (primary lambda-sensor signal) and engine operating data. This data is then compared with modeled calculations.

Diagnosis of lambda sensors

The lambda-sensor system generally consists of two sensors (upstream and downstream of the catalytic converter) and the lambda closed-loop control circuit. Upstream of the catalytic converter, there is normally a broad-band lambda sensor. It measures the λ level (i.e. the air/fuel mixture across the entire range from rich to lean) continuously by means of a variable voltage (Fig. 16a). It is also to controls this voltage. On older systems, a two-step lambda sensor was used upstream of the catalytic converter. The two-step sensor indicates only whether the mixture is lean ($\lambda > 1$) or rich ($\lambda < 1$) by an abrupt change in the voltage signal (Fig. 16b).

Present-day concepts incorporate a secondary lambda sensor – generally a two-step sensor – downstream of the primary catalytic converter and/or the main converter. It serves firstly to fine-tune the primary lambda sensor and secondly, for OBD functions. The lambda sensors not only measure the air/fuel mixture in the exhaust gas for the engine-management system, they also monitor the function of the catalytic converter(s).

Possible lambda-sensor faults include:
- Breaks or short circuits in the electrical circuit
- Aging (thermal, poisoning) of the sensor (leads to diminished sensor-signal dynamics)
- Corrupted signals caused by the sensor failing to reach operating temperature

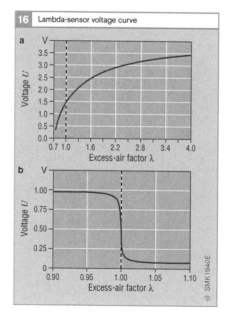

16 Lambda-sensor voltage curve

Fig. 16
a Broad-band lambda sensor
b Two-step lambda sensor

Primary sensor

The sensor upstream of the catalytic converter is referred to as the primary or upstream sensor. It is checked for:

- Plausibility (internal resistance, output voltage – the actual signal – and other parameters), and
- Dynamic response (rate of signal change at changeover from "rich" to "lean" and from "lean" to "rich", and period duration)

If the sensor has a heater, the heater function must also be checked. The tests are carried out while the vehicle is in motion at relatively constant operating conditions.

The broad-band lambda sensor requires different diagnostic procedures from the two-step sensor since it can also record levels other than $\lambda = 1$.

Secondary sensor

The secondary or downstream sensor(s) is/are used to monitor the catalytic converter as well as for other functions. They check the conversion efficiency of the catalytic converter and supply the most important data for converter diagnosis. The secondary-sensor signals can also be used to check the data provided by the primary sensor. In addition, the secondary sensor can help to ensure the long-term stability of emission levels by correcting the primary-sensor signals.

Apart from the period duration, all of the characteristics and parameters listed for the primary sensor are also checked for the secondary sensor.

Diagnosis of exhaust-gas recirculation system

The exhaust-gas recirculation system is an effective means of reducing the emissions of nitrogen oxides. Adding recirculated exhaust gas to the air/fuel mixture entering the engine lowers the peak combustion temperature. In turn, this reduces the formation of nitrogen oxides. The functional efficiency of

the exhaust-gas recirculation system therefore has to be monitored. There are two alternative methods that may be used.

The method based on intake-manifold pressure involves closing the exhaust-gas recirculation valve briefly while the engine is operating at medium power, and measuring the pressure change. A comparison between the measured intake-manifold pressure and the pressure calculated using a model allows diagnosis of the closing function of the exhaust-gas recirculation valve.

The method based on uneven engine running is used in systems without an air-mass sensor or without an additional intake-manifold pressure sensor. The exhaust-gas recirculation valve is slightly opened while the engine is idling. If the exhaust-gas recirculation system is functioning correctly, the slight increase in residual-gas mass causes the engine to run slightly more unevenly. This is used by the engine-smoothness monitoring function to diagnose exhaust-gas recirculation valve problems.

Diagnosis of crankcase ventilation

The so-called "blow-by gas" which enters the crankcase by escaping past the pistons, piston rings and cylinders has to be removed from the crankcase. This is the function of the Positive Crankcase Ventilation (PCV) system. The exhaust-gas enriched air has the soot removed from it by a cyclone separator and is then passed through the PCV valve into the intake manifold so that hydrocarbons can be returned to the combustion cycle.

One possible diagnosis method is based on measuring the idle speed which, when the PCV valve is opened, should exhibit specific characteristics that are predicted by a model. If the observed change in the idle speed differs too greatly from the modeled response, a leak is assumed to exist.

Diagnosis of engine cooling system

The engine cooling system consists of a large and a small circulation system which are connected by a thermostatic valve. The small circulation system is used during the starting phase to bring the engine quickly to operating temperature and is brought into action by closing the thermostatic valve. If the thermostat is defective or it has seized in the open position, the rate of coolant temperature rise is slowed down – particularly in cold weather conditions – and this results in higher emission levels. A thermostat monitoring function should therefore detect a slow rate of coolant temperature increase. The system's temperature sensor is first tested. Then the thermostatic valve is checked on this basis.

Diagnosis for monitoring heating measures

In order to obtain a high conversion rate, the catalytic converter needs to operate at a temperature of 400...800 °C. However, higher temperatures can damage the converter coating.

A catalytic converter operating at optimum temperature reduces engine emissions by more than 99 %. At low temperatures, its efficiency diminishes to the extent that a cold catalytic converter is almost totally ineffective. Therefore, in order to be able to meet emission-control requirements, a special strategy is required to bring the catalytic converter up to operating temperature as quickly as possible. The special converter-heating phase is terminated when the converter reaches a temperature of 200...250 °C (light-off temperature, approx. 50 % conversion efficiency). From this point on, the catalytic converter heats itself by exothermic conversion reactions.

When the engine is first started, two methods can be employed to heat up the catalytic converter more quickly:
- The ignition can be retarded to increase the exhaust-gas temperature
- The catalytic reactions produced in the catalytic converter by the incompletely burnt fuel in the exhaust manifold cause the converter to generate heat itself

The consequence of these effects is that the catalytic converter reaches its operating temperature more quickly so that exhaust-gas emissions are more rapidly reduced.

In order to ensure proper converter function, legislation requires monitoring of the temperature immediately upstream of the catalytic converter and monitoring of the converter warm-up phase. The warm-up phase can be monitored by checking and analyzing parameters such as ignition angle, engine speed and intake-air mass. In addition, other factors related to catalytic-converter warm-up are also specifically monitored during this phase (e.g., camshaft position/lock).

Air conditioner (components)

In order to cover the power requirements of the air conditioner, the engine is run in a different mode when given different requirements apply. If the optimized operating mode is not activated when the air conditioner is switched on (or if it is activated when the air conditioner is switched off), higher exhaust-gas emissions can result, which means that the function has to be monitored.

Diagnosis of variable valve timing (VVT)

Variable valve timing is used under certain conditions to reduce fuel consumption and exhaust-gas emissions. Whereas previously VVT was indirectly subject to legislative requirements as a "comprehensive component", the OBD II Update introduces explicit demands. Present diagnosis involves measuring the camshaft position and performing a setpoint/actual-value comparison.

Direct ozone-reduction system

A particular feature of the Californian emission-control legislation is the possibility of reducing not only exhaust gases and evaporative-loss emissions, but also the atmospheric concentration of the air pollutant, ozone. This is the purpose of the catalytic coating applied to the vehicle radiator to create a Direct Ozone Reduction (DOR) system. Ozone reduction is calculated on the basis of the surface area and the air through-put. A "credit" amount is then calculated. This can then be taken into account when calculating the total reduction of exhaust gases and evaporative-loss emissions (hydrocarbons only). The coated radiator is therefore an emission-reducing component and will have to be monitored by the OBD system with effect from model year 2006 (legal requirement under OBD II).

As yet, no cost-efficient testing method has become widespread. The possible methods of diagnosis currently under discussion are listed below.
- *Pressure sensor:* Contamination reduces the amount of ozone passing through the radiator. A pressure sensor can detect the resulting drop in pressure.
- *Measurement of resistance:* The coating has a specific electrical resistance. Corrosion of the coating changes this resistance.
- *Photodetectors:* The catalytic coating is impermeable to light. Detectors can identify if there are gaps in the coating.
- *Ozone sensors:* These are used to measure the ozone concentration levels in front of and behind the radiator.

Comprehensive components:
diagnosis of sensors
In addition to the specific diagnoses described above, which are explicitly required by Californian legislation and described individually in separate sections, all sensors and actuators (e.g., throttle valve or high-pressure pump) must also be monitored, as faults in these components could affect emission levels or prevent other diagnostic functions.

Sensors have to be monitored for:
- Electrical faults
- Range errors and – where possible –
- Plausibility errors

Electrical faults
The legislation defines electrical faults as short circuits to ground, power-supply short circuits or circuit breaks.

Checking for range errors ("range checks")
Normally, sensors have a specified output characteristic, often with a lower and an upper limit, i.e., the physical detection range of the sensor is mapped to an output voltage in the range of 0.5...4.5 V, for example. If the output voltage produced by the sensor is outside this range, a range error is present. This means that the limits for this test (the "range check") are fixed limits specific to each sensor that are not dependent on the momentary operating status of the engine.

Where certain types of sensor do not permit a distinction between electrical faults and range errors, the legislation allows for this.

Plausibility errors ("rationality checks")
As a means of achieving greater diagnosis sensitivity, the legislation demands that plausibility checks (so-called "rationality checks") be carried out in addition to the range checks. The characteristic feature of such plausibility checks is that the momentary sensor output voltage is not – as is the case with the range checks – compared with fixed limits but with narrower limits that are determined by the momentary operating status of the engine. This means that for this check current data from the engine-management system must be referred to. These checks may be implemented as comparisons between the sensor output voltage and a model by cross-reference with another sensor. The model defines an expected range for the modeled variable for all engine operating conditions.

In order to make repairs as straightforward and effective as possible, the defective component has to be identified as accurately as possible. In addition, the categories of fault referred to should be distinguishable from one another and – in the case of range and plausibility errors – as to whether the signal is outside the upper or lower limit. In the case of electrical faults, the problem can generally be assumed to be a wiring fault, whereas a plausibility error rather tends to indicate a fault in the component itself.

As testing for electrical faults and range errors must be continuous, checking for plausibility errors must take place at a specified minimum frequency during normal operation.

Among the sensors to be monitored in this way are the following:
- The air-mass sensor
- Various pressure sensors (intake manifold, atmospheric pressure, fuel tank)
- The engine-speed sensor
- The phase sensor
- The intake-air temperature sensor
- The exhaust-gas temperature sensor

Example
The diagnostic process is described below using the air-mass sensor as an example.

The air-mass sensor, which is used to detect the amount of air drawn in by the engine to enable calculation of the amount of fuel to be injected, measures the air-mass rate and sends that information to the Motronic system in the form of an output voltage signal. The air masses vary according to the throttle setting and/or engine speed.

The diagnostic function monitors whether the sensor output voltage is outside certain (specifiable, fixed) upper and lower limits, and if they output a range error.

By comparing the air mass indicated by the air-mass sensor with the current position of the throttle valve – taking account of the engine operating status – the sensor signal can be taken to be implausible if the divergence between the two signals is greater that a certain tolerance. Example: The throttle valve is fully open but the air-mass sensor indicates the contradiction of air mass equivalent to idle speed.

Comprehensive components: diagnosis of actuators
Actuators have to be monitored for electrical faults and – if technically possible – for correct function. Function monitoring in this case means that the execution of a given actuation command is monitored by observing whether the system responds in the appropriate manner. This means that – compared with sensor-signal plausibility checking – additional information has to be obtained from the system in order to assess component function.

Actuators include the following:
- All output stages
- The throttle valve
- The electronic throttle control system
- The canister-purge valve, and
- The carbon-canister shutoff valve

However, the majority of these components are already taken into account in the system diagnoses.

Example
The throttle valve is responsible for controlling the amount of air that mixes with each injection of fuel. In the electronic throttle control system, it is electronically controlled. The throttle-valve angle (aperture) for adjusting the rate of air intake is controlled by a digital position controller. In order to diagnose throttle-valve faults, the controller is monitored for discrepancies between the specified and the actual throttle-valve angle. If the divergence is too great, a throttle-valve positioner fault is registered. The same fault is recorded if the throttle-valve positioner response is too small.

With the electronic throttle control system, there is no longer a mechanical link between the accelerator pedal and the throttle valve. Instead, the throttle setting desired by the driver and indicated by the position of the accelerator pedal is detected by two identical potentiometers (for verification) and processed by the engine ECU.

▶ Global service

"When you ride in a motorcar you will discover that horses are really incredibly boring (…). But the car needs a conscientious mechanic (…)."

Robert Bosch wrote these lines to his friend Paul Reusch in 1906. In those days it was indeed possible for the hired chauffeur or mechanic to repair problems. Later on, however, in the period following the First World War, rising numbers of motorists driving their own vehicles led to a corresponding increase in the demand for service facilities. In the 1920's Robert Bosch launched a campaign aimed at creating a comprehensive service organisation. Within Germany these service centres received the uniform designation "Bosch-Dienst" in 1926

Today's Bosch operations bear the name "Bosch Car Service". They feature state-of-the-art electronic service equipment to meet the demands defined by the automotive technology and the customer expectations of today.

1 A repair operation in the year 1925 (Photo: Bosch)

2 Bosch Car Service in 2001, with state-of-the-art electronic test equipment

Diagnosis in the workshop

The function of this diagnosis is to identify the smallest, defective, replaceable unit quickly and reliably. The prompted troubleshooting procedure includes on-board information and off-board test procedures and testers. Help is provided, for example, by Electronic Service Information (ESI[tronic]) from Bosch. It contains instructions for further troubleshooting for many possible problems (e.g., engine bucks) and faults (e.g., short circuit in engine-temperature sensor).

Prompted troubleshooting

The main element is the prompted troubleshooting procedure. The workshop employee is prompted by means of a symptom-dependent, event-controlled procedure – starting from the symptom (vehicle symptom or fault-memory entry). On-board (fault-memory entry) and off-board facilities (actuator diagnosis and on-board testers) are used.

Prompted troubleshooting, fault-memory readouts, workshop diagnostic functions, and electrical communication with off-board testers take place using PC-based diagnostic testers. This may be a specific workshop tester from the vehicle manufacturer or a universal tester (e.g., KTS 650 from Bosch).

Reading out fault-memory entries

Fault information (fault-memory entries) stored during vehicle operation are read out via a serial interface during vehicle service or repair in the service workshop.

Fault entries are read out using a diagnostic tester. The workshop employee receives information about:
- Malfunctions (e.g., engine-temperature sensor)
- Fault codes (e.g., short circuit to ground, implausible signal, static fault)
- Ambient conditions (measured values on fault storage, e.g. engine speed, engine temperature, etc.)

Once the fault information has been retrieved in the workshop and the fault corrected, the fault memory can be cleared again using the tester.

A suitable interface must be defined for communication between the ECU and the tester.

---

1 Flowchart of a prompted troubleshooting procedure with CAS[plus]

Identification

Troubleshooting based on customer claim

Read out and display fault memory

Start component testing from fault-code display

Display SD actual values and multimeter actual values in component test

Setpoint/actual-value comparison allows fault definition

Perform repair, define parts, circuit diagrams, etc. in ESI[tronic]

Renew defective part

Erase fault memory

Fig. 1
The CAS[plus] system (Computer Aided Service) from Bosch combines ECU diagnosis with troubleshooting instructions for even more efficient troubleshooting.

Actuator diagnosis

The ECU contains an actuator diagnostic routine in order to activate individual actuators at the service workshop and test their functionality. This test mode is started using the diagnostic tester and only functions when the vehicle is at standstill below a specific engine speed, or when the engine is switched off. This allows an acoustic (e.g., valve clicking), visual (e.g., flap movement), or other type of inspection, e.g., measurement of electrical signals, to test actuator function.

Workshop diagnostic functions

Faults that the on-board diagnosis fails to detect can be localized using support functions. These diagnostic functions are implemented in the engine ECU and are controlled by the diagnostic tester.

Workshop diagnostic functions run automatically, either after they are started by the diagnostic tester, or they report back to the diagnostic tester at the end of the test, or the diagnostic tester assumes runtime control, measured data acquisition, and data evaluation. The ECU then executes individual commands only.

Signal testing

Electrical current, voltage and resistance can be measured with the diagnostic tester's multimeter function. In addition, an integrated oscilloscope allows the signal paths of the actuating signals for the actuators to be checked. This is especially relevant to actuators that are not checked in the actuator-diagnosis process.

Off-board tester

The diagnostic capabilities are expanded by using additional sensors, test equipment, and external . In the event of a fault detected in the workshop, off-board testers are adapted to the vehicle.

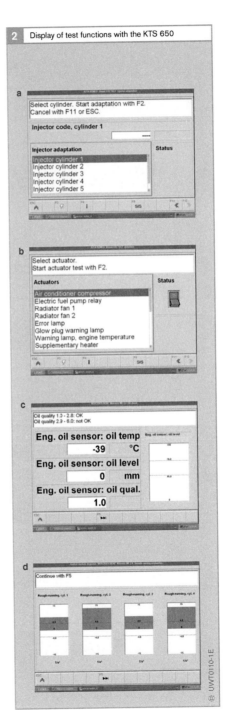

2 Display of test functions with the KTS 650

Fig. 2

a Multimeter function

b Selecting an actuator test

c Reading out engine-oil-specific data

d Evaluating smooth-running characteristics

ECU development

The Electronic Control Unit (ECU) is the central point from which an electronic system in the motor vehicle is controlled. High demands in terms of quality and reliability are therefore made of ECU development.

The tender specification contains the vehicle manufacturer's individual specifications and requirements. The performance specification is derived from the tender specification. This is the standard document for ECU development, which is made up of the individual fields of hardware development, function development, software development and application engineering.

Hardware development

The complexity of electronic systems has consistently increased over the years, and this trend is to continue in the future. Ongoing development can only be effectively managed by using highly integrated circuits. The demand for small dimensions in all system components in turn places ever more exacting demands on hardware development.

Efficient and economical development of hardware can only be achieved through the use of standard modules which are produced in large quantities.

Project start

All the functions which the ECU to be developed must fulfill are described in a block diagram. From this diagram, the following points can be clarified:
- Determination of the scope for the hardware
- Cost estimate for the hardware
- Development expenditure, and
- Tool costs

Hardware specimens

After the project start, hardware specimens are built and subjected to quality inspections. The specimens can be assigned as preliminary stages of the series-production ECU to four categories. Each specimen category builds on the preceding category and is suitable for the intended purpose in each case.

A-specimen

The A-specimen is built from an existing, if necessary modified ECU or a development printed-circuit board. Its functional scope is limited. Its technical function is to a large extent guaranteed; however, the A-specimen is not suitable for endurance testing. It is a *function specimen* which is used for preliminary tests and serves to verify the design.

B-specimen

The B-specimen contains all the circuit components. It is a *test specimen*, with which the entire functional scope and technical requirements are tested in preliminary test. It is already suitable for endurance testing in motor-vehicle prototypes.

The connection and installation dimensions correspond to series status. However, it is possible that not all the vehicle manufacturer's specifications have been fulfilled, because, for example, other materials have been used.

C-specimen

The C-specimen is the *release specimen*, with which the vehicle manufacturer's tests are carried out for "Technical Release". All the specifications are safely fulfilled with this ECU. The product release concludes the development phase. Where possible, the C-specimens are manufactured with standard tools and manufacturing processes closely approximating full-scale production.

D-specimen

The D-specimen is the *pilot-series specimen*, on which the series type plate showing the release number is incorporated. Pilot-series vehicles are equipped with D-specimens for large-scale vehicle production trials. These ECUs are assembled and tested with standard manufacturing processes and under series conditions. With these specimens, confirmation of manufacturing safety and reliability is provided.

Preparations

The preparations for the B-specimen begin at the project start:

- Determination of connector-pin assignment
- Determination of the housing
- Order and development for new circuit modules (function groups)
- Drawing up of the circuit diagram
- Determination of the components (only released components may be provided, or they must still be released)

When the circuit modules are selected (e.g., in an IC-integrated evaluation circuit for knock sensors), checks are made to ascertain whether already existing circuits – if necessary with modifications – can be used. Otherwise, new modules must be developed.

Circuit diagram and bill of materials

The circuit diagram (Fig. 1 a) complete with a bill of materials of the components used is created with a CAD system (Computer Aided Design). In the bill of materials, the following points are also established for each component:

- Component size
- Pin assignment
- Geometry in the layout
- Housing, and
- Supplier and terms of delivery

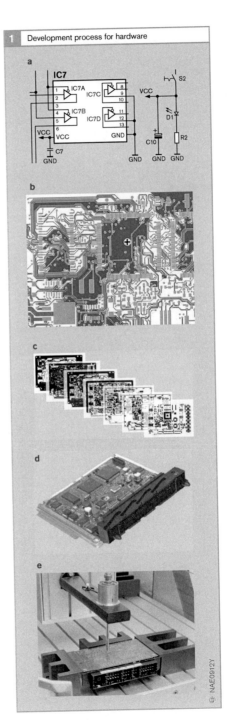

1   Development process for hardware

Fig. 1
a   Circuit diagram
b   Layout
c   Printed-circuit-board production
d   Specimen assembly
e   Testing

Layout

A layout (Fig. 1b) is required for printed-circuit board production. This reflects the printed conductors and the connector pins of the components.

The layout is created on a CAD system. The circuit-diagram data are adopted and converted. The network list now created (connection list of the components) provides information on how the components are interconnected. The layout can be created from this network list and from the CAD data for the components (component size and pin assignment).

The creation of the layout is subject to requirements, which must be taken into account. In terms of the positioning of components, these requirements may be as follows:
- The power loss created in components (possibility of heat dissipation)
- EMC influences (Electro-Magnetic Compatibility)
- Favorable arrangement of the components in relation to the connector
- Observance of blocked areas (component size)
- Insertability by automatic component machines
- Test points, and
- Space required for test adapters and visual inspection systems

Printed-circuit board

Films for printed-circuit-board production (Fig. 1c) can be made with the layout data. These films are used to expose, develop and etch the "raw" printed-circuit board's provided with a photosensitve layer. The individual layers of the multilayer printed-circuit board are laid one on top of the other and hardened.

Then a solder resist mask and a carbon lacquer are applied to the printed-circuit board.

Specimen assembly

The components must be inserted on the finished printed-circuit board (Fig. 1d). This is done in the case of specimen ECUs in specimen construction. Because of the miniaturization of the components and the high integration density on the printed-circuit board, it is essential that the components can be inserted even on the specimen units with an automatic insertion machine. The machine is controlled by the layout's CAD data.

The components are soldered after being inserted. Two methods are available:
- Wave soldering, or
- Reflow soldering

Testing the equipped printed-circuit board

Electrical testing

The equipped and soldered printed-circuit board must be tested. Electrical test routines, which run on a computer, are drawn up for this purpose. This automatic test checks the completeness of the equipped components and the functional capability of the circuit.

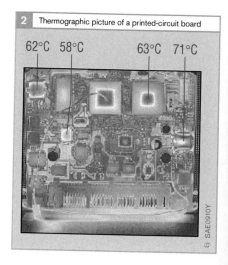

2 Thermographic picture of a printed-circuit board

62°C   58°C   63°C   71°C

SAE0910Y

Fig. 2
Engine ECU
Operating voltage
$U = 14$ V
Idle
$n = 1000$ rpm

Thermography

Thermographic pictures of the printed-circuit board show the buildup of heat by the components during operation (Fig. 2). The different temperature ranges are shown on the film in different colors. It is therefore possible to identify components which become too hot. The findings made here influence the change list from the B- to the C-specimen. Layout changes (e.g., heat through hole plating) can be used to reduce the buildup of heat.

Electromagnetic testing

The electromagnetic fields generated on the printed-circuit board can be sampled with a magnetic-field probe (Fig. 1e). The results are then evaluated on a PC. Different field strengths are identified by different colors. If necessary, layout changes must be made and additional components provided which prevent electromagnetic fields from being radiated.

These tests are also carried out as early as on the B-specimen so that the necessary changes can be taken into account in the C-specimen.

EMC tests

Tests conducted in an EMC test cell or EMC test chamber (Fig. 3) check the behavior of the ECU for electromagnetic irradiation and radiation. Tests are performed both on the installed ECU (vehicle tests) and in the laboratory (e.g., stripline procedure).

On the negative side, *vehicle tests* can only be carried out when the vehicle and the electronics are already at an advanced stage of development. The possibilities for intervention in the event of unsatisfactory EMC behavior are therefore extremely limited at this stage. For this reason, early *laboratory tests* are very important, because these allows tests at an early stage with the hardware specimens.

The EMC tests are carried out at different frequencies and different electrical field strengths. The output signals (e.g., ignition signals, injection signals) are analyzed to ascertain their immunity to interference from irradiation and their radiation behavior.

| 3 | Vehicle in an EMC test chamber |

SAE0911Y

Function development

Highly exacting demands regarding driver convenience and fuel consumption are placed on modern motor-vehicle engines.
In order that these demands can be met economically at the same time as compliance with increasingly strict emission-control legislation, the prime requirement is for optimized coordination between the engine management system and the entire drivetrain, numerous sensors and actuators and the fault-diagnosis functions for them.

As, in contrast with earlier solutions, direct mechanical adjustment of the actuators in response to changes in ambient conditions is not possible, the engine electronic control unit– and therefore the control software running on it – is the only link between the sensors and the actuators. Consequently, the fuel consumption characteristics of a vehicle are determined to a great extent by the algorithms implemented on the engine-management system and the quality of them.

It is the purpose of function development to provide these algorithms.

Requirements placed on functions
Modularity
Modern control systems have to be modular in terms of the structure of their functions and hardware in order to manage the enormous variety of engine-configuration parameters, such as number of cylinders, fuel-injection method, sensors used and type of exhaust-gas system. A hierarchical structure divided into subsystems (air system, fuel system, exhaust-gas system) with stable interfaces enables parallel and therefore rapid development which, at the rate of innovation in today's automotive industry, is becoming an increasingly important consideration.

Component packages
For every class of sensors and actuators, a generally applicable interface with a physical representation is created. Higher-level control functions connect to this interface without knowledge of the way in which the component is structured and leave the detailed handling of special component properties to the lower function level, the so-called component package. The latter provides for optimum interaction between the mechanical component – i.e. the sensor or actuator –, the processing or control hardware and the hardware-related software. In addition to essential functions such as protection against excessive temperatures or operating voltages, this is also where the correction of nonlinear characteristics and conversion into the selected physically based component-package interface format take place. It means, for example, that the changeover from a pneumatic to an electric actuator can be implemented without altering the higher-level control functions.

This concept is the basic requirement for the use of similar components of differing generations and manufacturers, with minor consequences for the overall system.

Development process
Requirement
When new requirements emerge in relation to a customer or platform project, for example with regard to the use of new components, a possible solution has to be found and the functions that are affected have to be identified. When doing so, the existing function structure should be retained as far as possible in order that the interfaces, and therefore the interaction with other functions, are not endangered. Once the necessary algorithms are known and capable of implementation on the ECU, an estimate is produced of the amount of development work, the cost and the completion date. Following discussions with internal and external clients, the project is commissioned by the customer and a definite delivery deadline agreed.

Concept

If the available algorithms are inadequate for performing the required task and if the task is sufficiently significant, a new development process is initiated involving basic calibration and testing on the vehicle or engine test bench. The resulting concept is tested for conformity with physical principles, absence of conflicts, capability of implementation in the available engine-management system and anticipated applicability in the process of a concept review involving representatives from Function Development, System Development and Application.

Function definition

There then follows the implementation of the concept by the creation and specification of each function. After offline simulation of critical components using measured vehicle data, these functions are entered as function definitions in a central database together with the associated documentation which encompasses the written descriptions and application instructions.

Function reviews with other function developers prevent the repetition of known errors.

Encoding

Next, the software developers convert the functions into program code either by automatic code generation or manually in the case of critical functions. As with function reviews, a code review minimizes error frequency.

Function test

On completion, all new software modules are integrated in a program version for a specific vehicle project. Only then can the system be tested under real conditions on the vehicle. This test involves three stages:

- The function developer checks that the specifications and the software implementation are consistent with one another
- The function developer compares the implemented solution with the customer/project requirements
- In the course of initial commissioning, Function Development and Application jointly assess whether the chosen solution is adequately applicable and also usable for other projects

These tests can lead to more or less extensive modifications to the selected solution. However, the aim is always to identify errors as early as possible in the development process in order to manage their impact on deadlines, costs and quality within acceptable limits.

Program version delivery

After the testing stage, the functions are delivered to the customer in the form of a program version. In the case of new concepts or extensive modifications, the function developer and the Application department assist the customer with commissioning by presenting the selected concept and discussing the calibration procedure on the engine test bench or in the course of summer or winter trials.

Software development

From Assembler...

When the first microcontrollers were used in control units, the programs they used required only 4 kilobytes of space or even less. At that time, it was all the memory that was available. For this reason, the programs had to be written in a space-saving language. The most commonly used programming language was Assembler. The commands in this language are mnemonics. They generally correspond directly to the instructions in the microcontroller's machine code. However, Assembler programs are generally difficult to read and maintain.

Over the course of time, the capacity of memory chips grew larger and larger, and the range of functions performed by the engine-management system became more and more complex. The expanding variety of functions made software modularization unavoidable. The ECU program is divided into modules, each of which incorporates a specific group of functions (e.g. lambda closed-loop control, idle-speed control). Of course, these modules have to be usable not just for one particular project but for large numbers of similar projects. For this reason, defined interfaces for the input and output variables of the functions are important. At this point, assembly language is reaching the limits of its capabilities.

...to high-level programming language

For the demands faced by present-day software developers, high-level programming languages are indispensable. The entire software for an engine-management system is nowadays written in a high-level language – the preferred choice being the C programming language. Programming in a high-level language ensures:

- Updatability of the software
- Modularity
- Interchangeability of software packages
- Independence of the software from the microcontroller used in the control unit

Software quality

Most innovations in automotive technology today are based on the use of electronics. In the past, the software was seen as an "appendage" to the hardware. Over time, however, the importance of the software has steadily grown. With the increasing complexity of microcontroller-controlled electronic systems, the quality of the software became a central pillar of software development – because problems caused by imperfections in the software damage the image of a manufacturer and drive up warranty-claim costs.

Software process improvement

The model used for improving processes in software development is CMM (Capability Maturity Model). It provides a framework for highlighting the elements of an effective software-development process. It describes an evolutionary path from a disorganized to a perfected, disciplined process and supports:

- Characterization of the degree of process maturity
- Definition of targets for process improvement and
- Setting of priorities for action to be taken

Distributed development

Bosch develops software not only in Germany but at a number of locations around the world. The same development process is applied throughout this international development network. Consequently, the same high standards of software quality are produced by Bosch worldwide.

Software sharing

As a result of software modularization and the use of defined interfaces, "external" software modules can also be incorporated in the ECU programs. This means that a vehicle manufacturer can use its own software for different vehicle models. The software then becomes a "distinguishing factor" on the competitive marketplace.

Creating program code

The basis for software development are the function definitions that are produced by the function developers. These documents describe the ECU functions (e.g. lambda closed-loop control) that are to be converted into a program by the software developers and then combined into an executable ECU program.

Creating source code

For every function, a separate module – a component of the overall program – is produced. The source code for the modules is written on a PC using a text editor (Fig. 1). The source code essentially contains the actual program instructions as well as documentation that facilitates the "readability" of the program (for program upgrading).

Compiler

Once created, the source code has to be translated into machine code so that it can be understood by the microcontroller. This involves the use of a compiler. It produces "object code" which contains relative rather than absolute memory addresses.

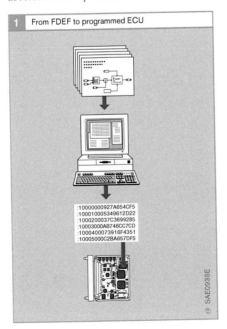

1   From FDEF to programmed ECU

```
:10000000927A654CF5
:10001005349612D22
:1000200037C3699285
:1000300AB746CC7CD
:1000400073916F4351
:10005000C2BA657DF5
```

SAE0938E

Linker

Once all the modules that make up the overall program have been created and compiled, all the object-code modules can be combined to form an executable program. This job is performed by a linker. The linker makes use of a file in which all the modules to be linked are listed. The file also details the memory addresses for the data and program memories. So all the relative addresses quoted in the object-code modules can be replaced by absolute addresses.

The result of the linking process is a machine-code program capable of running on the target system – the electronic control unit.

Module archiving

Software is subject to a rapid process of change. So that program versions supplied to customers can be reliably reproduced, it is essential to archive the modules. Archiving programs allow the tracking of every modification down to each individual module. For each archived program version, all the modules used can be listed and retrieved.

Testing station

The program code produced by the linker has to be tested in the laboratory before the ECU programmed with this code can be used on a vehicle. First of all, the new modules have to be individually tested in every detail. After this, the ability of all the modules to operate perfectly in combination with one another has to be thoroughly tested.

The tests are carried out on a testing station which is made up of a diverse range of testing equipment (Fig. 2 overleaf).

ECU with emulator module

A specially constructed laboratory version of the ECU is used for laboratory tests. It differs from the production version by virtue of an IC socket in place of the permanently soldered Flash-EPROM. Plugged into this socket is an emulator module which simulates the EPROM by means of a RAM.

This makes it possible to modify data and program code "online". The control operations are performed on a PC.

LabCar

In a real operating environment, the control unit receives input signals from sensors and desired-value generators. It then produces output signals to control the physical actuators on the vehicle. The sensor signals are simulated in a laboratory environment. The necessary signal generators (e.g. inductive speed sensor) or hardware circuits (e.g. resistor sequence for simulating the temperature sensor) are accommodated in a "black box" known as the LabCar.

The LabCar also emulates the connections with all of the actuators controlled by the control unit. One of the most important is the electronic throttle-valve device since feedback from this device is continuously monitored by the ECU. Without the throttle-valve link, it is not possible to operate the vehicle.

The immobilizer also has to be connected to the ECU in order to allow vehicle operation. The electrical signals are simulated.

The LabCar thus provides the means for simulating the vehicle for the purposes of testing the ECU program.

Connection adaptor

In the wiring harness which connects the LabCar to the ECU, there is a connection adaptor. Every lead in the wiring harness is plugged into a socket on the connection adaptor. This means this every signal traveling to or from the control unit is accessible for testing purposes (e.g. tracking the voltage curve of a control signal using an oscilloscope).

Fig. 2

1 LabCar
2 TRS 4.22 interface
3 INCA VME
  (calibration tool)
4 Throttle device
  (electronic throttle
  control)
5 Immobilizer
6 Connection adaptor
7 K-line (serial
  1-wire interface)
8 Engine ECU with
  emulator module
9 Serial interface
  (RS232)
10 Parallel interface
  (Centronics link
  cable with additional
  fiber-optic cores)
11 PC (engine model,
  LabCar control)
12 Serial interface
  (RS232)
13 PC (testing
  computer, auxiliary
  computer for control
  of automatic testing)

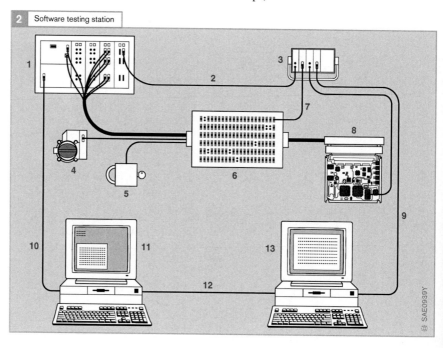

2 Software testing station

Emulator mode

The control unit (with microcontroller socket) has an emulator plugged into it. The emulator replaces the ECU's microcontroller. The ECU program then runs on the emulator.

The program code is loaded to the emulator from a PC. The emulator is also controlled from this PC. This means that:

- The program can be started from a specific memory address
- "Breakpoints" can be set (so this the program stops at defined points)
- The program sequence can be retraced from the breakpoint and the memory contents can be read and modified at each individual stage
- Trigger conditions can be defined and the program sequence can be analyzed before and after the trigger point
- Internal signals and registers (processor registers) that otherwise would not be accessible can be read
- In single-step mode, program sequences can be processed one step at a time and the processing sequence tracked
- Data and program code can be modified in order to manipulate the program sequence for testing purposes

The ability to set trigger conditions and record results makes it possible to analyze the program run in relation to specific input signals. As a result, every branch of the program can be tested against the function definitions.

Logic analyzer

Another means of tracking the program sequence is provided by the logic analyzer. It is connected via an adaptor in such a way that it can "listen in" to data traffic on the address and data buses. In this way, it can record the program sequence and also track read-and-write access to the external data memory. However, access to the data memory integrated in the microcontroller is not possible.

Trigger conditions (e.g. reaching a specific address, storage of a specific value in a memory cell) can be set on the logic analyzer. When the trigger condition has been met, the program sequence can be retraced.

The emulator and the logic analyzer offer similar capabilities. The advantage of the logic analyzer is that it does not interfere with the program sequence (real time) and can therefore be used in a real operating environment on the vehicle.

Automatic test

The LabCar testing station not only provides the means for manual testing of ECU programs, it also offers the facility for substantially reducing testing times, particularly where repeat tests are concerned, by using an automatic testing sequence. For such purposes, the ECU is operated in a closed-loop control circuit. At present, there are four different tests that can be used:

1. The most frequently used plausibility test is a "rough check" of the most important ECU functions. It performs an electrical and physical check on all input variables, and tests fuel injection and ignition, throttle-valve response, and maximum runtime load on the ECU.
2. The OBD (On-Board Diagnosis) test checks out the most important ECU diagnostic functions and fault management by means of fault simulation.
3. The CAN test checks communication signals with reference to range of values and reference signal.
4. The start/stop test analyzes fuel-injection and ignition response to ECU start-and-stop processes. Among other things, this involves varying the battery voltage.

Application-related adaptation

So that a car is able to meet the driver's expectations, extensive development work is required, particularly in respect of the engine.

As a rule, the vehicle manufacturer will center the development of a new vehicle around a basic engine. The most important operating parameters for this engine are known quantities, such as:

● Compression ratio and
● Valve timing (on engines with variable valve timing, this can be altered while the engine is running)

The engine peripherals have to be modified to meet installation space constraints. This applies in particular to:
● The air-intake system and
● The exhaust-gas system

Other important aspects (e.g. fitting location of knock sensors) have to be defined in consultation with the Application department at Bosch.

Parameter definition

The next stage involves adapting the electronic control system – the Motronic – to the engine. This is the function of application-specific adaptation. Application-specific adaptation means adapting an engine to suit a particular vehicle.

The ECU program consists of the program itself plus a large amount of data. The program meets the requirements set out in the specifications document (function framework), but the data still has to be adapted to the particular engine and vehicle-model variant.

In the adaptation phase, all data – also referred to as parameters – has to be adjusted to achieve optimum efficiency of operation. The main evaluation criteria include:
● Low-emission exhaust gas (compliance with prevailing emission standards)
● High torque and power output
● Low fuel consumption and
● High level of user-friendliness

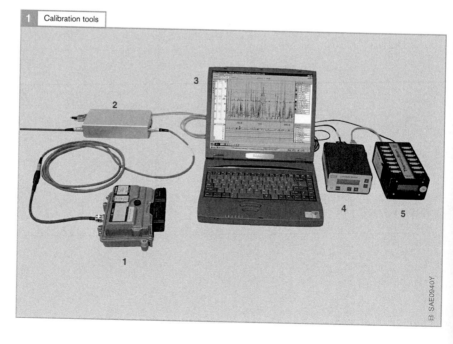

1   Calibration tools

Fig. 1
1   Engine ECU
2   MAC
    (compact testing
    and calibration tool)
3   Laptop computer
4   Lambda Scan
    (test interface for
    broadband lambda
    sensors)
5   Thermo Scan
    (test interface for
    temperature sensors)

The aim of adaptation is to ensure that the objectives outlined above are achieved as fully as possible, i.e. the best possible compromise is reached between competing demands.

To this end, there is a data record with up to 5500 "labels" that are capable of adjustment. These labels are subdivided into:
- Individual parameter values (e.g. temperature thresholds at which specific functions are activated)
- Data curves (e.g. engine speed vs. temperature curve for threshold at which a function is activated, temperature vs. ignition-timing curve) and
- Data maps (e.g. ignition timing as a function of engine load and speed)

Initially, work has to be carried out on the engine test bench. In this basic adaptation phase, parameters such as the ignition-timing map are defined. When basic adaptation has established the foundation for the initial vehicle trials, all parameters that affect engine response and dynamic characteristics are adapted. This work is for the greater part performed with the engine fitted in the vehicle.

The scope for optimization on Motronic systems has become so extensive and complex that many of the functions are now only possible with the aid of automated optimization methods and powerful tools.

Calibration tools

A large part of the adaptation work is carried out using PC-based calibration tools. Such programs allow developers to modify the engine-management software. One such calibration tool is the INCA (Integrated Calibration and Acquisition System) program. INCA is an integrated suite of several tools. It is made up of the following components:
- The Core System incorporates all measurement and adjustment functions
- The Offline Tools (standard specification) comprise the software for analyzis of measured data and management of adjustment data, and the programming tool for the Flash-EPROM

The ECU used for adaptation purposes has an emulator module instead of the program memory (EPROM) which emulates the ECU's EPROM and RAM. The INCA system has access to the data in these memories.

This memory emulator represents the most powerful ECU interface currently available for connecting calibration tools.

A simpler method of linking calibration tools (laptop) to the ECU is offered by the MAC (compact testing and calibration device). It connects via the K-line of the diagnosis interface or – if present – the emulator module (Fig. 1).

The use and function of the calibration tools can be illustrated by the description below of a typical calibration process (Fig. 2 overleaf).

Software calibration process

Defining the desired characteristics

The desired characteristics (e.g. dynamic response, noise output, exhaust-gas composition) are defined by the engine manufacturer and the (exhaust-gas emissions) legislation. The aim of calibration is to alter the characteristics of the engine to meet these requirements. This requires testing on the engine test bench and in the vehicle.

Preparations

Special engine ECUs with emulator modules are used for calibration. Compared with the ECUs used on production models, they allow parameters that are fixed for normal operation to be altered. An important aspect of preparations is choosing and setting up the appropriate hardware and/or software interface.

Additional testing equipment (e.g. temperature sensors, flow meters) permit detecting other physical variables for special tests using INCA hardware components such as:

- Thermo Scan (for measuring temperatures)
- Lambda Scan (for measuring exhaust-gas oxygen content)
- Baro Scan (for measuring pressures) and
- A/D Scan (for other analog signals)

Determining and documenting the actual system responses

The detection of specific measured data is carried out using the INCA kernel system. The information concerned can be displayed on screen and analyzed in the form of numerical values or graphs (Fig. 3).

The measured data can not only be viewed after the measurements have been taken but while measurement is still in progress. In this way, engine response to changes (e.g. in the exhaust-gas recirculation rate) can be investigated. The data can also be recorded for subsequent analyzis of transient processes (e.g. engine starting).

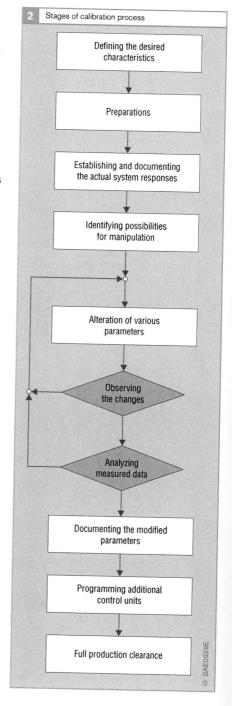

2   Stages of calibration process

- Defining the desired characteristics
- Preparations
- Establishing and documenting the actual system responses
- Identifying possibilities for manipulation
- Alteration of various parameters
- Observing the changes
- Analyzing measured data
- Documenting the modified parameters
- Programming additional control units
- Full production clearance

SAE0929E

Identifying possibilities for manipulation
With the help of the ECU software documentation (function framework), it is possible to identify which parameters are best suited to altering system response in the manner desired.

Altering selected parameters
The parameters stored in the ECU software can be displayed as numerical values (in tables) or as graphs (curves) on the PC and altered. Each time an alteration is made, system response is observed.

All parameters can be altered while the engine is running so that the impacts are immediately observable and measurable.

In the case of short-lived or transient processes (e.g. engine starting), it is effectively impossible to alter the parameters while the process is in progress. In such cases, therefore, the process has to be recorded during the course of a test. The measured data is saved to file and then the parameters that are to be altered are identified by analyzing the recorded data.

Further tests are performed in order to evaluate the success of the adjustments made or to learn more about the process.

because several people will be involved in the process of engine optimization at different times.

Documenting the modified parameters
The changes to the parameters are also compared and documented. This is done with the offline tool ADM (Application Data Manager), sometimes also called CDM (Calibration Data Manager).

The calibration data obtained by various technicians is compared and merged into a single data record.

Programming additional control units
The new parameter settings obtained can also be used on other engine ECUs for further calibration. This requires reprogramming of the Flash-EPROMs in the ECUs. This is carried out using the INCA kernel system tool PROF (Programming of Flash-EPROM).

Depending on the extent of the calibration and design innovations, multiple looping of the steps described above may take place.

Analyzing measured data
Analyzis and documentation of the measured data is performed with the aid of the offline tool MDA (Measure Data Analyzer). This stage of the calibration process involves comparing and documenting system response before and after parameter alteration. Such documentation encompasses improvements as well as problems and malfunctions. Documentation is important

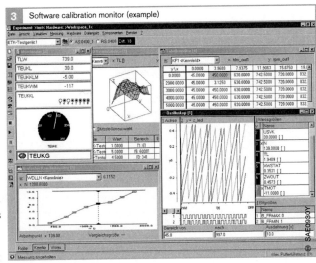

3 Software calibration monitor (example)

Example of calibration

Exhaust-gas temperature control

The following aspects play an essential role in operating a modern gasoline engine:

- Exhaust-gas emissions
- Fuel consumption and
- Thermal stresses on engine components

In order to minimize the stresses to which the engine is subjected, a low exhaust-gas temperature has to be aimed at (e.g. peak temperature below 1050 °C and continuous temperature below 970 °C). This is a particularly important consideration in the case of turbocharged engines because the turbocharger has to be protected against thermal damage.

The exhaust-gas temperature can be reduced by enriching the air-fuel mixture by manipulating the optimization parameters, for example. Unfortunately, the negative side of enriching the mixture is that both fuel consumption and exhaust-gas emissions (CO and HC) are raised. This means that the mixture need only be enriched by the absolute minimum amount required.

Engine operation generally follows a dynamic pattern, which is why the exhaust-gas temperature is subject to fluctuations. Therefore, a physical model to take account of heat capacities, heat transfer and response times is required to determine exhaust-gas temperature. Such complex models can generally only be configured with the aid of optimization tools. This involves performing tests under all relevant operating conditions in which all essential input and output variables are recorded. In the case of exhaust-gas temperature control, the optimizer then adapts the optimization parameters until the modeled temperature matches the measured temperature as closely as possible. The accuracy of the modeled temperature is equivalent to that achieved with a temperature sensor if the choice of parameters is appropriate.

The advantages are obvious – the temperature sensor, sensor wire and installation position layout can be dispensed with on the production model. In addition, this removes the risk of the component failing over the life of the vehicle. It means that the fault-diagnosis function for the component is not required either.

Other adjustments

Safety-related adaptation

As well as the functions that determine emission levels, performance and user-friendliness, there are also numerous safety functions that require adaptation (e.g. response to failure of a sensor or actuator).

Such safety functions are primarily intended to restore the vehicle to a safe operating condition for the driver and/or to ensure the safe operation of the engine (e.g. to prevent engine damage).

Communication

The engine ECU is normally part of a network of several ECUs. The exchange of data between vehicle, transmission, and other systems takes place via a data bus (usually a CAN). Correct interaction between the various ECUs involved cannot be fully tested and optimized until they are installed in the vehicle, as the process of basic configuration on the engine test bench usually involves only the engine-management module on its own.

A typical example of the interaction between two vehicle ECUs is the process of changing gear with an automatic transmission. The transmission ECU sends a request via the data bus to reduce engine torque at the optimum point in the gear shifting operation. The engine ECU then initiates actions independently of the driver to reduce engine torque output and thus facilitate a smooth and judder-free gear change. The data that result in torque reduction has to be adapted.

Electromagnetic compatibility

The large number of electronic vehicle systems and the wide use of other electronic communications equipment (e.g. radio telephones, two-way radios, GPS navigation systems) make it necessary to optimize the Electro-Magnetic Compatibility (EMC) of the engine ECU and all its connecting leads in terms of both immunity to external interference and of emission of interference signals. A large proportion of this optimization work is carried out during the development of ECUs and the sensors concerned, of course. However, the dimensioning (e.g. length of cable runs, type of shielding) and routing of the wiring harnesses in the actual vehicle has a major impact on immunity to and creation of interference. As a result, testing and, if necessary, optimization of the complete vehicle inside an EMC room is absolutely essential.

Fault diagnosis

Due to legal requirements, the capabilities demanded of fault-diagnosis systems are very extensive. The engine ECU constantly checks that the signals from all connected sensors and actuators are within specified limits. It also tests for loose contacts, short circuits to ground or to the battery terminal, and for plausibility with other signals. The signal range limits and plausibility criteria must be defined by the application developer. These limits must firstly be sufficiently broad to ensure that extreme conditions (e.g. hot or cold weather, high altitudes) do not produce false diagnoses, but secondly, sufficiently narrow to provide adequate sensitivity to real faults. In addition, fault response procedures must be defined to specify whether and in what way the engine may continue to be operated if a specific fault is detected. Finally, detected faults have to be stored in a fault memory so that service technicians can quickly locate and rectify the problem.

Testing under extreme climatic conditions

Testing procedures include trials under extreme climatic conditions that are normally only encountered under exceptional circumstances during the service life of the vehicle. The conditions that are encountered during these trials can only be simulated to a limited degree on a test bench because the subjective judgement of the test driver and long experience play an important part in such tests. Temperature itself can easily be simulated on a test bench, but using a chassis dynamometer to assess a vehicle's response when pulling away, for example, is very difficult compared to making the judgement under real driving conditions.

In addition, road tests generally involve longer distances and several vehicles. This enables testing of calibration parameters across the spread of the vehicles tested and, therefore, allows wider conclusions to be drawn than with calibration based on a single test subject.

Another essential aspect is the impact of variations in fuel grade from one part of the world to another. The chief effect of such variations in fuel grade is on the engine's starting characteristics and warm-up phase. Vehicle manufacturers go to great lengths to ensure this a vehicle will run properly on all the fuels on the market.

Cold-weather trials

Cold-weather trials cover the temperature range from approx. 0 °C to −30 °C. Preferred locations for cold-weather trials are places such as northern Sweden and Canada. The primary function is to assess starting and pulling away.

During the starting sequence, every individual combustion process is analyzed and the appropriate parameters optimized where necessary. Correct configuration of parameters for every individual injection sequence is a decisive factor in engine start time and smooth increase of engine speed from starter speed to idle speed. Even a single imperfect combustion process with resulting reduced torque development during the startup phase is perceived as a deficiency – even by inexperienced customers.

Hot-weather trials

Hot-weather trials cover the temperature range from approx. +15 °C to +40 °C. These trials are carried out at locations such as southern France, Spain, Italy, the U.S., South Africa and Australia. Despite the great distances involved and the corresponding high cost of equipment transportation, South Africa and Australia are of interest because they offer hot-weather conditions during the European winter. Due to the ever increasing demands to shorten development times, such possibilities have to be considered.

Hot-weather trials test such things as hot starting, tank ventilation, tank leakage detection, knock control, exhaust-gas temperature control and a wide variety of diagnostic functions.

Altitude trials

Altitude trials involve testing at altitudes between 0 and approx. 4000 meters. It is not only the absolute altitude that is of importance to the tests but, in many cases, a rapid change in altitude within a short space of time. Altitude trials are generally carried out in combination with hot or cold-weather trials.

Once again, an important component is testing the start characteristics. Other aspects examined include mixture adaptation, tank ventilation, knock control and a range of diagnostic functions.

4   Recording cold-starting response in the cold room

Quality management

Quality assurance measures accompany
the entire development process, and subsequently the production process as well.
Only in that way can consistent quality of the
end product be guaranteed. The quality requirements placed on safety-related systems
(e. g. ABS) are particularly strict.

Quality assurance systems

All elements of a quality management system
and all quality assurance measures have to be
systematically planned. The various tasks, authorities and responsibilities are defined in
writing in the quality management handbook. International standards such as ISO
9001 to 9004 are also adopted.

In order to regularly monitor all elements of a
quality management system, quality audits
are carried out. Their purpose is to assess the
extent to which the requirements of the quality management system are being followed
and the effectiveness with which the quality
requirements and objectives are being met.

Quality assessment

On completion of specific stages in the development process, all information available up
to that point about quality and reliability is
subjected to a quality assessment and any
necessary remedial action initiated.

FMEA

FMEA (Failure Mode and Effects Analyzis) is
an analytical method for identifying potential
weaknesses and assessing their significance.
Systematic optimisation results in risk and
fault cost reduction and leads to improved reliability. FMEA is suitable for analyzing the
types of fault occurring on system components and their effects on the system as a
whole. The effect of a fault can be described
by a causal chain from point of origin (e.g.
sensor) to system (e.g. vehicle).

The following types of FMEA are distinguished:
- Design FMEA: assessment of the design of
systems for compliance with the specifications. It also tests how the system reacts in
the event of design faults
- Process FMEA: assessment of the production process
- System FMEA: assessment of the interaction of system components

FMEA assessments are based on theoretical
principles and practical experience.

Example: a direction indicator fails. The
effects in terms of road safety are serious. The
likelihood of discovery by the driver is small,
however, since the indicator is not visible
from inside the vehicle. As a means of making
the fault obvious, the rate at which the indicators flash must be made to change if an indicator fails. The higher flashing rate is discernible both visually on the instrument cluster and audibly. As a result of this modification, the effect of the fault can be reduced.

Review

The review is an effective quality assurance
tool in software development in particular.
Reviewers check the compliance of the work
produced with the applicable requirements
and objectives.

The review can be usefully employed as
a means of checking progress made even
at early stages of the development process.
Its aim is to identify and eliminate any faults
at as early a point as possible.

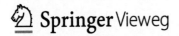

Made in the USA
San Bernardino, CA
05 July 2015